Air Quality: Monitoring, Measuring and Modeling

Air Quality: Monitoring, Measuring and Modeling

Edited by Elliot Franco

SYRAWOOD
PUBLISHING HOUSE
New York

Published by Syrawood Publishing House,
750 Third Avenue, 9th Floor,
New York, NY 10017, USA
www.syrawoodpublishinghouse.com

Air Quality: Monitoring, Measuring and Modeling
Edited by Elliot Franco

International Standard Book Number: 978-1-64740-425-3 (Hardback)

Cataloging-in-publication Data

Air quality : monitoring, measuring and modeling / edited by Elliot Franco.
p. cm.
Includes bibliographical references and index.
ISBN 978-1-64740-425-3
1. Air quality. 2. Air quality--Measurement. 3. Air quality management. 4. Air--Pollution.
5. Environmental monitoring. 6. Environmental impact analysis. I. Franco, Elliot.
TD883 .A37 2023
363.739 2--dc23

TABLE OF CONTENTS

PREFACE

In my initial years as a student, I used to run to the library at every possible instance to grab a book and learn something new. Books were my primary source of knowledge and I would not have come such a long way without all that I learnt from them. Thus, when I was approached to edit this book; I became understandably nostalgic. It was an absolute honor to be considered worthy of guiding the current generation as well as those to come. I put all my knowledge and hard work into making this book most beneficial for its readers.

Air quality monitoring refers to a process which involves taking samples and performing tests of air emissions. It aids in measuring the quality of air in relation to pollutants in a given time duration. Furthermore, it also helps to evaluate whether the quality of air is deteriorating or improving through the assessment of patterns and trends. It involves a systematic and long-term evaluation of pollutant levels in the surroundings, by determining the quantity and types of contaminants in the air. Air quality is measured using an index called the air quality index (AQI). An AQI of 50 or less is considered good whereas AQI above this level is unhealthy and may result in health issues. AQI helps in measuring the level of different pollutants that are present in the atmosphere, namely, particulate matter (PM10), particulate matter (PM2.5), nitrogen dioxide (NO_2), sulfur dioxide (SO_2), carbon monoxide (CO), ozone (O_3), ammonia (NH_3), and lead (Pb). The modeling of air quality is done using computer programs and mathematics for estimating pollutant concentrations in the air. Systems for assessing and forecasting air quality help the decision makers in improving public health and air quality. This book contains some path-breaking studies on air quality monitoring, measuring and modeling. It will serve as a reference to a broad spectrum of readers.

I wish to thank my publisher for supporting me at every step. I would also like to thank all the authors who have contributed their researches in this book. I hope this book will be a valuable contribution to the progress of the field.

Editor

1

Mobile DOAS Observations of Tropospheric NO_2 using an Ultralight Trike and Flux Calculation

Daniel-Eduard Constantin [1,*], Alexis Merlaud [2], Mirela Voiculescu [1], Carmelia Dragomir [1], Lucian Georgescu [1], Francois Hendrick [2], Gaia Pinardi [2] and Michel Van Roozendael [2]

1 European Center of Excellence for the Environment, Faculty of Sciences and Environment, "Dunarea de Jos" University of Galati, Str. Domneasca 111, Galati 800008, Romania; mirela.voiculescu@ugal.ro (M.V.); carmelia.dragomir@ugal.ro (C.D.); lucian.georgescu@ugal.ro (L.G.)

2 Royal Belgian Institute for Space Aeronomy, Ringlaan-3-Avenue Circulaire, B-1180 Brussels, Belgium; alexis.merlaud@aeronomie.be (A.M.); francois.hendrick@aeronomie.be (F.H.); gaia.pinardi@aeronomie.be (G.P.); michel.vanroozendael@aeronomie.be (M.V.R.)

* Correspondence: daniel.constantin@ugal.ro

Academic Editors: Pius Lee, Rick Saylor and Jeff McQueen

Abstract: In this study, we report on airborne Differential Optical Absorption Spectroscopy (DOAS) observations of tropospheric NO_2 using an Ultralight Trike (ULT) and associated flux calculations. The instrument onboard the ULT was developed for measuring the tropospheric NO_2 Vertical Column Density (VCD) and it was operated for several days between 2011 and 2014, in the South-East of Romania. Collocated measurements were performed using a car-DOAS instrument. Most of the airborne and mobile ground-based measurements were performed close to an industrial platform located nearby Galati city (45.43° N, 28.03° E). We found a correlation of $R = 0.71$ between tropospheric NO_2 VCDs deduced from airborne DOAS observations and mobile ground-based DOAS observations. We also present a comparison between stratospheric NO_2 Slant Column Density (SCD) derived from the Dutch OMI NO_2 (DOMINO) satellite data product and stratospheric SCDs obtained from ground and airborne measurements. The airborne DOAS observations performed on 13 August 2014 were used to quantify the NO_2 flux originating from an industrial platform located nearby Galati city. Measurements during a flight above the industrial plume showed a maximum tropospheric NO_2 VCD of $(1.41 \pm 0.27) \times 10^{16}$ molecules/cm^2 and an associated NO_2 flux of $(3.45 \pm 0.89) \times 10^{-3}$ kg/s.

Keywords: mobile DOAS; airborne observations; nitrogen dioxide; emission flux

1. Introduction

Nitrogen dioxide (NO_2) is a chemical gaseous compound with an important role in the Earth's atmosphere. NO_2 is a key trace element in the chemistry of ozone, since it is involved in the catalytic destruction of ozone in the stratosphere [1], while in the troposphere its photolysis leads directly to the formation of ozone (O_3) in the presence of VOCs (volatile organic compounds). NO_2 is released in the atmosphere from natural sources (soil, lightning, solar cosmic rays) and anthropogenic emissions (fossil fuels and biomass burning, industrial activities). Long-term exposure to NO_2 may affect the respiratory system and lead to coronary diseases. NO_2 can lead to acidification of the aquatic ecosystem following the oxidation to HNO_3.

The Differential Optical Absorption Spectroscopy (DOAS) technique [2] has been used for NO_2 atmospheric measurements since the early 1970s [3,4]. Nowadays, besides ground-based zenith sky measurements, DOAS techniques have developed into Multi-Axis Differential Optical Absorption Spectroscopy (MAX-DOAS) observations [5]. The mobile DOAS technique was recently used on several platforms such as: cars [6,7], airplanes [8–10], Unmanned Aerial Vehicles (UAVs) [11] or satellites [12–16].

Airborne observations have a number of important advantages for atmospheric research such as: the flexibility during the flights and the possibility to access remote areas such as oceans, deserts, rural or areas without roads.

The ULMs (Ultralight Motorized) are airborne platforms with an important scientific potential for atmospheric research. So far, ULMs have been used to study the ultraviolet actinic radiation flux [17], formaldehyde distribution [18], aerosol profiles [19], SO_2, NO_2 and ozone distribution [20–23].

This work highlights the capability of a low-cost system (ULT-DOAS) used for measurements of tropospheric NO_2 VCD and associated flux calculations. This study presents airborne DOAS observations of tropospheric NO_2 using an Ultralight Trike (ULT) and associated flux calculations. The work presented here was motivated by the need to further assess the intrapixel variability of NO_2 detected by UV-VIS DOAS instruments onboard satellites. This work comes in the context of a validation programme of the future Atmospheric Sentinels, starting with the Sentinel-5 Precursor to be launched in summer 2017. A similar system based on DOAS onboard the ULM was used during the Airborne Romanian Measurements of Aerosols and Trace gases (AROMAT) campaign, held in Romania in August 2015 [24]. The AROMAT campaign was conducted under the aegis of the European Space Agency (ESA) in the framework of a series of ESA field campaigns.

2. Methodology

2.1. Experimental and Instrumental Descriptions

The mobile DOAS observations were performed onboard of an Ultralight Trike (ULT), in the South-East of Romania (Figure 1) during several days between 2011 and 2014. All measurements were performed under clear-sky conditions (see Table 1). The Mobile DOAS system used for measurements will be described in the following as the ULT-DOAS system. The measurements took place in an area around Galati (located at 45.43° N, 28.03° E), Braila (45.26° N, 27.95° E), and close to the industrial areas of Slobozia (44.56° N, 27.35° E). Note that an operational steel mill factory is located in the vicinity of Galati, while Slobozia was chosen due to the presence of a fertilizer factory. Due to security concerns, direct flights above the cities or industrial platforms were not performed. Most of the airborne DOAS observations were performed in nadir geometry. Details about the ULT-DOAS measurements are presented in Table 1.

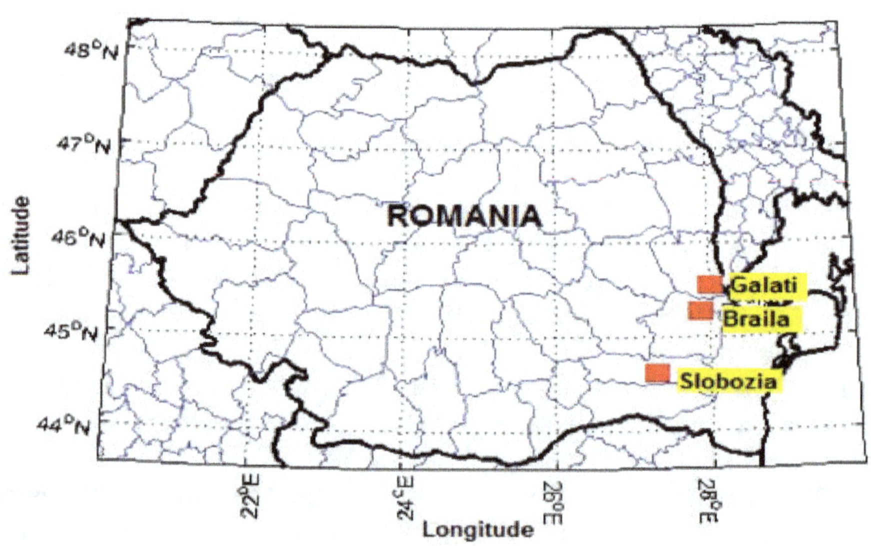

Figure 1. The main locations where airborne and/or ground-based mobile Differential Optical Absorption Spectroscopy (DOAS) observations were performed.

Table 1. Coordinates and temporal coverage of the mobile DOAS measurements.

Day	Time Interval UTC *	Route of ULT-DOAS ** Measurements	NO_2 Source Target
1 September 2011	8.31–9.45	Galati–Braila	Braila urban area
25 August 2012	7.53–8.89	Galati–Braila	Braila urban area
21 July 2014	9.51–10.96	Galati–Slobozia	Slobozia industrial area
13 August 2014	7.32–8.19	Galati	Galati industrial area

* Coordinated Universal Time. ** ULT-DOAS = Ultralight Trike-Differential Optical Absorption Spectroscopy.

Airborne-DOAS measurements were accompanied by car-DOAS measurements and static twilight observations on 21 July 2014. Static twilight DOAS measurements, used for determination of the NO_2 content from the reference spectra, were performed in a rural area close to Galati city. The car-DOAS measurements were performed right before or after the experimental flights.

The aircraft used for all the flight experiments presented in this paper is a double-seated, open-cockpit ultralight aircraft, trike type (model Fanagoria 21, produced by Plovdiv Air Bulgaria). The flexible wing (Atlant-21, produced by Plovdiv Air Bulgaria) has an area of 16 m^2. The ULT is powered by a Subaru EA81 engine with 75 HP. The cruise speed is 75 km/h and the maximum speed is 100 km/h relative to the ground. The aircraft has a maximum total weight at take-off of 450 kg.

The ULT-DOAS instrument consists of a compact Czerny-Turner spectrometer (AvaSpec-ULS2048XL-USB2, of 175 × 110 × 44 mm dimensions and 855 g weight) placed in the Ultralight Trike. Figure 2 presents the instrumental DOAS set-up. The spectral range of the spectrometer is 280–550 nm with 0.7 nm resolution (FWHM—Full Width at Half Maximum) with a focal length of 75 mm. The entrance slit is 50 μm and the grating is 1200 L/mm, blazed at 250 nm. The spectrometer is connected to the telescope through a 400 μm chrome-plated brass optical fiber. The telescope achieves a 2.3° field-of-view with fused silica collimating lenses. Each spectrum is recorded by a laptop and georeferenced by a GPS receiver. The spectrometer and the GPS receiver are powered by the laptop USB ports. The entire set-up is powered by 12 V of the ULT through an inverter. Each measurement is a 10-second average of 10 scans accumulations at an integration time between 50 and 150 ms.

This work is mostly based on nadir-DOAS observations but we also present zenith-sky observations onboard ULT for stratospheric NO_2 measurements. The same DOAS system was used in the case of the zenith sky car-DOAS observations.

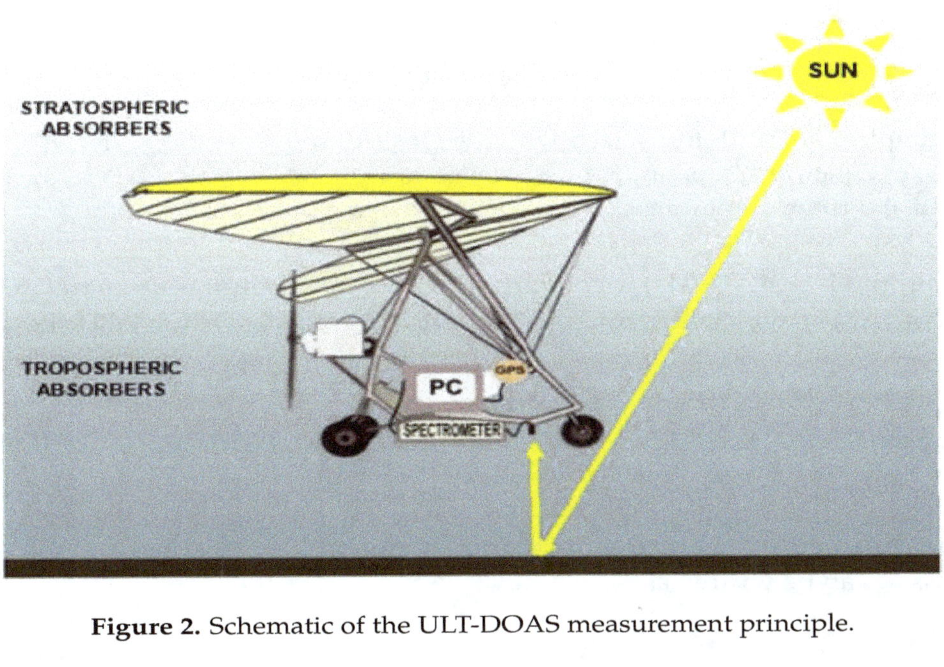

Figure 2. Schematic of the ULT-DOAS measurement principle.

2.2. *Determination of the NO_2 Tropospheric Vertical Column and Flux Calculation*

2.2.1. Retrieval of NO_2 Slant Column

The analysis of the measured spectra was performed using the QDOAS software [25]. For the NO_2 fit, the spectral window of 425–490 nm was used. The NO_2 spectral analysis included five absorption cross sections: the NO_2 cross sections at 298 K and 220K [26], the O_3 cross section at 223 K [27], the O_4 cross section [28] and a Ring spectrum [29]. A fifth-degree polynomial to account for scattering processes and broad-band absorption in the atmosphere was used in the DOAS analysis. The direct result of the spectral analysis is a differential slant column density (DSCD), which is the integrated trace gas concentration along the light path through the atmosphere. The DSCD is the difference between the slant column densities in the measured spectra (SCD_{meas}) and the Fraunhofer reference spectrum (SCD_{ref}). The NO_2 amount in the Fraunhofer reference spectrum is unknown and its retrieval is important for the determination of the SCD_{meas} (Equation (1)).

$$SCD_{meas} = DSCD + SCD_{ref} \tag{1}$$

Figure 3 presents a typical DOAS fit using a spectrum recorded during the experiment close to the Galati steel factory, on 13 August 2014.

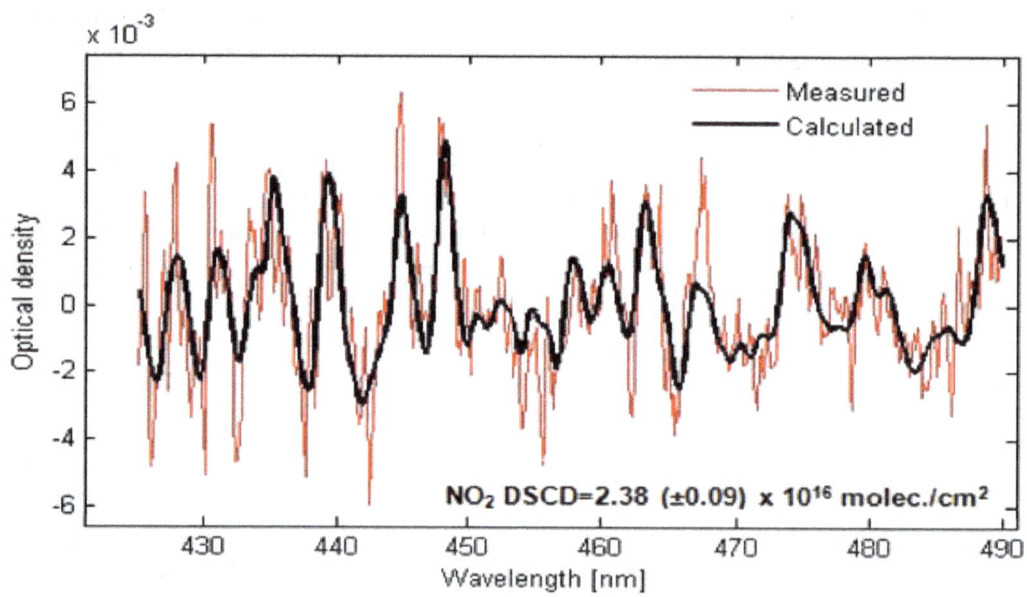

Figure 3. Example of a DOAS fit realized with the QDOAS software; the analyzed spectrum was recorded close to Galati, on 13 August 2014. Black line corresponds to molecular cross sections scaled to the detected absorptions in the measured spectrum (red line).

The Slant Column Density (SCD) is converted to a Vertical Column Density (VCD) by means of an Air Mass Factor (AMF), which is defined as the ratio between SCD and VCD (Equation (2))

$$AMF = \frac{SCD}{VCD} = \frac{\tau_{SCD}}{\tau_{VCD}} \tag{2}$$

where τ_{SCD} and τ_{VCD} are the optical thickness for the slant column (SCD) and vertical column (VCD), respectively.

Since the measured spectra contain information about both stratospheric and tropospheric NO_2 content, the SCD_{meas} can be written as:

$$SCD_{meas} = AMF_{tropo} \cdot VCD_{tropo} + AMF_{strato} \cdot VCD_{strato} - SCD_{ref} \tag{3}$$

The above equations can be further simplified assuming that SCD_{ref} is dominated by stratospheric NO_2. Using this assumption, the stratospheric contributions can be canceled by the NO_2 amount in the reference spectrum (Equation (4)) [8].

$$VCD_{tropo} = DSCD/AMF_{tropo} \tag{4}$$

The assumption presented above is valid if the reference spectrum (needed for the spectral evaluation) is recorded at noon, in an area with a very low NO_2 content and if SCD_{strato} does not vary in time. The Fraunhofer reference spectrum could also be a zenith-sky spectrum recorded at high altitude over the boundary layer [30]. However, in this work the tropospheric NO_2 VCD is based on calculations using Equation (3).

2.2.2. Deduction of the SCD_{ref} and VCD_{strato}

To avoid systematic errors due to the use of multiple reference spectra, only one spectrum will be used for the spectral analysis of all DOAS observations presented in this paper.

The NO_2 amount in the reference spectrum was calculated using a photo-chemically modified Langley plot [6,31]. The SCD_{ref} corresponds to a zenith spectrum recorded at noon, in a clean rural area close to Galati city. The spectrum was recorded on 13 August 2014 (9.70 UTC and solar zenith angle (SZA) = 31.55°).

The photo-chemically modified Langley plot was applied for the twilight sunrise observations performed on 21 July 2014. By applying the Langley plot method, we calculated the SCD_{ref} as 4.1×10^{15} molecules/cm^2 of NO_2.

The stratospheric contribution used for the retrieval of the VCD_{tropo} is derived from the assimilated vertical stratospheric columns simulated by Dutch OMI NO_2 (DOMINO). Table 2 shows the satellite overpass data sets that were used for the retrieval algorithm presented in this work.

Table 2. OMI satellite overpasses data sets.

Day	Orbit Nr.	Overpass Time UTC	Stratospheric VCD [$\times 10^{15}$ molecules/cm^2]
1 September 2011	37,923	11:04:28	3.76
25 August 2012	43,151	11:11:07	3.75
21 July 2014	53,272	11:17:36	4.14
13 August 2014	53,607	11:23:47	3.74

VCD = Vertical Column Density.

Figure 4A presents the SCD determined at twilight sunrise on 21 July 2014 compared with the SCD_{strato} derived from the DOMINO Level 2 product [32] scaled with a chemically modified AMF calculated by PSCBOX [33,34]. More details about the retrieval of the SCD_{strato} using twilight observations and model simulations are presented in [6]. A good agreement between the two types of SCD_{strato} determinations is obtained, which gives confidence in the stratospheric SCD measured by our static DOAS observations.

Figure 4B shows the SCDstrato derived from DOMINO compared with the SCDs determined by car-DOAS zenith-sky observations and ULT-DOAS measurements performed in the zenith geometry, for the same day of 21 July 2014. From this plot, one can see that the car-DOAS measurements are dominated by tropospheric NO_2 while the zenith-sky ULT-DOAS observation presents a low amount of NO_2 close to the stratospheric NO_2 SCD derived from OMI. This is due to the fact that zenith-sky ULT-DOAS observations are performed above the NO_2 plume or above the planetary boundary layer.

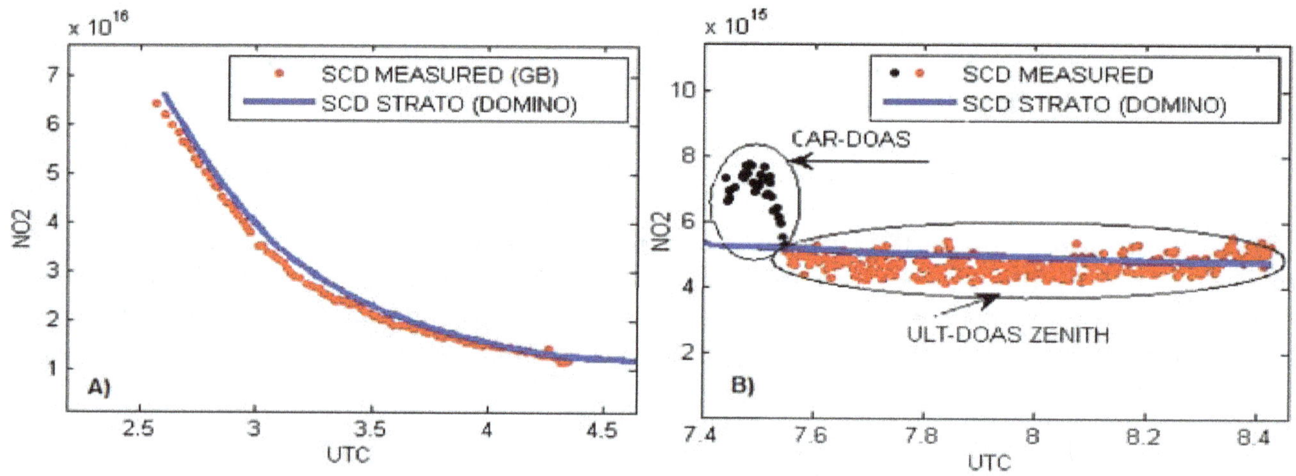

Figure 4. Comparisons between measured SCD using various methods of determination and stratospheric SCD derived from OMI (21 July 2014); (**A**) Comparison between SCD determined from ground-based (GB) observations and stratospheric SCD derived from DOMINO; (**B**) Comparison between SCD determined from CAR-DOAS (black dots) and ULT-DOAS (red dots) and stratospheric SCD derived from DOMINO.

2.2.3. Radiative Transfer Calculation

In order to determine the VCD, the SCD retrieved with the DOAS method has to be converted using an appropriate AMF. The geometric approximation for the airborne DOAS observations assumes a simple reflection of the sunlight on the earth's surface. In this case (neglecting the earth's sphericity) the nadir AMF can be described as a function of the solar zenith angle (SZA) as:

$$AMF_{geo} = 1 + 1/\cos(SZA) \tag{5}$$

Since the ULT-DOAS measurements were performed in the open atmosphere using scattered sunlight radiation, the radiative transfer during the observations needs to be modeled to interpret the retrieved data. In this work, the AMF was calculated using the radiative transfer model (RTM) UVspec/DISORT [35], which is a fully spherical model. This RTM has been validated using six other different codes [34].

The general assumptions made for the radiative transfer calculation using the RTM UVspec/DISORT are introduced in Table 3.

Table 3. Input parameters used for the radiative transfer model (RTM) calculations.

Trace Gas	Nitrogen Dioxide		
wavelength		440 nm	
flight altitude	1000 m		1500 m
albedo		0.1	
visibility	1 km	5 km	20 km
line of sight		0°	

Results of AMF simulations for the nadir flight performed on 13 August 2014, using the input parameters presented in Table 3, are displayed in Figure 5. The geometric AMF for the nadir view is also shown. The visibility parameter accounts for the effect of aerosols.

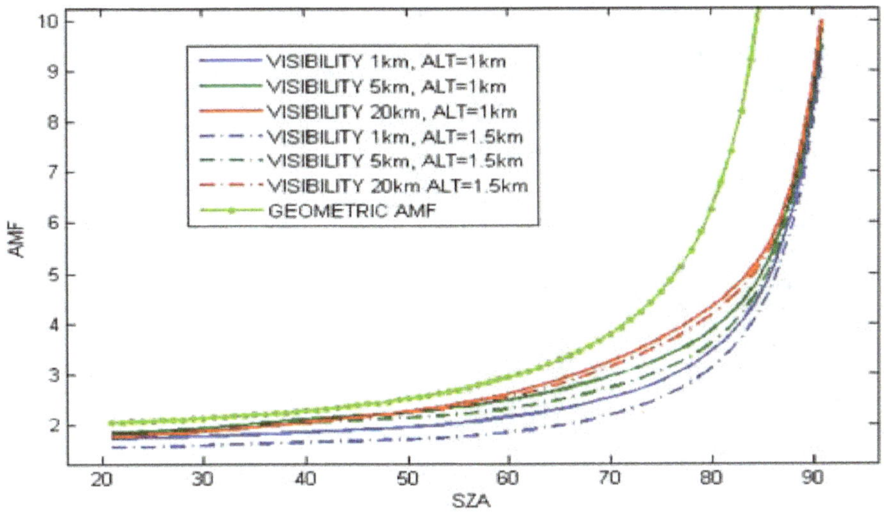

Figure 5. AMF simulations obtained from RTM calculations using UVspec/DISORT for various input parameters, for 13 August 2014. AMF = Air Mass Factor.

3. Results and Discussions

The airborne DOAS observations were designed to determine the distribution of tropospheric NO_2 from the South-East of Romania from urban, industrial and rural areas and associated flux.

The first flights were performed on 1 September 2011 and 25 August 2012, and aimed at measuring the NO_2 around the industrial area of Galati city and from Braila city. The flight performed on 21 July 2014 aimed to measure the NO_2 emitted by a fertilizer factory located nearby Slobozia city; unfortunately, during the DOAS flight the fertilizer factory was not operational. Figure 6 presents the horizontal distribution of the tropospheric NO_2 determined in nadir geometry for 1 September 2011, while Figure 7 depicts the results during a similar flight, but on 25 August 2012. During this experiment, the plume from the industrial platform was not fully crossed by the optical instrument onboard the ULT. The wind was northerly resulting in local increases of the NO_2 amount detected by the spectrometer. The maximum tropospheric NO_2 VCD detected during this experiment was $(1.1 \pm 0.24) \times 10^{16}$ molecules/cm^2 while the minimum tropospheric NO_2 VCD was $(2.1 \pm 0.81) \times 10^{15}$ molecules/cm^2.

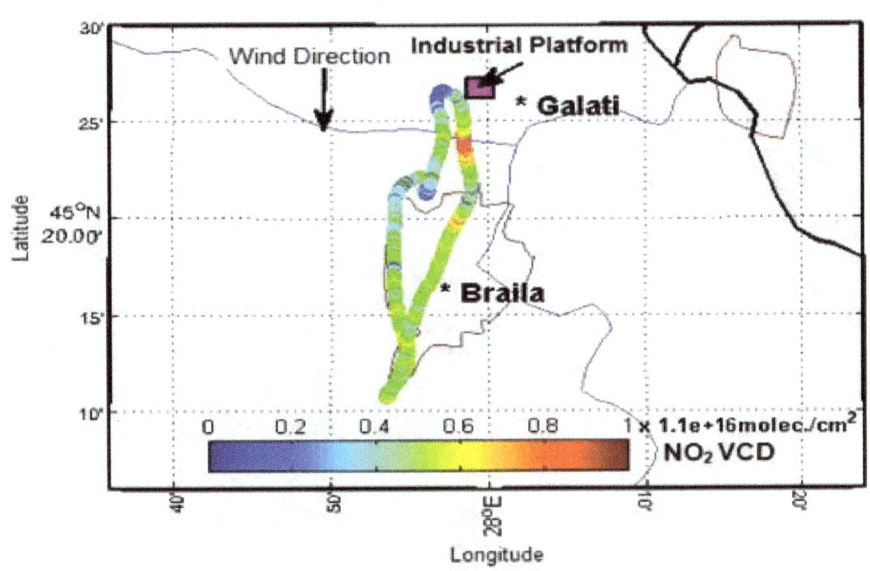

Figure 6. Map of tropospheric NO_2 VCD determined on 1 September 2011 using ULT-DOAS observations.

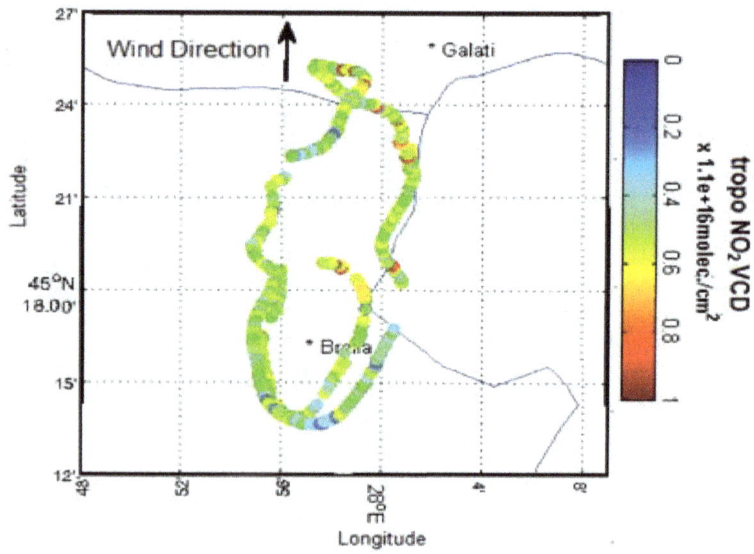

Figure 7. Map of tropospheric NO_2VCD determined on 25 August 2012 using ULT-DOAS observations.

The trajectory of the flight on 13 August 2014 gave us the opportunity of calculating the NO_2 flux emissions around the industrial area of Galati city. This was not possible for the other flights because encircling the NO_2 source was not authorized.

On 1 September 2011, the NO_2 amount was low relative to the other day. The flight comprised almost two complete circles around Braila; however, the NO_2 displayed no clear variation. The horizontal distribution of NO_2 was quite homogenous over Braila city on this day.

A double experiment was performed on 13 August 2014, using both a ULT-DOAS and a car-based DOAS system. The mobile ground-based DOAS observations were performed using the same equipment during 9.75–10 UTC, while the ULT-DOAS observations were performed during 7.30–8.15 UTC.

Figure 8 shows the tropospheric NO_2 VCD derived along the trajectory of the ULT-DOAS measurements. The right plot shows a photograph of the plume crossed by the ULT flights. During the same day, approximately 1 h after the acquisition of the ULT-DOAS measurements, a zenith-sky car-DOAS system was used to sample the NO_2 plume at the same location as the ULT-DOAS observations. We assume that the quantity of the NO_2 emitted by the steel factory was almost constant during the airborne and car-DOAS system.

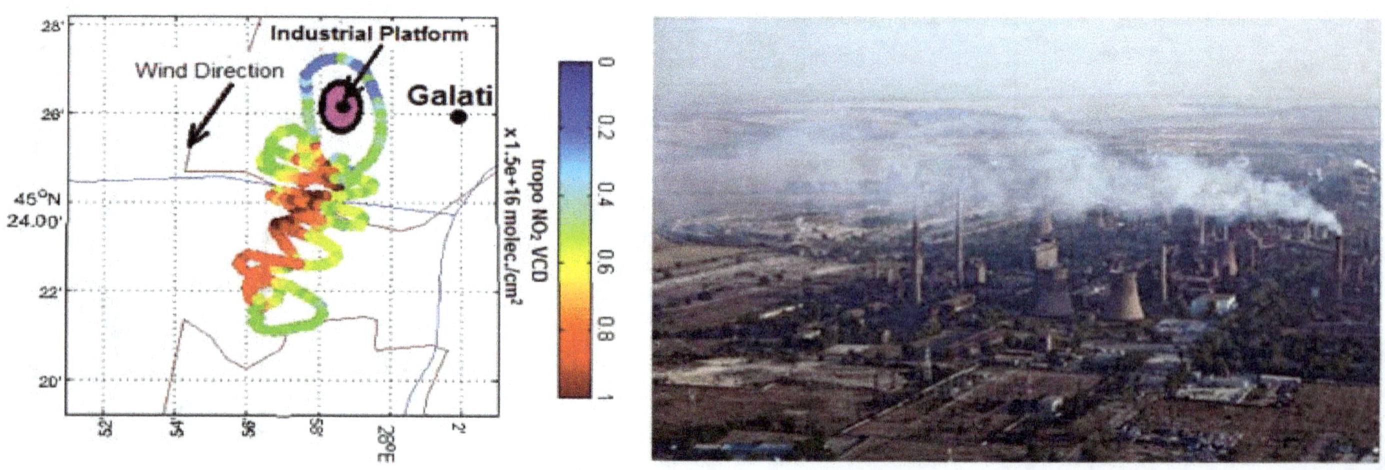

Figure 8. The tropospheric NO_2 VCD along the flight trajectory using the ULT-DOAS system on 13 August 2014 (**left**); Photography of the NO_2 plume determined on the same day (**right**).

Figure 9 presents the NO_2 VCD derived from the nadir airborne DOAS observations performed over the industrial area of Galati city compared with zenith-sky ground-based mobile DOAS measurements performed over the same area in the same day (13 August 2014). In these figures, we show the original SCDs (A) and the retrieved tropospheric NO_2 VCD (B). Figure 9A shows that the DSCDs determined from the ULT-DOAS system are ~30% higher than the DSCDs determined using car-DOAS observations. This difference is attributed to the different observation geometries. After appropriate AMF calculation (see Section 2.2.2), both observations show close results.

The ULT flight above the industrial plume led to the detection of a maximum tropospheric NO_2 VCD of $(1.41 \pm 0.27) \times 10^{16}$ molecules/cm^2 while the car-DOAS observation shows a maximum tropospheric NO_2 VCD of $(1.36 \pm 0.21) \times 10^{16}$ molecules/cm^2. Figure 10 displays the correlation between the tropospheric NO_2 VCD retrieved by the ULT-DOAS and the car-DOAS instrument, where closest spatially coincident data were selected. A Pearson correlation coefficient R = 0.71 was obtained between ground mobile DOAS observations and airborne DOAS measurements.

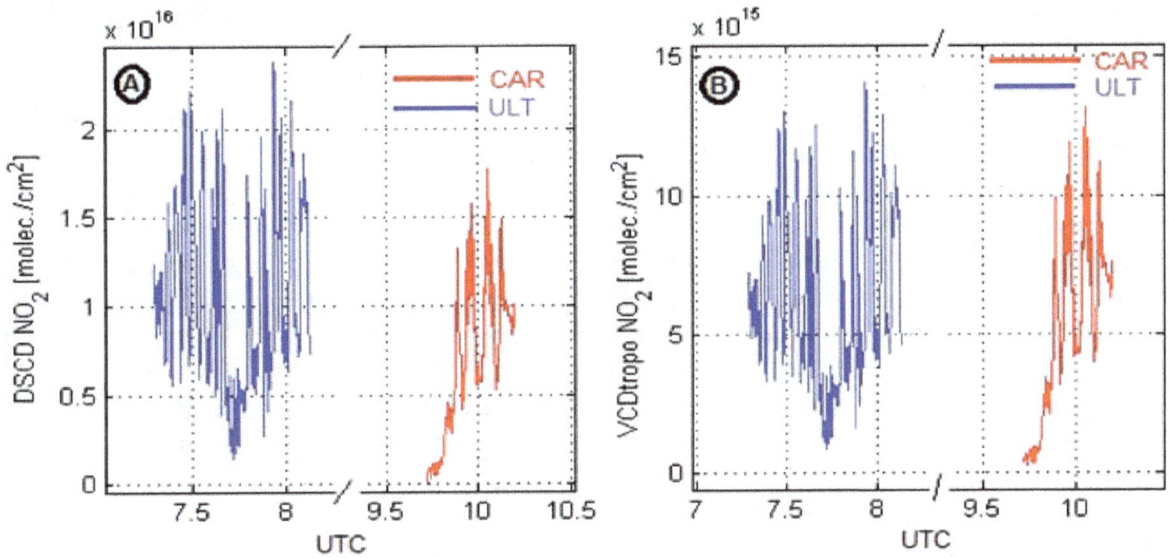

Figure 9. Comparisons between ULT-DOAS and car-DOAS observations performed on 13 August 2014. (**A**) The results of the preliminary DOAS analysis (DSCDs) and (**B**) after determination of the vertical columns (VCDs).

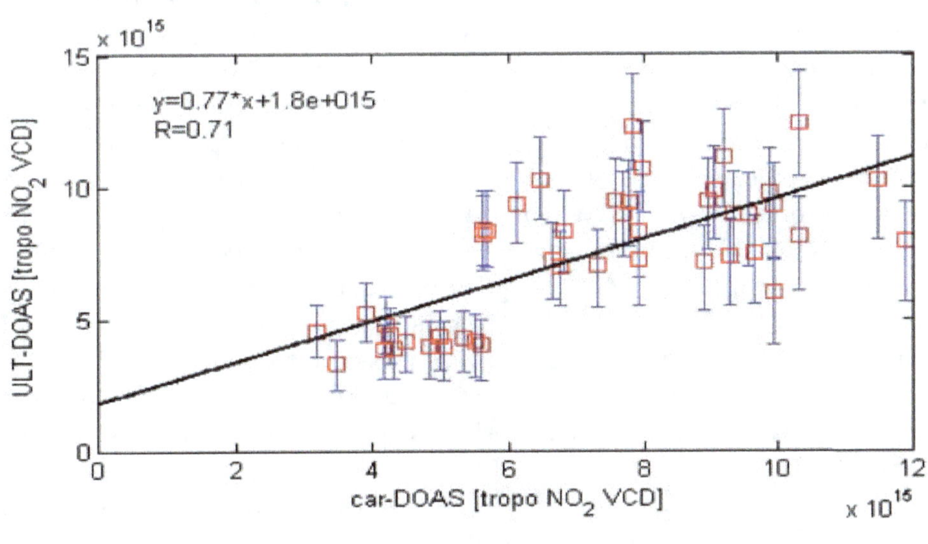

Figure 10. Correlation between tropospheric VCDs measured by the ULT-DOAS and the car-DOAS instrument on 13 August 2014.

NO_2 Flux Calculation

The NO_2 flux above the industrial platform located nearby Galati city was calculated by performing upwind and downwind measurements around the point source using ULT-DOAS observations on 13 August 2014. The calculation of the emission flux is based on the following parameters: the NO_2 VCD determined from the transect over the plume, the wind speed and the wind factor correction, taking into account the angle between the flight direction and wind direction (Equation (6)), [20,36,37]:

$$\mathrm{Flux}_{\mathrm{NO_2}} = \sum_{\mathrm{i}} \mathrm{VCD}_{\mathrm{NO_2}}(\mathrm{s_i}) \cdot \mathrm{v} \cdot \sin(\alpha_{\mathrm{i}}) \cdot \Delta \mathrm{s_i} \tag{6}$$

where VCD_{NO2} is the NO_2 tropospheric vertical column, v is the wind speed, α the angle between wind direction and driving route, i is the observations index, and Δs_i is the distance between two successive spectra.

The wind data used for the NO_2 flux calculation rely on measurements of the automatic weather station (Davis Vantage Pro2) located in the campus university of Galati city (45.44° N, 28.05° E), while vertical wind profiles come from the Hybrid Single Particle Lagrangian Integrated Trajectory Model (HYSPLIT) [38] model using archived dataset GDAS 0.5° × 0.5°. The weather station is located at 30 m height and acquires data every 30 min. Since no atmospheric sounding was possible during the experiments, the wind measured on the ground was scaled to the output of HYSPLIT model simulation at 1000 m altitude. During the period of the ULT-DOAS measurements, the mean NO_2 emission flux was determined to be $(3.45 \pm 0.89) \times 10^{-3}$ kg/s. A local environmental report indicates ~600 tons/year NO_x emissions emitted by the steel factory [39]. Using a Leighton ratio (L = [NO]/[NO_2]) of 0.3, we calculated that the steel factory has emitted a mean of ~10×10^{-3} kg/s of NO_2. The difference between the two types of estimation may be attributed to the fact that the NO_x emissions from the steel factory are dependent on the quantity of the steel produced, which can vary from one day to another. Also, the derived NO_2 fluxes must be dealt with some care because of the probably incomplete NO to NO_2 conversion [40].

4. Conclusions

Ultralight-trike DOAS observations were performed in the South-East of Romania during four days between 2011 and 2014. The first two flights were focused over Braila city, the third aimed at measurements of the NO_2 plume emitted by a fertilizer factory near Slobozia city. Unfortunately, during the DOAS flight the fertilizer factory was not operational. The last flight, performed on 13 August 2014, was focused over the industrial area of Galati city. Nadir observations were performed around the industrial platform of Galati city aiming at measuring the tropospheric NO_2 VCD around the source and at evaluating the associated NO_2 flux. To retrieve the tropospheric NO_2 VCD from ULT-DOAS observations, complementary ground- and space-based measurements were used.

We showed that the tropospheric NO_2 VCD deduced from the ULT-DOAS observations are consistent with measurements performed from the ground using a zenith-sky car-DOAS system. Although two hours separated the two experiments, a correlation coefficient of R = 0.71 was found between the two results, a tropospheric NO_2 VCD of $(1.41 \pm 0.27) \times 10^{16}$ molecules/cm^2 and an estimated associated flux of $(3.45 \pm 0.89) \times 10^{-3}$ kg/s was measured close to the industrial area of Galati city on 13 August 2014, the only day that it was possible to determine the NO_2 flux.

Also, we showed that the stratospheric SCD derived from ground-based and airborne measurements correlates well with stratospheric NO_2 derived from observations by the OMI satellite sensor.

Based on this study, we conclude that the ULT is an efficient tool which allows determining with a high resolution the NO_2 distribution around urban or industrial sources. Also, the ULT-DOAS system is a very efficient tool for measuring fluxes due to its flexibility during the flights and the possibility to access remote areas. The ULT-DOAS system might also constitute a promising tool for satellite validation and calibration under clear-sky conditions, especially for upcoming high-resolution sensors such as the TROPOMI/Sentinel-5 Precursor instrument to be launched in summer 2017.

Acknowledgments: The work of D.E. Constantin was supported by Project PN-II-RU-TE-2014-4-2584, a grant of the Romanian National Authority for Scientific Research and Innovation, CNCS-UEFISCDI. The DOMINO data product was taken from the ESA TEMIS archive (www.temis.nl) maintained at KNMI, The Netherlands.

Author Contributions: Daniel-Eduard Constantin and Alexis Merlaud conceived and designed the study; Daniel-Eduard Constantin analyzed the data; and Mirela Voiculescu, Carmelia Dragomir, Lucian Georgescu, Gaia Pinardi, Francois Hendrick and Michel Van Roozendael improved the paper.

References

1. Solomon, S.; Portmann, R.W.; Sanders, R.W.; Daniel, J.S.; Madsen, W.; Bartram, B.; Dutton, E.G. On the role of nitrogen dioxide in the absorption of solar radiation. *J. Geophys. Res.* **1999**, *104*, 12047–12058. [CrossRef]
2. Platt, U. Differential optical absorption spectroscopy (DOAS). *Chem. Anal. Ser.* **1994**, *127*, 27–83.
3. Brewer, A.W.; McElroy, C.T.; Kerr, J.B. Nitrogen dioxide concentration in the atmosphere. *Nature* **1973**, *246*, 129–133. [CrossRef]
4. Noxon, J.F. Nitrogen dioxide in the stratosphere and troposphere measured by ground-based absorption spectroscopy. *Science* **1975**, *189*, 547–549. [CrossRef] [PubMed]
5. Hendrick, F.; Müller, J.-F.; Clémer, K.; Wang, P.; De Mazière, M.; Fayt, C.; Gielen, C.; Hermans, C.; Ma, J.Z.; Pinardi, G.; et al. Four years of ground-based MAX-DOAS observations of HONO and NO_2 in the Beijing area. *Atmos. Chem. Phys.* **2014**, *14*, 765–781. [CrossRef]
6. Constantin, D.-E.; Merlaud, A.; van Roozendael, M.; Voiculescu, M.; Fayt, C.; Hendrick, F.; Pinardi, G.; Georgescu, L. Measurements of Tropospheric NO_2 in Romania Using a Zenith-Sky Mobile DOAS System and Comparisons with Satellite Observations. *Sensors* **2013**, *13*, 3922–3940. [CrossRef]
7. Dragomir, C.; Constantin, D.-E.; Voiculescu, M.; Georgescu, L.; Merlaud, A.; Roozendael, M.V. Modeling results of atmospheric dispersion of NO_2 in an urban area using METI–LIS and comparison with coincident mobile DOAS measurements. *Atmos. Pollut. Res.* **2015**, *6*, 503–510. [CrossRef]
8. Merlaud, A.; van Roozendael, M.; van Gent, J.; Fayt, C.; Maes, J.; Toledo-Fuentes, X.; Ronveaux, O.; de Mazière, M. DOAS measurements of NO_2 from an ultralight aircraft during the Earth Challenge expedition. *Atmos. Meas. Tech.* **2012**, *5*, 2057–2068. [CrossRef]
9. Meier, A.C.; Schönhardt, A.; Bösch, T.; Richter, A.; Seyler, A.; Ruhtz, T.; Constantin, D.-E.; Shaiganfar, R.; Wagner, T.; Merlaud, A.; et al. High-resolution airborne imaging DOAS-measurements of NO_2 above Bucharest during AROMAT. *Atmos. Meas. Tech. Discuss.* **2016**. [CrossRef]
10. Tack, F.; Merlaud, A.; Iordache, M.-D.; Danckaert, T.; Yu, H.; Fayt, C.; Meuleman, K.; Deutsch, F.; Fierens, F.; van Roozendael, M. High resolution mapping of the NO_2 spatial distribution over Belgian urban areas based on airborne APEX remote sensing. *Atmos. Meas. Tech. Discuss.* **2017**. [CrossRef]
11. Merlaud, A.; Constantin, D.; Fayt, C.; Maes, J.; Mingireanu, F.; Mocanu, I.; Georgescu, L.; Roozendael, M. Small whiskbroom imager for atmospheric composition monitoring (SWING) from an unmanned areal vehicle (UAV). In Proceedings of the 21st ESA Symposium on European Rocket and Balloon Programmes and related Research, Thun, Switzerland, 9–13 June 2014; pp. 1–7.
12. Bovensmann, H.; Burrows, J.P.; Buchwitz, M.; Frerick, J.; Nöel, S.; Rozanov, V.V.; Chance, K.V.; Goede, A.H.P. SCIAMACHY—Mission objectives and measurement modes. *J. Atmos. Sci.* **1999**, *56*, 127–150. [CrossRef]
13. Levelt, P.; van den Oord, G.; Dobber, M.; Malkki, A.; Visser, H.; de Vries, J.; Stammes, P.; Lundell, J.; Saari, H. The ozone monitoring instrument. *IEEE T. Geosci. Remote.* **2006**, *44*, 1093–1101. [CrossRef]
14. Munro, R.; Eisinger, M.; Anderson, C.; Callies, J.; Corpaccioli, E.; Lang, R.; Lefebvre, A.; Livschitz, Y.; Albinana, A.P. GOME-2 on MetOp. In Proceedings of the 2006 EUMETSAT Meteorological Satellite Conference, Helsinki, Finland, 12–16 June 2006; p. 48.
15. van der A, R.J.; Eskes, H.J.; Boersma, K.F.; van Noije, T.P.C.; van Roozendael, M.; de Smedt, I.; Peters, D.H.M.U.; Meijer, E.W. Trends, seasonal variability and dominant NO_x source derived from a ten year record of NO_2 measured from space. *J. Geophys. Res.* **2008**, *113*, D04302. [CrossRef]
16. Varotsos, C.; Christodoulakis, J.; Tzanis, C.; Cracknell, A.P. Signature of tropospheric ozone and nitrogen dioxide from space: A case study for Athens, Greece. *Atmos. Environ.* **2014** *89*, 721–730. [CrossRef]

17. Junkermann, W. An ultralight aircraft as platform for research in the lower troposphere: System performance and first results from radiation transfer studies in stratiform aerosol layers and broken cloud conditions. *J. Atmos. Ocean. Technol.* **2001**, *18*, 934. [CrossRef]
18. Junkermann, W. On the distribution of formaldehyde in the western Po-Valley, Italy, during FORMAT 2002/2003. *Atmos. Chem. Phys.* **2009**, *9*, 9187–9196. [CrossRef]
19. Chazette, P.; Sanak, J.; Dulac, F. New Approach for Aerosol Profiling with a Lidar Onboard an Ultralight Aircraft: Application to the African Monsoon Multidisciplinary Analysis. *Environ. Sci. Technol.* **2007**, *41*, 8335–8341. [CrossRef] [PubMed]
20. Wang, P.; Richter, A.; Bruns, M.; Burrows, J.P.; Scheele, R.; Junkermann, W.; Heue, K.-P.; Wagner, T.; Platt, U.; Pundt, I. Airborne multi-axis DOAS measurements of tropospheric SO_2 plumes in the Po-valley, Italy. *Atmos. Chem. Phys.* **2006**, *6*, 329–338. [CrossRef]
21. Grutter, M.; Basaldud, R.; Rivera, C.; Harig, R.; Junkerman, W.; Caetano, E.; Delgado-Granados, H. SO_2 emissions from Popocatépetl volcano: Emission rates and plume imaging using optical remote sensing techniques. *Atmos. Chem. Phys.* **2008**, *8*, 6655–6663. [CrossRef]
22. General, S.; Pöhler, D.; Sihler, H.; Bobrowski, N.; Frieß, U.; Zielcke, J.; Horbanski, M.; Shepson, P.B.; Stirm, B.H.; Simpson, W.R.; et al. The Heidelberg Airborne Imaging DOAS Instrument (HAIDI)—A novel imaging DOAS device for 2-D and 3-D imaging of trace gases and aerosols. *Atmos. Meas. Tech.* **2014**, *7*, 3459–3485. [CrossRef]
23. Liu, L.; Flatøy, F.; Ordóñez, C.; Braathen, G.O.; Hak, C.; Junkermann, W.; Andreani-Aksoyoglu, S.; Mellqvist, J.; Galle, B.; Prévôt, A.S.H.; et al. Photochemical modelling in the Po basin with focus on formaldehyde and ozone. *Atmos. Chem. Phys.* **2007**, *7*, 121–137. [CrossRef]
24. Airborne Romanian Measurements of Aerosols and Trace Gases (AROMAT). Available online: http://uv-vis.aeronomie.be/aromat/index.php (accessed on 15 January 2017).
25. Danckaert, T.; Fayt, C.; van Roozendael, M.; de Smedt, I.; Letocart, V.; Merlaud, A.; Pinardi, G. QDOAS Software user manual Version 2.111-April 2016, UV-Visible DOAS Research Group of the Royal Belgian Institute for Space Aeronomy Web Site. Available online: http://uv-vis.aeronomie.be/software/QDOAS/QDOAS_manual.png (accessed on 15 December 2016).
26. Vandaele, A.; Hermans, C.; Simon, P.; Carleer, M.; Colin, R.; Fally, S.; Mérienne, F.; Jenouvrier, A.; Coquart, B. Measurements of the NO_2 absorption cross-section from 42000 cm^{-1} to 10000 cm^{-1} (238–1000 nm) at 220 K and 294 K (220 K). *J. Quant. Spectrosc. Radiat. Transf.* **1998**, *59*, 171–184. [CrossRef]
27. Bogumil, K.; Orphal, J.; Homann, T.; Voigt, S.; Spietz, P.; Fleischmann, O.C.; Vogel, A.; Hartmann, M.; Kromminga, H.; Bovensmann, H.; et al. Measurements of molecular absorption spectra with the SCIAMACHY pre-flight model: Instrument characterization and reference data for atmospheric remote-sensing in the 230–2380 nm region. *J. Photochem. Photobiol. A Chem.* **2003**, *157*, 167–184. [CrossRef]
28. Thalman, R.; Volkamer, R. Temperature dependent absorption cross-sections of O_2-O_2 collision pairs between 340 and 630 nm and at atmospherically relevant pressure. *Phys. Chem. Chem. Phys.* **2013**, *15*, 15371–15381. [CrossRef] [PubMed]
29. Chance, K.V.; Spurr, R.J.D. Ring effect studies: Rayleigh scattering, including molecular parameters for rotational Raman scattering, and the Fraunhofer spectrum. *Appl. Opt.* **1997**, *36*, 5224–5230. [CrossRef] [PubMed]
30. Baidar, S.; Oetjen, H.; Coburn, S.; Dix, B.; Ortega, I.; Sinreich, R.; Volkamer, R. The CU Airborne MAX-DOAS instrument: Vertical profiling of aerosol extinction and trace gases. *Atmos. Meas. Tech.* **2013**, *6*, 719–739. [CrossRef]
31. Vaughan, G.; Quinn, P.T.; Green, A.C.; Bean, J.; Roscoe, H.K.; van Roozendael, M.; Goutail, F. SAOZ measurements of stratospheric NO_2 at Aberystwyth. *J. Environ. Monit.* **2006**, *8*, 353–361. [CrossRef] [PubMed]
32. Dirksen, R.J.; Boersma, K.F.; Eskes, H.J.; Ionov, D.V.; Bucsela, E.J.; Levelt, P.F.; Kelder, H.M. Evaluation of stratospheric NO_2 retrieved from the Ozone Monitoring Instrument: Intercomparison, diurnal cycle and trending. *J. Geophys. Res.* **2011**, *116*, D08305. [CrossRef]
33. Hendrick, F.; Barret, B.; van Roozendael, M.; Boesch, H.; Butz, A.; De Mazière, M.; Goutail, F.; Hermans, C.; Lambert, J.-C.; Pfeilsticker, K.; et al. Retrieval of nitrogen dioxide stratospheric profiles from ground-based zenith-sky UV-visible observations: Validation of the technique through correlative comparisons. *Atmos. Chem. Phys.* **2004**, *4*, 2091–2106. [CrossRef]

34. Hendrick, F.; van Roozendael, M.; Kylling, A.; Petritoli, A.; Rozanov, A.; Sanghavi, S.; Schofield, R.; von Friedeburg, C.; Wagner, T.; Wittrock, F.; et al. Intercomparison exercise between different radiative transfer models used for the interpretation of ground-based zenith-sky and multi-axis DOAS observations. *Atmos. Chem. Phys.* **2006**, *6*, 93–108. [CrossRef]
35. Mayer, B.; Kylling, A. Technical note: The libRadtran software package for radiative transfer calculations—Description and examples of use. *Atmos. Chem. Phys.* **2005**, *5*, 1855–1877. [CrossRef]
36. Rivera, C.; Sosa, G.; Wöhrnschimmel, H.; de Foy, B.; Johansson, M.; Galle, B. Tula industrial complex (Mexico) emissions of SO_2 and NO_2 during the MCMA 2006 field campaign using a mobile mini-DOAS system. *Atmos. Chem. Phys.* **2009**, *9*, 6351–6361. [CrossRef]
37. Johansson, M.; Rivera, C.; de Foy, B.; Lei, W.; Song, J.; Zhang, Y.; Galle, B.; Molina, L. Mobile mini-DOAS measurement of the outflow of NO_2 and HCHO from Mexico City. *Atmos. Chem. Phys.* **2009**, *9*, 5647–5653. [CrossRef]
38. Stein, A.F.; Draxler, R.R.; Rolph, G.D.; Stunder, B.J.B.; Cohen, M.D.; Ngan, F. NOAA's HYSPLIT atmospheric transport and dispersion modeling system. *Bull. Am. Meteorol. Soc.* **2015**, *96*, 2059–2077. [CrossRef]
39. *Studiu Privind Calitatea Aerului in Municipiul Galati*; Municipiul Galati: Galati, Romania, 2016.
40. Frins, E.; Bobrowski, N.; Osorio, M.; Casaballe, N.; Belsterli, G.; Wagner, T.; Platt, U. Scanning and mobile multi-axis DOAS measurements of SO_2 and NO_2 emissions from an electric power plant in Montevideo, Uruguay. *Atmos. Environ.* **2014**, *98*, 347–356. [CrossRef]

2

Comparing CMAQ Forecasts with a Neural Network Forecast Model for $PM_{2.5}$ in New York

Samuel D. Lightstone *, Fred Moshary and Barry Gross

Optical Remote Sensing Lab, City College of New York, New York, NY 10031, USA; moshary@ccny.cuny.edu (F.M.); gross@ccny.cuny.edu (B.G.)

* Correspondence: slights01@citymail.cuny.edu

Abstract: Human health is strongly affected by the concentration of fine particulate matter ($PM_{2.5}$). The need to forecast unhealthy conditions has driven the development of Chemical Transport Models such as Community Multi-Scale Air Quality (CMAQ). These models attempt to simulate the complex dynamics of chemical transport by combined meteorology, emission inventories (EI's), and gas/particle chemistry and dynamics. Ultimately, the goal is to establish useful forecasts that could provide vulnerable members of the population with warnings. In the simplest utilization, any forecast should focus on next day pollution levels, and should be provided by the end of the business day (5 p.m. local). This paper explores the potential of different approaches in providing these forecasts. First, we assess the potential of CMAQ forecasts at the single grid cell level (12 km), and show that significant variability not encountered in the field measurements occurs. This observation motivates the exploration of other data driven approaches, in particular, a neural network (NN) approach. This approach makes use of meteorology and $PM_{2.5}$ observations as model predictors. We find that this approach generally results in a more accurate prediction of future pollution levels at the 12 km spatial resolution scale of CMAQ. Furthermore, we find that the NN is able to adjust to the sharp transitions encountered in pollution transported events, such as smoke plumes from forest fires, more accurately than CMAQ.

Keywords: air quality model; Air Quality System (AQS); Community Multi-Scale Air Quality (CMAQ) model; fine particulate matter ($PM_{2.5}$); Aerosol Optical Depth (AOD)

---

1. Introduction

Fine particulate matter air pollution ($PM_{2.5}$) is an important issue of public health, particularly for the elderly and young children. The study by Pope et al. suggests that exposure to high levels of $PM_{2.5}$ is an important risk factor for cardiopulmonary and lung cancer mortality [1,2]. Furthermore, increased risk of asthma, heart attack and heart failure have been linked to exposure to high $PM_{2.5}$ concentrations [3].

$PM_{2.5}$ levels are dynamic and can fluctuate dramatically over different time scales. In addition to local emission sources, pollution events can be the result of aerosol plume transport and intrusion into the lower troposphere. When there is a potential high pollution event, the local air quality agencies must alert the public, and advise the population on proper safety measures, as well as direct the reduction of emission producing activities. Therefore, accurately measuring and predicting fine particulate levels is crucial for public safety.

The U.S. Environmental Protection Agency (EPA) established the National Ambient Air Quality Standards (NAAQS), which regulate levels of pollutants such as fine particulate matter. The New York State Department of Environment Conservation (NYSDEC) operates ground stations for monitoring $PM_{2.5}$ and speciation throughout NY State [4]. However, surface sampling is expensive and existing networks are limited and sparse. This results in data gaps that can affect the ability to forecast $PM_{2.5}$ over a 24-h period. The EPA developed the Models-3 Community Multi-scale Air Quality system (CMAQ), to provide 24–48 h air quality forecasts. CMAQ provides an investigative tool to explore proper emission control strategies. CMAQ has been the standard for modeling air pollution for nearly two decades because of its ability to independently model different pollutants while describing the atmosphere using "first-principles" [5].

In their studies, McKeen et al. and Yu et al. evaluate the accuracy of CMAQ forecasts [6,7]. To do so, they use the CMAQ 1200 UTC (Version 4.4) forecast model. They observe the midnight-to-midnight local time forecast and compare the hourly and daily average forecasts to the ground monitoring stations. McKeen et al. [6] observed minimal diurnal variations of $PM_{2.5}$ at urban and suburban monitor locations, with a consistent decrease of PM values between 0100 and 0600 local time. However, the CMAQ model showed significant diurnal variations, leading McKeen et al. to conclude that aerosol loss during the late night and early morning hours has little effect on $PM_{2.5}$ concentrations, while the CMAQ model does not account for this. Therefore, in addition to testing the hourly CMAQ forecast for a 24-h period, we focus on the daytime window for two reasons: (1) to assess the accuracy of CMAQ when aerosols do not play a reduced roll in forecasting; (2) the forecast should predict the air quality during the time of maximum human exposure.

While these studies make a distinction between rural and urban locations, they take the average results for all rural and urban locations respectively; thereby, their assessment of the CMAQ model was as at a regional scale, rather than a localized one. In addition to regional emissions, these studies also considered extreme pollution events such as the wildfires in western Canada and Alaska, which occurred during the observation period for the studies by Yu et al. and McKeen et al. The results of this assessment concluded that due to insufficient representation of transport pollution associated with the burning of biomass, CMAQ significantly under predicted the $PM_{2.5}$ values for these events.

In the study by Huang et al. [8], the bias corrected CMAQ forecast was assessed for both the 0600 and 1200 UTC release times. The study revealed a general improvement of forecasting skill for the CMAQ model. However, it was observed that the bias correction was limited in predicting extreme events, such as wildfires, and new predictors must be included in the bias correction to predict these events. In this study, CMAQ was assessed as a regional forecasting tool, taking 551 sites, and evaluating the average results in six sub-regions.

In our present assessment of the current operational CMAQ forecast model (Version 4.6), we differ from the regional studies above in the following ways: Firstly, in addition to the 1200 UTC forecast, we evaluated the 0600 UTC forecast for the same period to determine if release time affects the CMAQ forecast. Second, we focused on specific locations, both rural and urban, to assess the potential of CMAQ as a localized forecasting tool. In addition, we revisited the forecast potential of CMAQ for high pollution events, to determine if these events are generally caused by transport, or by local emissions. Finally, we tailor the forecast comparisons to focus on the potential of providing next day forecasts using data prior to 5 p.m. of the previous day, since this is an operational requirement for the state environmental agencies.

In focusing on both rural and urban areas in New York State, previous studies have shown anomalies in $PM_{2.5}$ from CMAQ forecasts. For example, in [9], using CMAQ (Version 4.5) with various planetary boundary layer (PBL) parameterizations, $PM_{2.5}$ forecasts during the summer pre-dawn and post-sunset periods were often highly overestimated in New York City (NYC). Further analysis of these cases demonstrated that the most significant error was the retrieval of the PBL height, which was often compressed by the CMAQ model, and did not properly take into account the Urban Heat Island mechanisms that expand the PBL layer [10]. This study showed the importance of PBL height dynamics and meteorological factors that motivated the choice of meteorological forecast inputs used during the NN development.

The objective of this paper is to determine the best method to forecast $PM_{2.5}$ by direct comparison with CMAQ output products. In particular, using the CMAQ forecast model, as a baseline, we explore the performance of a NN based data driven approach with suitable meteorological and prior $PM_{2.5}$ input factors.

Paper Structure

Our present paper is organized in the following manner: In Section 2, we analyze CMAQ as the baseline forecaster. We briefly describe the CMAQ model and the forecast schedules that are publically available, as well as the relevant ground stations we use for comparison. We then describe and perform a number of statistical tests using both the direct, as well as the bias compensated, CMAQ outputs. In this section, we show the large dispersion in using the direct results without bias correction.

In Section 3, we present our NN data driven strategy. This includes a description of all the relevant input factors used, including a combination of present and predicted meteorology, as well as diurnal trends of prior $PM_{2.5}$ levels. We present our first statistical results for the comparisons between CMAQ and the NN for a variety of experiments in order to highlight the conditions in which the NN results are generally an improvement. Then we explore the forecast performance for high pollution multiday transport events, which result in the highest surface $PM_{2.5}$ levels during the observed time period. In this comparison, analyzed by combining a sequence of next-day forecasts together, we find that the neural network seems to follow the trends in $PM_{2.5}$ more accurately than the CMAQ model.

In Section 4, we summarize our results and describe potential improvements.

2. CMAQ Local and Regional Assessment

2.1. Datasets

2.1.1. Models

The CMAQ V4.6 (CB05 gas-phase chemistry) with 12 km horizontal resolution was used for this paper. The CMAQ product for meteorology predictions used is the North American Model Non-hydrostatic Multi-scale Model (NAM-NMMB). This version was made available starting February 2016. The CMAQ data used for this paper is from 1 February 2016 until 31 October 2016. The station names and locations are listed in Table A1. The data can be accessed from [11], and the model description can be found in [12,13].

The CMAQ model used has a few different configurations: release times of 0600 UTC and 1200 UTC, and each release time has a standard forecast as well as a bias corrected forecast. The analog ensemble method is used for bias corrections. The idea is to look at similar weather patterns for the forecast period, and statistically correct the numerical $PM_{2.5}$ forecast based on historical errors. The analog ensemble method is described in detail in Huang et al. [8]. For each release time, CMAQ provides a 48-h forecast. The release time of 0600 UTC and 1200 UTC (2 a.m. and 8 a.m. EDT) does not give the public enough time to react to the forecast on the same day as the release. Consequently, for the 0600 UTC release time, the forecast hours 22–45 were used, and for the release time of 1200 UTC the forecast hours of 16–39 were used. This allowed us to construct a complete 24-h diurnal period for the forecast time window, which facilitated comparison with the field station data.

2.1.2. Ground-Based Observations

$PM_{2.5}$ ground data is collected from the EPA's AirNow, which collects NYSDEC monitoring station measurements in real time. The station data used for the forecast experiments in this article are from the New York State stations listed in Table A2, from 1 January 2011 until 31 December 2016. To assess the accuracy of CMAQ model forecasts, matching the model to the ground monitoring station is necessary. To do this, we use the ground NYSDEC stations that lay within the CMAQ grid cell only. Ground stations that are not found in a CMAQ grid cell were not used for comparison; therefore, no spatial interpolation was done on the model results while mapping the model or meteorological data to the AirNow ground stations. This matching method is widely used for comparing the CMAQ model to ground monitoring stations [6,7,14]. The locational data-points are depicted in Figure A1, the CMAQ grid cell information can be found in Table A1, and the NYSDEC station information can be found in Table A2.

2.2. Methods

Assessing Accuracy of CMAQ Forecasting Models

The forecasting skill of the different models were evaluated by computing the R^2 and the root mean square (RMSE) values from a regression analysis comparing the model to the AirNow observations. High R^2 values and low RMSE values indicated a good match between the prediction and the observations. Finally, to directly assess potential biases in the regression assessment, residual plots are provided to show significant concentration bias.

2.3. Results

2.3.1. Effects of Bias and Release Time

Figure 1 shows the regression plots for the hourly CMAQ model output compared to the ground station data for the City College of New York (CCNY Station) to illustrate the general behavior of the CMAQ model, and how the forecast is affected by different forecast release times, and by the bias corrections applied. The results of the R^2 analysis for all ground stations can be found in the supplementary materials.

All forecasts from the CMAQ model over CCNY have a positive correlation to the ground data. The effect on the forecast for different release times, if any, is minimal.

As seen in Figure 1a,c, the standard model generally overestimates the ground. While the bias correction improves the over-prediction, the results are more dispersed. This can be verified from the fact that the bias correction decreases the root mean square error (RMSE), but it also decreases the R^2 value for both release times.

In Figure 1 we assess the overall skill for a 24-h CMAQ forecast. In Figure 2, we determine if the CMAQ model could be improved by simply moving the forecast release time to a later point in the day, thereby including the most up-to-date inputs in the model. To do this, we make a direct comparison between CMAQ forecasts with different release times. In Figure 2, the R^2 value is computed for each hour of the day. The release time of 0600 UTC, with forecast hours of 22–45, is compared to the 1200 UTC release time, with forecast hours 16–39, to determine if the lower number of forecast hours yields more accurate predictions. It is clear from Figure 2 that the later release time does not lead to a significant improvement in the accuracy of the forecast, and this is true for both urban and non-urban test sites.

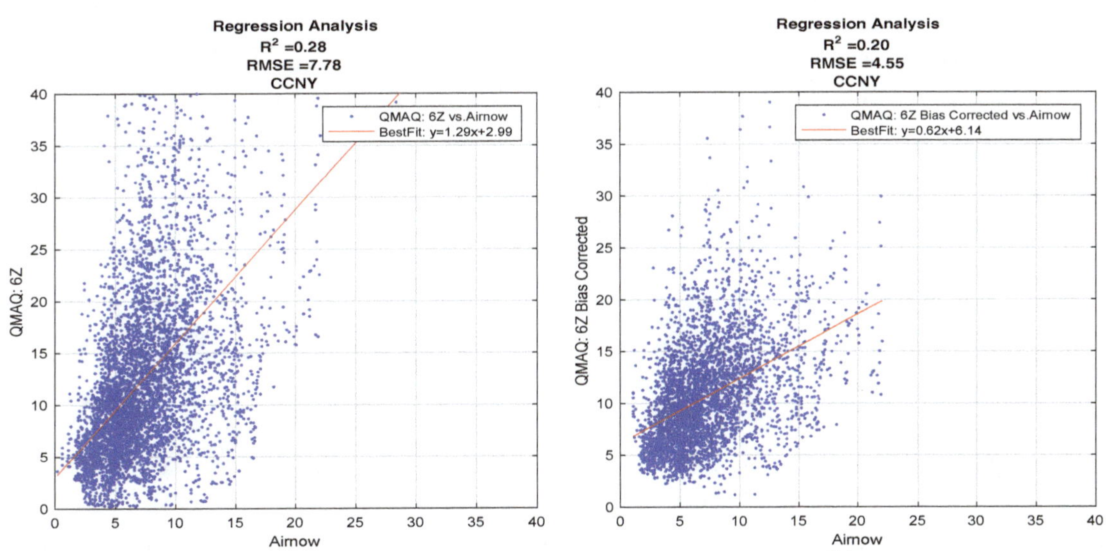

(a) (b)

Figure 1. *Cont.*

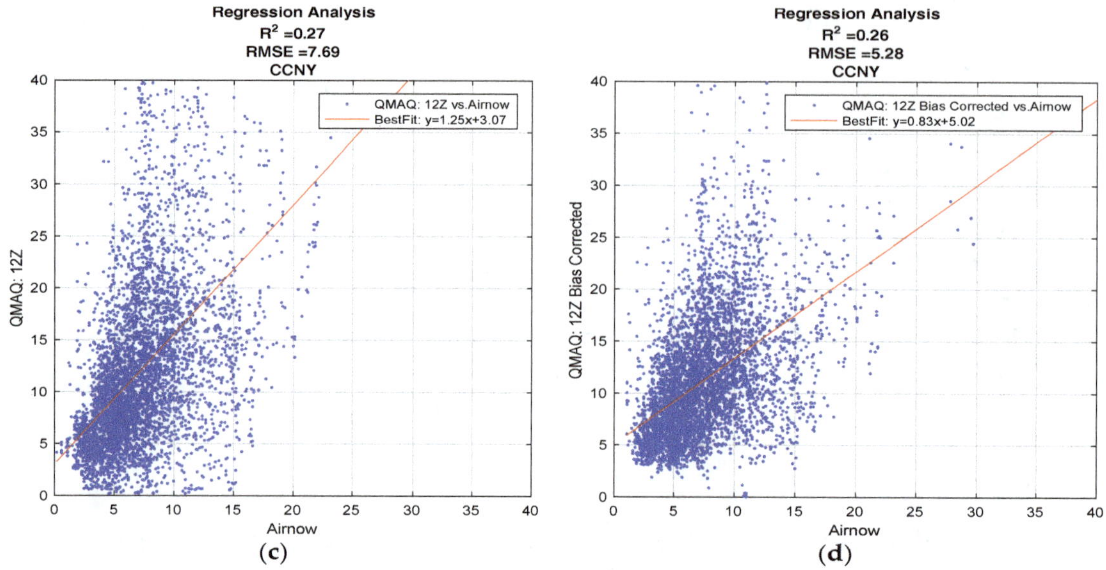

Figure 1. Community Multi-Scale Air Quality (CMAQ) regression analysis. (**a**) Standard, 06Z release time; (**b**) Bias Corrected, 06Z release time; (**c**) Standard, 12Z release time; (**d**) Bias Corrected, 12Z release time.

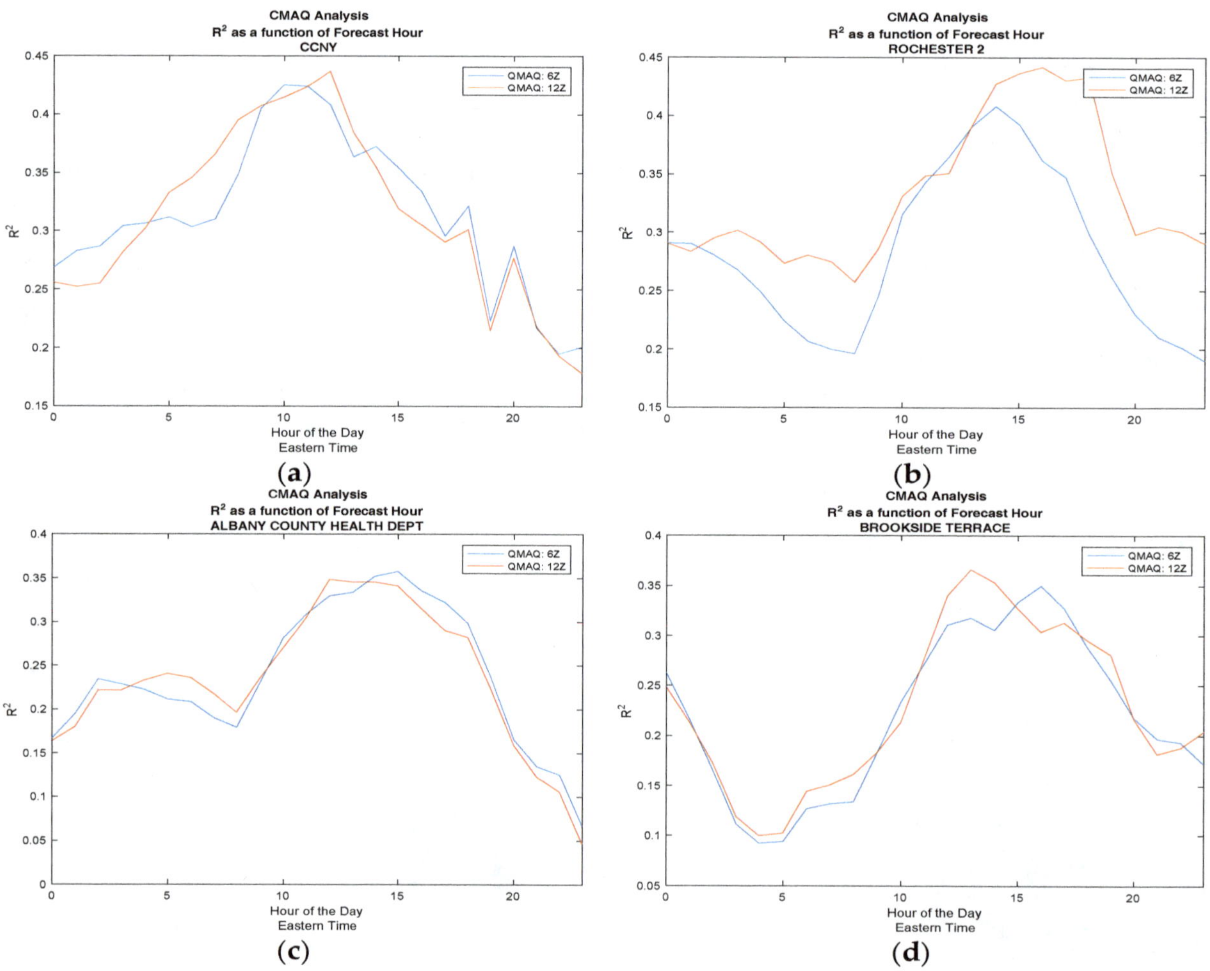

Figure 2. Comparing the effect of different release times for CMAQ by plotting the R^2 value as a function of time of day. (**a**) City College of New York (CCNY); (**b**) Rochester; (**c**) Albany; (**d**) Brookside Terrace.

It can be seen from this analysis that the CMAQ model performs best for midday hours, which is reasonable, since this is the period when convective mixing is most dominant. As discussed in reference [9], PBL modeling is very complex during the predawn/post-sunset period and errors in the PBL height clearly are a significant concern for further model development.

2.3.2. Differences between Urban and Non-Urban Locations

To get a better understanding of the spatial performance of the model, a multi-year time-series of daily averaged $PM_{2.5}$ observations from ground monitoring is used to compare the relationship between $PM_{2.5}$ values in New York City to the rest of New York State. Figure 3a is the regression analysis for this time period, and shows how the $PM_{2.5}$ values for NYC are strongly correlated to non-NYC areas, R^2 ~0.6. This indicates that while $PM_{2.5}$ values in NYC are generally higher than the rest of the state, the $PM_{2.5}$ level in NYC are still correlated to the levels in the rest of the state.

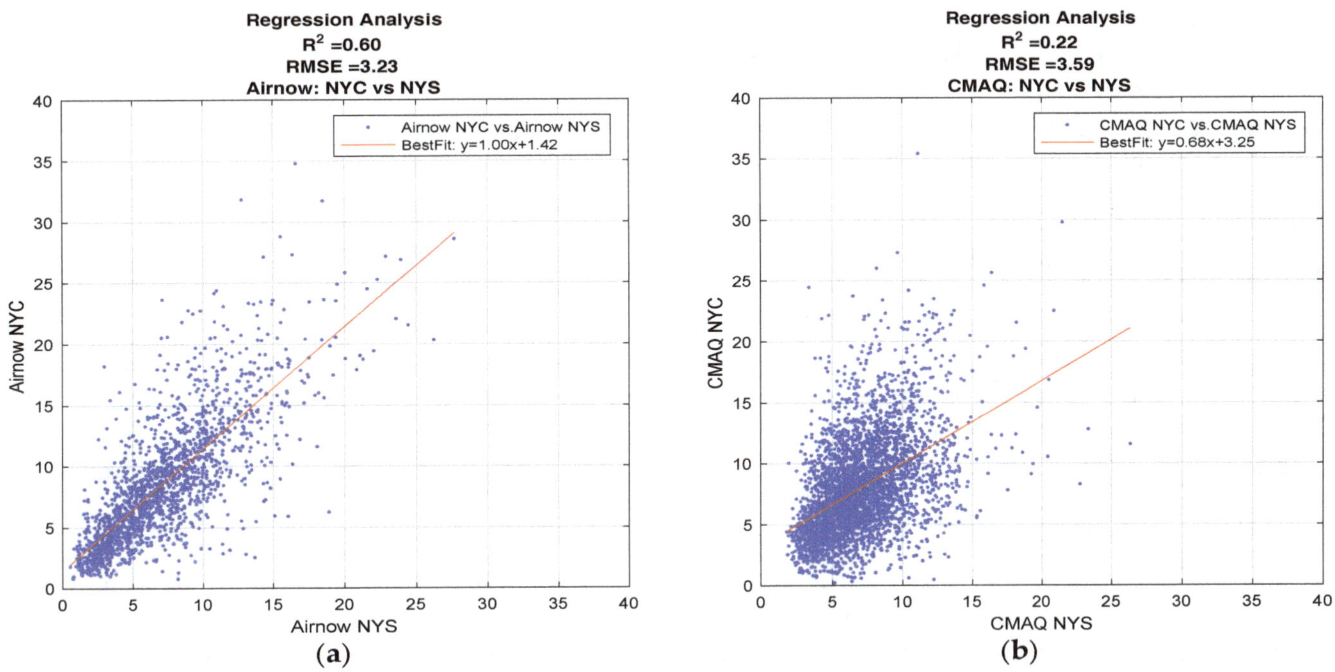

Figure 3. Regression analysis comparing $PM_{2.5}$ levels between NYC and the rest of NYS (non-NYC sites). (**a**) Multi-year day-averaged $PM_{2.5}$ analysis from NYSDEC ground observations; (**b**) CMAQ model comparison between NYC and NYS.

The same analysis comparing NYC to the rest of NYS was done with CMAQ forecast values as seen in Figure 3b. In this case, the correlation between NYC and NYS is not so strong, R^2 ~0.2. From this analysis alone, we can only speculate the reason for a low correlation between CMAQ forecasts for NYC and the rest of NYS is due to strong spatial differences in the National Emission Inventory (NEI) entries. However, the strong correlation in ground observations between NYC and NYS shows that while urban source emission may be a significant cause for somewhat higher levels of $PM_{2.5}$, there is still a strong correlation between NYC and NYS, and an accurate forecasting model must take this into account.

The limitations of CMAQ forecasting on a local pixel level indicate that other approaches should be explored. In particular, we explore the potential of data-driven models for localized forecasting in the next section.

3. Data Driven (Neural Network) Development

3.1. Datasets

3.1.1. Ground-Based Observations

$PM_{2.5}$ data collected from NYSDEC ground-monitoring stations is used for inputs in the neural network. These are the same ground stations listed above, in Section 2.1.2.

3.1.2. Models

The meteorological data was collected from the National Centers for Environmental Prediction (NCEP) North American Regional Reanalysis (NARR). NARR has high-resolution reanalysis of the North American region, 0.3 degrees (32 km) at the lowest latitude, including assimilated precipitation. The NARR makes available 8-times-daily and monthly means respectively. The data collected for this paper is the 8-times-daily means for the duration 1 January 2011 until 31 December 2016. Figure A1 shows the proximity of the meteorological data and the CMAQ model outputs to the ground stations.

The NN network was created and tested using historical data. In this paper, meteorology "forecast" data refers to NARR data that was observed the day of the $PM_{2.5}$ forecast. "Observed" or "measured" meteorology refers to NARR data that was observed before the forecast release time.

3.2. Methods

3.2.1. Development of the Neural Network

As stated above, the accurate prediction of $PM_{2.5}$ values is crucial for air quality agencies, so that they could alert the public of the severity and duration of a high pollution event. Therefore, it is imperative that the forecast predictions are released to the public the day before the event. For this paper, we chose 5 p.m. as a target for the forecast release time. Therefore, we ensure that all the methods tested, utilize factors that are available to the state agency prior to 2100 UTC (5 p.m. EDT).

Input Selection Scenarios

The NN input includes the following NARR meteorological data: surface air temperature, surface pressure, planetary boundary layer height (PBLH), relative humidity, and horizontal wind (10 m). To account for the seasonal variations, the month is also used as an input in the neural network. The PM input variables for the NN are the $PM_{2.5}$ measurements averaged over a three-hour frequency to match the meteorological dataset. The NN output is the next day $PM_{2.5}$ values.

In order to optimize the performance of the neural network, preliminary tests were done to determine the optimum utilization of the meteorological input variables. These test were done to determine if the "forecast" or the "observed" meteorology, or a combination of the two, should be used as input variables.

The forecast time window is midnight-to-midnight EDT for the forecast day, while the time window with the observed data is midnight to 5 p.m. EDT the day the forecast is released.

For the PBLH, the forecast value is always used as the input. One NN design employed only the forecast meteorological values as inputs. The second design utilized a combination of the forecast and the observed data, by subtracting the eight observation datasets from the eight forecast datasets. This first NN architecture uses the meteorological values as predictors, while the second design uses metrological trends as predictors. We note that this comparison does not affect the number of inputs used, allowing for a direct comparison of information content.

In scenario 1, where only the MET forecasts are used, we use the following inputs, where i represents the indices for time windows for the observation day, and j represents the indices for time windows for the forecast day (from the NARR forecasts), the NN inputs design is:

$PM_{2.5}(i)$	$i = 1:5$	$time\ window\ (i) = (i-1) \times 3\ : i \times 3$	(Field measurements)
$MET_{forecast}(j)$	$j = 1:8$	$time\ window\ (j) = (j-1) \times 3\ : j \times 3$	(NARR Forecasts)
$PBLH(j)$	$j = 1:8$	$time\ window\ (j) = (j-1) \times 3\ : j \times 3$	(NARR Forecasts)

In scenario 2, where the differential between the observation day and forecast day of the MET variables are used, the architecture for the NN inputs is:

$PM_{2.5}(i)$	$i = 1:5$	$time\ window\ (i) = (i-1) \times 3\ : i \times 3$	(Field measurements)
$MET_{forecast}(j) -$	$j = 1:8$	$time\ window\ (j) = (j-1) \times 3\ : j \times 3$	(NARR Forecasts)
$MET_{observed}(i)$	$i = 1:8$	$time\ window\ (i) = (i-1) \times 3\ : i \times 3$	(NARR Observations)
$PBLH(j)$	$j = 1:8$	$time\ window\ (j) = (j-1) \times 3\ : j \times 3$	(NARR Forecasts)

To show the robustness of the NN, the data used for training the neural networks came from 2011–2015 alone, while the network was tested with data from 2016. In both scenarios, the targets for the NN were taken to be the complete set of $PM_{2.5}$ over all time windows of the forecast day:

Targets: $PM_{2.5}(j)$	$j = 1:8$	$time\ window\ (j) = (j-1) \times 3\ : j \times 3$	(Field measurements)

Neural Network Training Approach

In developing a NN $PM_{2.5}$ forecast for all of New York State (NYS), we needed to take into account the very different emission sources, and to a lesser extent the meteorological conditions, between New York City (NYC) and the other sites in NYS. We found that the best solution is to design two different neural networks. The first is trained only over NYC sites, while the second is trained for the rest of NYS. It is important to note that we do not try to build a unique NN for every station, since this is not a useful approach for local agencies. PM and Meteorological data from 2011–2015, were used for training.

For NYC, since the stations are very close to each other, the NN was trained with spatial mean values of the ground PM monitors and NARR meteorological datasets. For NYS, all the PM and meteorological data from each site outside of NYC were used. Some site-specific information was implicitly included by using the surface pressure as inputs, which provides some indicator of surface elevation.

The neural network was developed using the MATLAB Neural Network Toolbox [15]. The Levenberg-Marquardt network was deployed using 10 hidden nodes. The break down for the NN input data is: 70% training, 15% validation, and 15% testing. Because the sample set of training, validation, and testing is divided randomly over the entire dataset, accuracy of the NN was determined by testing each network over 2016 data only, a time window that was not included in training. Once the NN function was created, the 2016 meteorological and PM data was passed through the network, and the outputs were stored with the date-time and station location as indices.

Neural Network Scenario Results

Figure 4a shows the performance of the NN using the forecast metrological data as inputs, while Figure 4b shows the performance of the NN using the difference between the forecast and the current days measurements. The NN utilizing the difference configuration is clearly better, with a higher R^2 value, 0.44 compared to 0.36, and a lower root-mean-square value, 3.09 compared to 4.59. In addition, there are substantially less anomalous high $PM_{2.5}$ forecasts. Since this improvement was seen in all test cases, we only used scenario 2, (differential meteorology) NN configuration. From these results, we see that meteorological trends are better indicators of $PM_{2.5}$ than meteorology alone. This appears to us to be a reasonable result since the meteorology trend better isolates particular mesoscale conditions, which is known to be a significant factor in boundary layer dynamics.

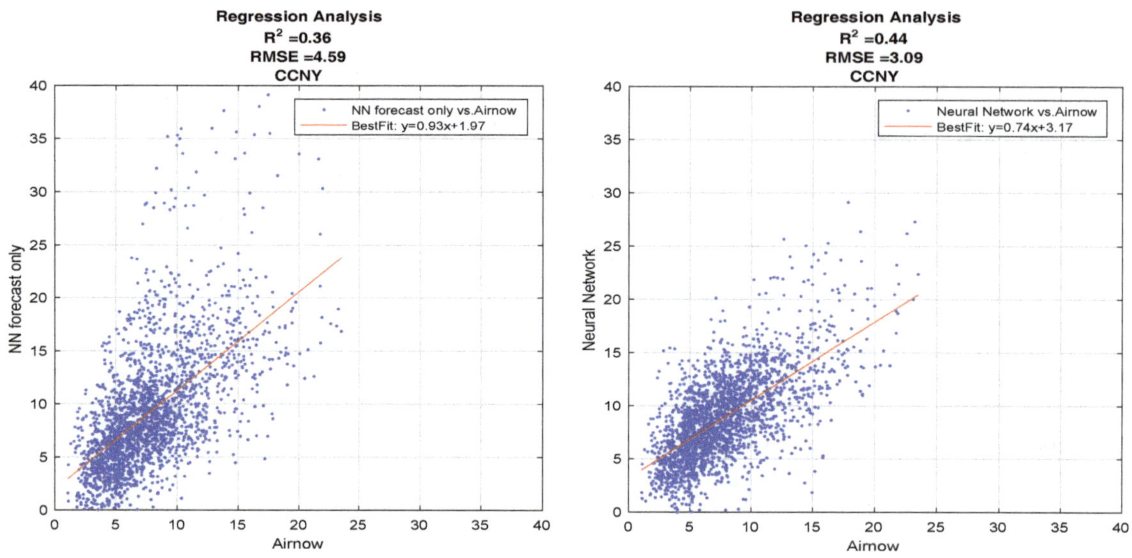

(a) (b)

Figure 4. Results from the regression analysis to maximize Neural Network performance for the different scenarios. (**a**) NN designed with the forecast meteorological data (Scenario 1); (**b**) NN designed by taking the difference between the forecast and the current days measurements (Scenario 2).

3.3. Results

3.3.1. Neural Network and CMAQ Comparison

The R^2 value for CMAQ and the NN, both compared to AirNow observations, is computed for each forecast model and for each location. As a representative example of the overall performance, the R^2 value for NYC, represented by CCNY, is compared to NYS, represented by Brookside Terrace, a non-NYC, non-urban station, and these results are displayed in Figure 5. The individual results for each location can be found in the supplementary materials.

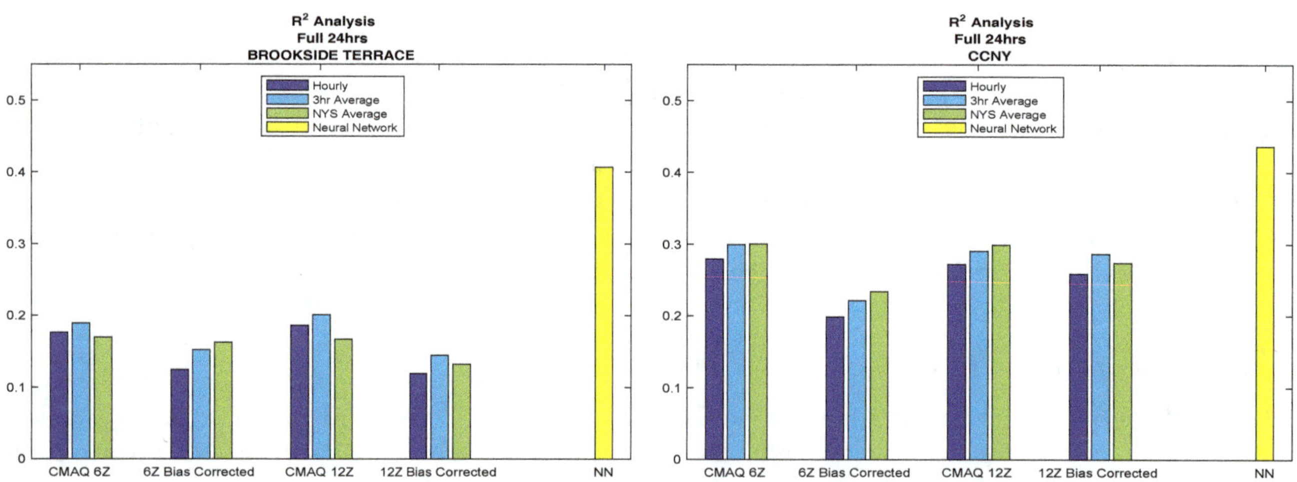

(a) (b)

Figure 5. Regression analysis is computed for the comparison between AirNow observations and the various prediction models, for the complete CMAQ database time period (February 2016, through October 2016). The R^2 value for each model is plotted in the figure above to compare CMAQ to the NN. The CMAQ model includes the different release times as well as bias compensated vs. uncompensated runs. In addition, different time and spatial averaging of CMAQ is considered at each location. (**a**) Brookside Terrace, representative of non-NYC; (**b**) CCNY, representative of NYC.

From Figure 5 above, it can be seen that the most accurate forecast model is the neural network for both NYS and NYC over any of the CMAQ forecasts studied. Regarding CMAQ, we note better performance for NYC than for non-urban areas. This is in contrast to the neural network, where there is very little variation in the results for locations that are urban versus non-urban, indicating that locational inputs in the model, such as the surface pressure, improves forecasting skill.

In addition, for all cases, it can be seen that taking the time average improves the CMAQ results. Furthermore, the spatial averaging over NYS (with 1-h time sampling) shows more improvement in most NYC cases and some non-NYC cases as well. These results indicate the possibility that the best use for CMAQ forecasting is on a regional level. This is supported from the 12 km grid cell resolution for CMAQ, a cell size typical for regional analysis.

We note again that the different release times for CMAQ has almost no effect on the forecast accuracy. In Figure 6, we compared the diurnal performance of the NN to the CMAQ model. The most apparent result is the dramatic improvement of the NN during the night and morning hours, where the CMAQ model has the most difficulty. This is clearly due to the machine learning approach where the time differences, the inputs, and forecast periods have a dramatic effect on output performance.

This also explains the general downward trend, where performance tails off in the late afternoon and becomes closer to the CMAQ performance. This can be expected, because larger time delays should lead to more dispersion between the outputs and input PM levels.

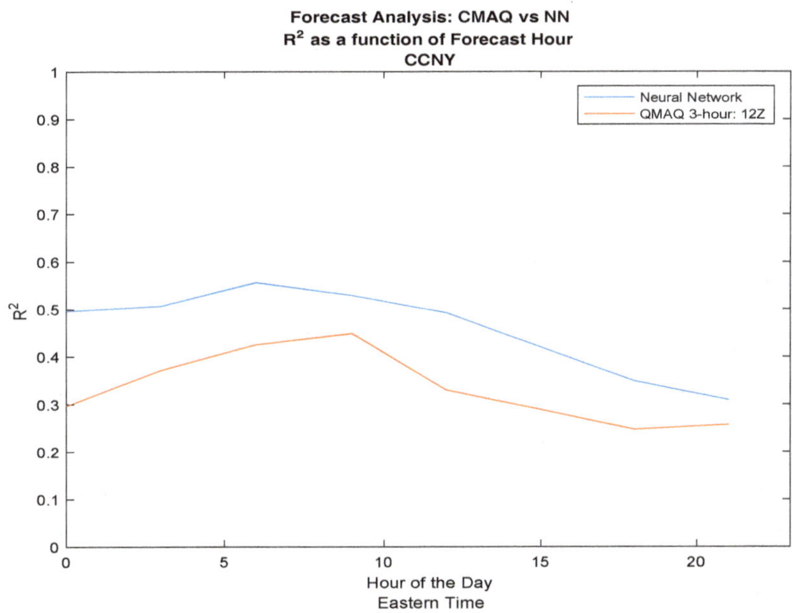

Figure 6. Comparing the effect of different release times for the NN in comparison to CMAQ by plotting the R^2 value as a function of time of day.

Figure 7 below shows the residual results for CMAQ in comparison to the neural network. For CMAQ, as noted above, there seems to exist a non-random bias pattern, where CMAQ generally over predicts for low and high PM values, and under predicts for medium values. This pattern seems to indicate that the CMAQ model may not capture all of the underlying variability factors. On the other hand, for the neural network, the behavior of the residuals is clearly stochastic in nature.

We find that an optimized NN approach generally results in a more accurate prediction of future pollution levels, as compared to CMAQ, for a single grid cell (resolution 12 km).

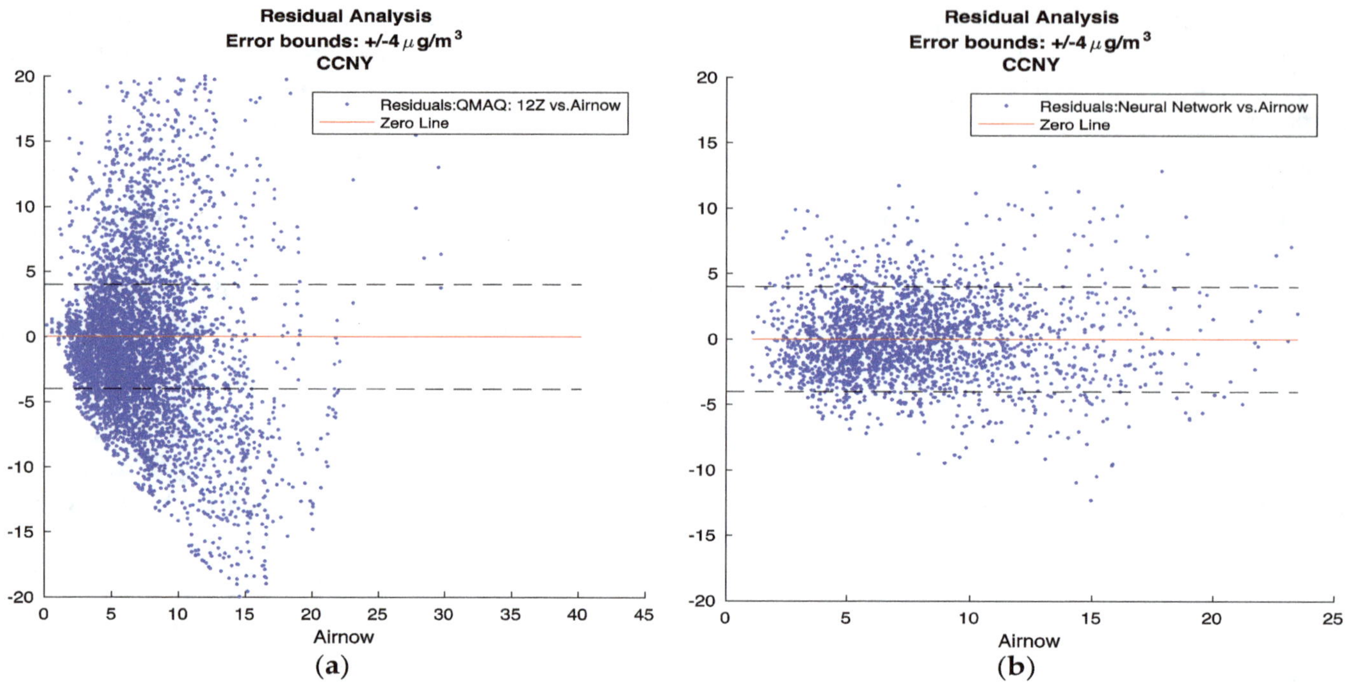

Figure 7. Residual analysis. The standard deviation of $PM_{2.5}$ from the AirNow ground monitoring sites was calculated to be 4 μg/m^3, therefore, +/−4 μg/m^3 was used for the error bounds (**a**) CMAQ; (**b**) Neural Network.

3.3.2. Heavy Pollution Transport Events

Because the neural network is data-driven, the network performs better when the most up-to-date inputs are used. This explains the degradation of performance with time, as seen in Figure 6. In the current design of the neural network, we only used five $PM_{2.5}$ inputs, instead of maximum possible in a 24-h period, eight. In the training of the NN, there were very few extreme event cases, $PM_{2.5} > 25$ μg/m^3. The lack of suitable training statistics for these events causes the NN approach to have difficultly in adjusting to the sharp contrast with the onset of the event.

Therefore, a second neural network was trained with the same design as the neural network illustrated above; however, this neural network produces a 24-h forecast at 5 p.m. for the time period, 5 p.m.–5 p.m. (instead of a next day 24-h midnight-to-midnight forecast). This neural network uses all eight PM measurements, because there is no lag time between the release time and the first forecast hour. This neural network, referred to as NN Continuous, was not used in the statistical analysis for the different forecast models (because the 24-h forecast period is different than the forecast analysis above), but is being explored in the extreme event cases. The reason for developing this continuous neural network is to determine if the continuous nature of the network produces better results in extreme pollution events.

To explore the behavior of the different models under high pollution transport conditions, the forecasts coinciding with the wildfires of Fort McMurray in Alberta, Canada were analyzed. The wildfire started on 1 May 2016, and was declared under control on 5 July 2016. Although the wildfire lasted for over two months, evidence of increased $PM_{2.5}$ surface levels in NYC resulting from the wildfire were detected on 9 May, and on 25 May. On these dates, instances of aloft plume intrusions and the mixing down into the planetary-boundary layer were observed by a ceilometer and a Raman-Mie Lidar [16]. In Figure 8, we plot the CMAQ and NN model forecasts, focusing on the transport intrusions into NYC on 25 May.

The first thing to notice in Figure 8a, is the oscillations in the CMAQ model, and to notice how these oscillations smooth out in Figure 8b,c, where the three-hour time average and the New York State spatial average are tested respectively. It is logical that for heavy transport cases, domain averaging helps decrease oscillations; however, we still see significant underestimation of the event.

This is the first case where we analyze the behavior of the continuous neural network. Looking at Figure 8c,d, it is clear that the continuous neural network is able to respond to the trend of the high pollution event faster, and more accurately, then the standard neural network.

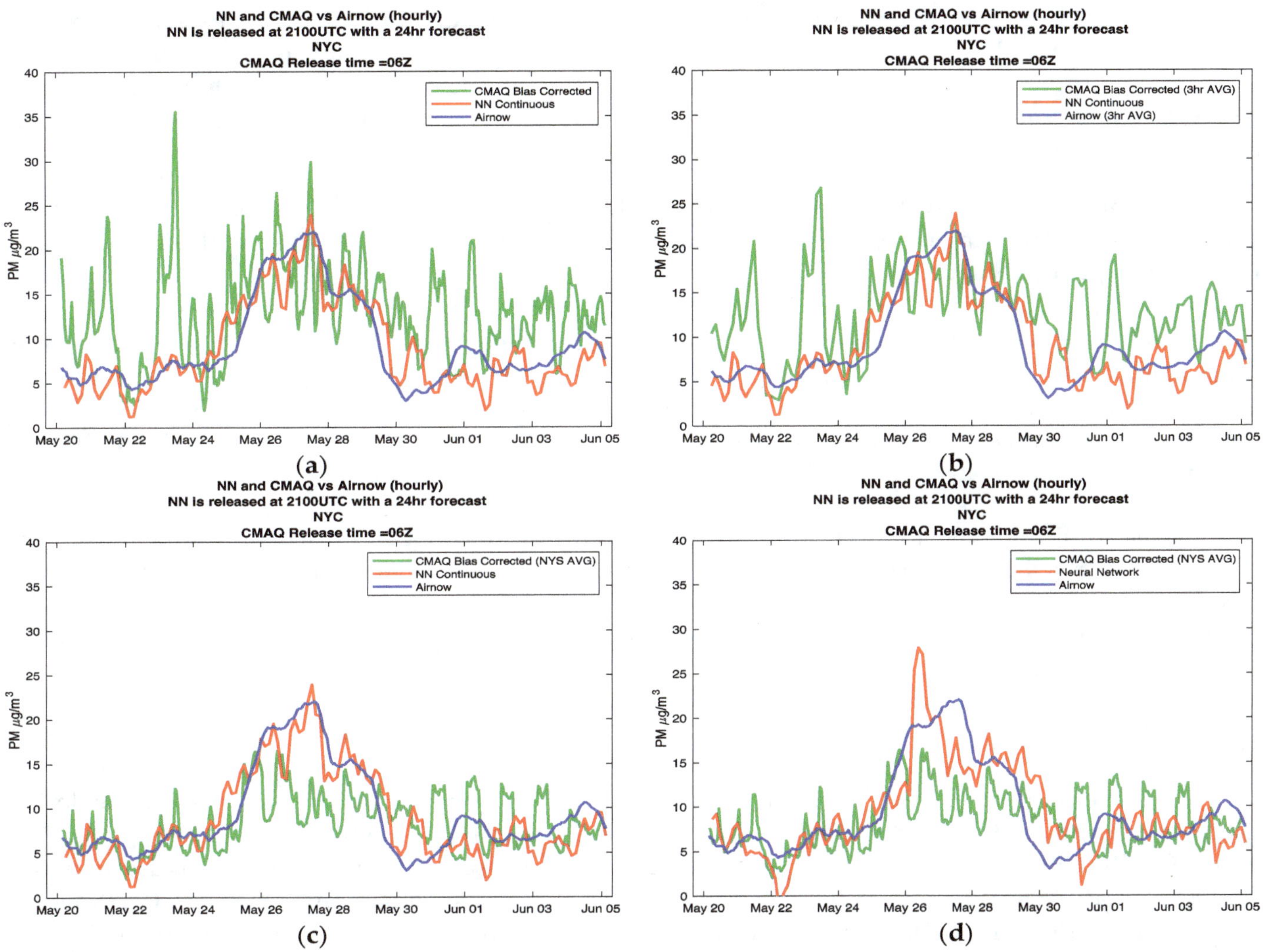

Figure 8. NYC surface $PM_{2.5}$ levels affected from the wildfires of Alberta Canada 2016. The plots focus on the aloft plumes mixing down into the PBL on May 25. The plots show different models vs. AirNow observation (**a**) CMAQ Biased hourly, NN continuous; (**b**) CMAQ Biased 3-h average, NN continuous; (**c**) CMAQ Biased state average, NN continuous; (**d**) CMAQ Biased state average, standard Neural Network.

4. Conclusions

In this paper, we first made a baseline assessment of the V4.6 CMAQ forecasts, and found significant dispersion as well as a tendency for the model to overestimate the ground truth field measurements. Even in the bias corrected case, the residuals error in the model was found to have significant bias patterns, indicating that there are predictors not included in the model that could significantly improve the results.

These results motivated the development of data driven approaches such as a NN. In developing a data driven NN next day forecast model, we found a general improvement of performance when using prior $PM_{2.5}$ inputs together with the difference between present and next day meteorological parameter forecasts. This "differential NN" approach performed significantly better than if we used only the future forecast variables, indicating that meteorological pattern trends are important indicators.

Using this NN architecture, we then made extensive regression based comparisons between CMAQ next day forecast models and regionally trained NN next day forecasts for the NYS and NYC regions. In general, we found that the NN results are a significant improvement over the CMAQ forecasts in all cases. These comparisons were made to be consistent with state agencies where forecasts should be available by 5 p.m. In addition, we also made a diurnal comparison, which illustrated that; the NN approach had superior forecasting skills during the early part of the day but degraded smoothly as the forecast time increased. By mid-day, the differences between the two approaches was much closer.

To improve the CMAQ forecasts, we found limited improvement when spatial averaging is extended beyond the single pixel 12 km resolution to all of New York State. Even in this case, the NN results were generally more accurate.

Finally, we focused on forecast performance for transported high pollution events such as Canadian wildfires. In these cases, we found that the CMAQ forecasts had large temporal fluctuations, which could hide most of the event. In this case, significant improvement was obtained when using state averaged bias corrected outputs; however, in general, the smoothed results underestimate the local $PM_{2.5}$ measurements.

In this application, we found the neural network approach provides a reasonably smooth forecast, although the transition from a clean state to a polluted state is very poor. Nevertheless, the standard NN performed better than CMAQ in this scenario. Further improved results for the NN were obtained in the transition period when the forecast time of the NN was reduced (NN continuous), making the transition from training to testing continuous.

Future Work

While the continuous NN does adjust quickly to the sharp contrast in transport events, this design limits the scope of the forecast period. Clearly, local data alone is not ideal for this application. Non-local data that can identify high pollution events and assesses their potential mixing with our region is needed. As a preliminary analysis, we explored the use of a combination of HYSPLIT Air Parcel Trajectories with GOES satellite Aerosol Optical Depth (AOD) retrievals to improve the NN. In particular, we analyzed the use of these tools to quantify the relative AOD levels for all air parcels that reach our target area. We found that by properly counting the trajectories weighted by the AOD, a good correlation was seen between the relative AOD and the $PM_{2.5}$ levels. Therefore, we believe that using the relative AOD metric as an additional input factor can make improvements in the NN approach. When GOES-R AOD retrievals, with high data latency and multispectral inversion capabilities [17,18], become available, we plan to incorporate these AOD metrics as predictors in the NN.

Acknowledgments: The authors gratefully acknowledge the NOAA Air Resources Laboratory (ARL) for the provision of the HYSPLIT transport and dispersion model used in this publication. We further gratefully acknowledge Jeff McQueen for providing the V4.6 CMAQ forecasts. Finally, S. Lightstone would like to acknowledge partial support of this project through a NYSERDA EMEP Fellowship award.

Author Contributions: Samuel Lightstone was responsible for the NN experiments and the CMAQ assessment; Lightstone also wrote the paper. Barry Gross was responsible for design of the different experiments as well as conceiving the use of the GOES AOD and HYSPLIT trajectories to estimate transported AOD as a predictor of $PM_{2.5}$. Fred Moshary provided significant critical assessment and suggestions of all results.

Appendix A. Datasets

Table A1. CMAQ Grid Cell Information.

Name	Abbreviation	Latitude	Longitude	Land Type
Amherst	AMHT	42.99	−78.77	Suburban
CCNY	CCNY	40.82	−73.95	Urban
Holtsville	HOLT	40.83	−73.06	Suburban
IS 52	IS52	40.82	−73.90	Suburban
Loudonville	LOUD	42.68	−73.76	Urban
Queens College 2	QC2	40.74	−73.82	Suburban
Rochester Pri 2	RCH2	43.15	−77.55	Urban
Rockland County	RCKL	41.18	−74.03	Rural
S. Wagner HS	WGHS	40.60	−74.13	Urban
White Plains	WHPL	41.05	−73.76	Suburban

Table A2. NYSDEC Station Information.

NYSDEC ID	Station Name	Latitude	Longitude	Land Type
360010005	Albany County Health Dept	42.6423	−73.7546	Urban
360050112	IS 74	40.8155	−73.8855	Suburban
360291014	Brookside Terrace	42.9211	−78.7653	Suburban
360551007	Rochester 2	43.1462	−77.5482	Urban
360610135	CCNY	40.8198	−73.9483	Urban
360810120	Maspeth Library	40.7270	−73.8931	Suburban
360850055	Freshkills West	40.5802	−74.1983	Suburban
360870005	Rockland County	41.1821	−74.0282	Rural
361030009	Holtsville	40.8280	−73.0575	Suburban
361192004	White Plains	41.0519	−73.7637	Suburban

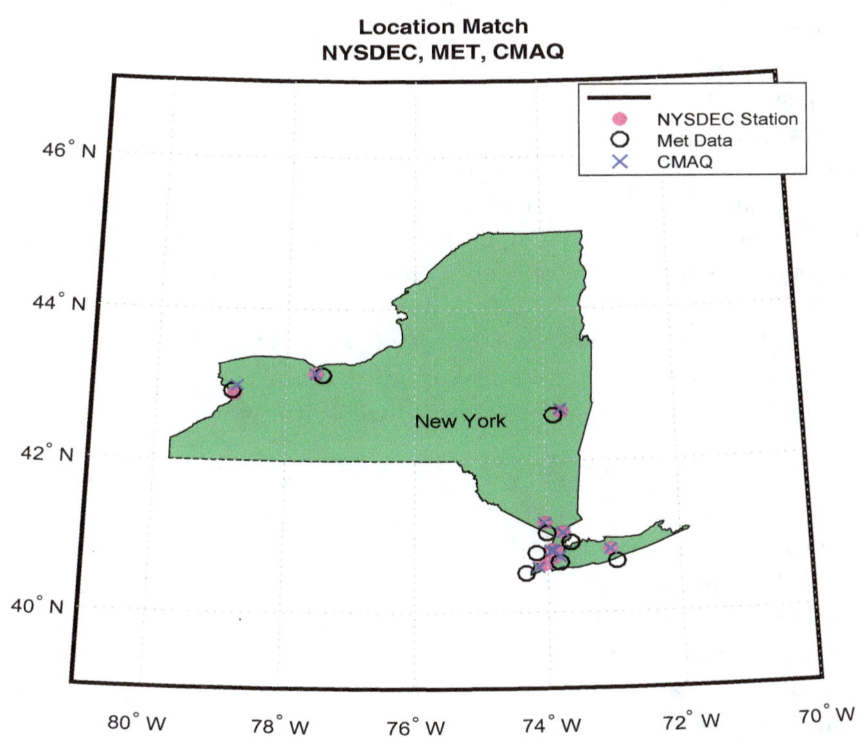

Figure A1. This map shows the proximity of the ground NYSDEC stations to the NARR meteorological data, and the CMAQ forecast data.

References

1. Pope, C.A., III; Dockery, D.W. Health effects of fine particulate air pollution: Lines that connect. *J. Air Waste Manag. Assoc.* **2006**, *56*, 709–742. [CrossRef] [PubMed]
2. Pope, C.A., III; Burnett, R.T.; Thun, M.J.; Calle, E.E.; Krewski, D.; Ito, K.; Thurston, G.D. Lung cancer, cardiopulmonary mortality, and long-term exposure to fine particulate air pollution. *JAMA* **2002**, *287*, 1132–1141. [CrossRef] [PubMed]
3. Weber, S.A.; Insaf, T.Z.; Hall, E.S.; Talbot, T.O.; Huff, A.K. Assessing the impact of fine particulate matter ($PM_{2.5}$) on respiratory-cardiovascular chronic diseases in the New York City Metropolitan area using Hierarchical Bayesian Model estimates. *Environ. Res.* **2016**, *151*, 399–409. [CrossRef] [PubMed]
4. Rattigan, O.V.; Felton, H.D.; Bae, M.; Schwab, J.J.; Demerjian, K.L. Multi-year hourly $PM_{2.5}$ carbon measurements in New York: Diurnal, day of week and seasonal patterns. *Atmos. Environ.* **2010**, *44*, 2043–2053. [CrossRef]
5. Byun, D.; Schere, K.L. Review of the governing equations, computational algorithms, and other components of the Models-3 Community Multiscale Air Quality (CMAQ) modeling system. *Appl. Mech. Rev.* **2006**, *59*, 51–77. [CrossRef]
6. McKeen, S.; Chung, S.H.; Wilczak, J.; Grell, G.; Djalalova, I.; Peckham, S.; Gong, W.; Bouchet, V.; Moffet, R.; Tang, Y.; et al. Evaluation of several $PM_{2.5}$ forecast models using data collected during the ICARTT/NEAQS 2004 field study. *J. Geophys. Res. Atmos.* **2007**, *112*. [CrossRef]
7. Yu, S.; Mathur, R.; Schere, K.; Kang, D.; Pleim, J.; Young, J.; Tong, D.; Pouliot, G.; McKeen, S.A.; Rao, S.T. Evaluation of real-time $PM_{2.5}$ forecasts and process analysis for $PM_{2.5}$ formation over the eastern United States using the Eta-CMAQ forecast model during the 2004 ICARTT study. *J. Geophys. Res. Atmos.* **2008**, *113*. [CrossRef]
8. Huang, J.; McQueen, J.; Wilczak, J.; Djalalova, I.; Stajner, I.; Shafran, P.; Allured, D.; Lee, P.; Pan, L.; Tong, D.; et al. Improving NOAA NAQFC $PM_{2.5}$ Predictions with a Bias Correction Approach. *Weather Forecast.* **2017**, *32*, 407–421. [CrossRef]
9. Doraiswamy, P.; Hogrefe, C.; Hao, W.; Civerolo, K.; Ku, J.Y.; Sistla, G. A retrospective comparison of model-based forecasted $PM_{2.5}$ concentrations with measurements. *J. Air Waste Manag. Assoc.* **2010**, *60*, 1293–1308. [CrossRef] [PubMed]
10. Gan, C.-M.; Wu, Y.; Madhavan, B.L.; Gross, B.; Moshary, F. Application of active optical sensors to probe the vertical structure of the urban boundary layer and assess anomalies in air quality model $PM_{2.5}$ forecasts. *Atmos. Environ.* **2011**, *45*, 6613–6621. [CrossRef]
11. Files in /mmb/aq/sv/grib. Available online: http://www.emc.ncep.noaa.gov/mmb/aq/sv/grib/ (accessed on 1 December 2016).
12. NCEP Operational Air Quality Forecast Change Log. Available online: http://www.emc.ncep.noaa.gov/mmb/aq/AQChangelog.html (accessed on 1 May 2017).
13. Lee, P.; McQueen, J.; Stajner, I.; Huang, J.; Pan, L.; Tong, D.; Kim, H.; Tang, Y.; Kondragunta, S.; Ruminski, M.; et al. NAQFC developmental forecast guidance for fine particulate matter ($PM_{2.5}$). *Weather Forecast.* **2017**, *32*, 343–360. [CrossRef]
14. McKeen, S.; Grell, G.; Peckham, S.; Wilczak, J.; Djalalova, I.; Hsie, E.; Frost, G.; Peischl, J.; Schwarz, J.; Spackman, R.; et al. An evaluation of real-time air quality forecasts and their urban emissions over eastern Texas during the summer of 2006 Second Texas Air Quality Study field study. *J. Geophys. Res. Atmos.* **2009**, *114*. [CrossRef]
15. Demuth, H.; Beale, M. *MATLAB and Neural Network Toolbox Release 2016a*; The MathWorks, Inc.: Natick, MA, USA, 1998.
16. Wu, Y.; Pena, W.; Diaz, A.; Gross, B.; Moshary, F. Wildfire Smoke Transport and Impact on Air Quality Observed by a Multi-Wavelength Elastic-Raman Lidar and Ceilometer in New York City. In Proceedings of the 28th International Laser Radar Conference, Bucharest, Romania, 25–30 June 2017.
17. Anderson, W.; Krimchansky, A.; Birmingham, M.; Lombardi, M. The Geostationary Operational Satellite R Series SpaceWire Based Data System. In Proceedings of the 2016 International SpaceWire Conference (SpaceWire), Yokohama, Japan, 25–27 October 2016.
18. Lebair, W.; Rollins, C.; Kline, J.; Todirita, M.; Kronenwetter, J. Post launch calibration and testing of the Advanced Baseline Imager on the GOES-R satellite. In Proceedings of the SPIE Asia-Pacific Remote Sensing,

3

Transport Pathways and Potential Source Region Contributions of PM 2.5 in Weifang: Seasonal Variations

Chengming Li, Zhaoxin Dai *, Xiaoli Liu and Pengda Wu

Chinese Academy of Surveying and Mapping, Beijing 100830, China; cmli@casm.ac.cn (C.L.); liuxl@casm.ac.cn (X.L.); wupd@casm.ac.cn (P.W.)

* Correspondence: daizx@lreis.ac.cn

Abstract: As air pollution becomes progressively more serious, accurate identification of urban air pollution characteristics and associated pollutant transport mechanisms helps to effectively control and alleviate air pollution. This paper investigates the pollution characteristics, transport pathways, and potential sources of $PM_{2.5}$ in Weifang based on $PM_{2.5}$ monitoring data from 2015 to 2016 using three methods: Hybrid Single-Particle Lagrangian Integrated Trajectory (HYSPLIT), the potential source contribution function (PSCF), and concentration weighted trajectory (CWT). The results show the following: (1) Air pollution in Weifang was severe from 2015 to 2016, and the annual average $PM_{2.5}$ concentration was more than twice the national air quality second-level standard (35 $\mu g/m^3$). (2) Seasonal transport pathways of $PM_{2.5}$ vary significantly: in winter, spring and autumn, airflow from the northwest and north directions accounts for a large proportion; in contrast, in summer, warm-humid airflows from the ocean in the southeastern direction dominate with scattered characteristics. (3) The PSCF and CWT results share generally similar characteristics in the seasonal distributions of source areas, which demonstrate the credibility and accuracy of the analysis results. (4) More attention should be paid to short-distance transport from the surrounding areas of Weifang, and a joint pollution prevention and control mechanism is critical for controlling regional pollution.

Keywords: $PM_{2.5}$; spatiotemporal characteristics; back-trajectory clustering; potential source contribution; concentration weighted trajectory

---

1. Introduction

With rapid socioeconomic development, accelerated industrialization, and continuously increasing energy consumption, particulates have become major urban air pollutants in China [1–3]. In particular, fine particulates $PM_{2.5}$ can not only reduce atmospheric visibility but also increase mortality and incidence of diseases such as respiratory diseases [4–6], and these particulates have aroused great public concern and attention. Studies have indicated that urban air pollution levels and their spatiotemporal distributions are not only associated with local emissions but also influenced by cross-regional transport from sources in surrounding areas [7–9]. Accurate identification of urban atmospheric pollution characteristics and transport mechanisms is critical to control and mitigate air pollution [10].

Existing studies on the spatiotemporal characteristics of $PM_{2.5}$ and transport mechanisms have mostly focused on geo-statistics analysis [11], air quality models (e.g., large-scale Weather Research and Forecasting (WRF) model) [12], and back-trajectory clustering-based mechanism analysis. For back-trajectory clustering-based transport analysis, previous research has mostly focused on hotspots [13,14], while there is limited research on small but seriously polluted cities in China.

Li et al. (2018, 2019) indicated that $PM_{2.5}$ concentration and its influencing factors varied in different seasons, and they also preliminarily showed that the air pollution in Weifang City was affected by meteorological conditions and pollution of surrounding cities [6,15]. The above-mentioned studies are based on qualitative analysis. To quantitatively analyze the cross-regional influences of Weifang in different seasons, this paper describes an in-depth study of the seasonal variations in internal and external potential sources and provides targeted pollution control measures for government over different seasons. In addition, to improve the credibility and accuracy of the results, different trajectory analysis methods are used to analyze the potential sources of Weifang $PM_{2.5}$ in this paper, which can be used to evaluate the results through mutual verification.

Therefore, based on hourly high-frequency $PM_{2.5}$ data from 38 provincial monitoring stations in Weifang from 2015 to 2016, this paper examines the transport characteristics and potential sources of $PM_{2.5}$ in Weifang City in different seasons by using a back-trajectory clustering method. This research helps to better understand the causes of $PM_{2.5}$ pollution and the sources in Weifang as well as provides important scientific references for joint atmospheric pollution control in Weifang City.

This paper includes five sections. Section 2 summarizes relevant studies on pollutant transport mechanisms and potential sources. Section 3 describes the data sources and methods used. The analytical results of the $PM_{2.5}$ pollution characteristics and potential source contributions in different seasons are presented in Section 4, followed by the conclusions in Section 5.

2. Literature Review

The causes and sources of urban air pollution are closely related to the characteristics of airflow transport trajectories. Methods such as the back-trajectory model, clustering analysis, and the potential source contribution function (PSCF) method have become important ways to address these characteristics [16–18]. For example, Yan et al. (2015) conducted a trajectory analysis on a smog process in Beijing in February 2014 and discovered that Baoding, Hengshui, and Handan are important potential source areas in the region [19]. Donnelly et al. (2015) analyzed the effects of various long-range transport pathways of the concentrations of particulate matter with diameter less than 10 μm (PM_{10}) in Ireland by using the HYSPLIT4 model and showed that air quality in Ireland is heavily dependent on air mass origin and the inherent characteristics of the air mass [20]. Lee et al. (2011) examined the pathways of PM_{10} by back-trajectory analysis and showed that the transboundary pollutants of high-PM_{10} levels are more than twice as high as those from internal sources, especially in winter and spring; in addition, the local pollutants contributing to high-PM_{10} concentrations have decreased through the efforts to reduce emissions, but the transboundary pollutants have not decreased [21]. By analyzing the transport pathways of particulate matter over Guwahati, located in the Brahmaputra River Valley (BRV), Tiwari et al. (2017) found that the turbid air masses transported over Guwahati mostly from the western and southwestern directions contribute to higher PM concentrations, either carrying anthropogenic pollution from the Indo-Gangetic Plains or locally and LRT (long-range transported) dust from BRV and western India, respectively [22]. Li et al. (2017) employed the PSCF and concentration weighted trajectory (CWT) methods to analyze the transport trajectories and potential sources of $PM_{2.5}$ and PM_{10} in Beijing and demonstrated that, in summer and autumn, the impacts of air pollution are mostly from the south and southeast, while those in spring and winter are influenced from the southeast and north [23]. By analyzing the transport pathways and potential sources of PM_{10} pollution in Shanghai, Li et al. (2014) reported that there are two potential sources of PM_{10} pollution in Shanghai: one is located in the northwest (for example, Hebei and Shandong), while the other is in the southwest (for example, Zhejiang and Jiangxi) [24]. Yang et al. (2017) examined the spatiotemporal characteristics and trajectories of a smog process in Beijing in December 2015 and discovered that the $PM_{2.5}$ concentration in Beijing was high in the south but low in the north [25]. The major potential pollutant sources are the deserts in the northwest and the built-up areas in the Beijing-Tianjin-Hebei region.

Based on eight monitoring locations in Chengdu and meteorological data over three months, Liao et al. (2017) analyzed the spatiotemporal characteristics and sources of $PM_{2.5}$ in Chengdu. The results reveal that the major potential sources of $PM_{2.5}$ in Chengdu are located along the western margin of the Sichuan Basin and in the southeastern cities [26]. Xin et al. (2016) adopted daily average PM_{10} concentration data and Global Data Assimilation System (GDAS) data to study the transport trajectories that significantly influence PM_{10} in Xining and found that atmospheric pollution is easily affected by inland trajectories [27]. Based on nine air quality monitoring locations in Qingdao in winter, Li et al. (2017) analyzed the characteristics of atmospheric pollution and pollutant sources, the results revealed that $PM_{2.5}$ is a major urban atmospheric pollutant in Qingdao, and the greatest contributions are from Shanxi, the southern part of Hebei, and the western part of Shandong [23]. Additionally, pollutants such as aerosols from the deserts in Inner Mongolia and the Yellow Sea are also a cause of atmospheric pollution. Lv et al. (2015) discovered that the $PM_{2.5}$ concentration in Guangzhou is more sensitive to the velocities of air mass movements than their directions. The regional $PM_{2.5}$ contributions in spring, summer, autumn, and winter are 15%, 28%, 16%, and 22%, respectively [28].

However, previous studies have shortcomings with respect to air pollution transport characteristics. Most studies: (1) have focused on typical large-scale regions, while limited research has focused on small yet heavily polluted cities, such as Weifang City [24]; and (2) have adopted low temporal resolution data (with 6- or 24- h resolution); however, high time resolution data have been shown to contribute to improved resolutions of source areas in PSCF calculations [23]. Moreover, in previous studies, several air quality monitoring data have mostly been used only to obtain average pollutant concentrations in PSCF models, whereas sparse and unevenly distributed monitoring data cannot truly represent the concentrations over the study area.

3. Data and Methodology

3.1. *Study Area and Data Sources*

Weifang is located in the middle of the Shandong Peninsula (Figure 1), with Zibo to the west, Linyi to the south, Qingdao to the east and Laizhou Bay and the Bohai Sea to the north, covering a total area of 16,000 square kilometers. Weifang City is high in the south (with ground elevation 100–1032 m) and low in the north (with ground elevation under 7 m). The south is mainly covered with hills and low mountains, while the northeast is mainly characterized by plains and lakes. Weifang City has been experiencing industrial and economic advancements and was one of the most rapidly developing cities in Shandong Province. As of 2019, the number of registered vehicles had exceeded 2 million. This growth has resulted in damage to the Weifang City environment. Existing studies on the characteristics of air pollution in Weifang have mainly focused on spatiotemporal patterns, while few reports have described the transport trajectories and mechanisms as well as the potential sources of air pollution. This paper utilizes hourly $PM_{2.5}$ monitoring data and a back-trajectory model to analyze the transport pathways and potential source contributions of $PM_{2.5}$ pollution in Weifang and can provide important scientific support for joint atmospheric pollution control and management.

The $PM_{2.5}$ monitoring data used in this paper were acquired from 5 national monitoring stations, 4 provincial monitoring stations, and 29 city monitoring stations in Weifang (Table A1 in Appendix A). The data were obtained by automatic air quality monitors through 24-h continuous monitoring and cover the period from 1 March 2015 to 29 February 2016. The data were acquired from the Data Center of the Ministry of Environmental Protection of the People's Republic of China (http://datacenter.mep.gov.cn/index). Thermo Fisher 1405F monitoring devices were used to measure the $PM_{2.5}$ concentrations, and this instrument operates on the principle of measuring $PM_{2.5}$ concentrations by a filter dynamic measurement system (FDMS) with the tapered element oscillating microbalance (TEOM) and the beta-attenuation method [6].

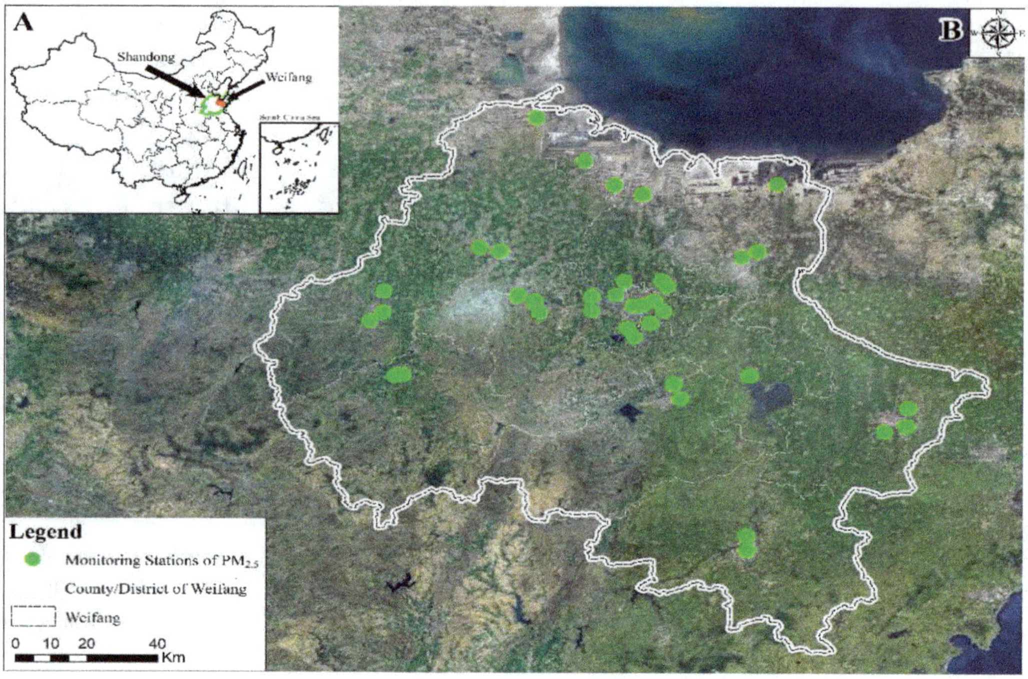

Figure 1. Study area and spatial distribution of monitoring sites.

The major data structure is illustrated in Figure 2a. The meteorological data used in the back-trajectory model were obtained from GDAS data provided by the National Centers for Environmental Prediction (NCEP). The 1.0° resolution global reanalysis data are adopted. These data are recorded every 6 h, namely, at 00:00, 06:00, 12:00 and 18:00 (UTC). Their structure is shown in Figure 2b.

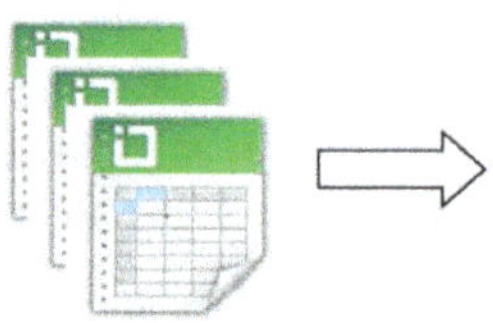

StationID	Date	Longitude	Latitude	PM2.5
W001	2015-03-01 00:00	119.078	36.7336	85.29
W001	2015-03-01 01:00	119.078	36.7336	86.48
W001	2015-03-01 02:00	119.078	36.7336	85.55
………	………	………	………	
W206	2016-02-29 08:00	118.520	36.500	101.09
W206	2016-02-29 09:00	118.520	36.500	103.17
………	………	………	………	

(**a**) Structure of PM2.5 monitoring data

Date	Press	TMP (K)	Relative humidity (%)	PWAT (kg/m2)	………	u wind (m/s)	………
2015-03-01 00:00	1019.6	305.69	62%	0	………	6.9	………
2015-03-01 01:00	1019.5	305.51	62%	0	………	6.9	………
2015-03-01 02:00	1019.2	306.05	61%	0	………	6.9	………
………	………	………	………	………	………	………	………
2016-02-29 00:00	1023.3	301.37	63%	0	………	5.4	………
2016-02-29 01:00	1023.3	301.73	63%	0	………	5.4	………
………	………	………	………	………	………	………	………

(**b**) Structure of GDAS data

Figure 2. Structures of monitoring station data and GDAS data.

3.2. Methodology

(1) HYSPLIT model

The Hybrid Single-Particle Lagrangian Integrated Trajectory (HYSPLIT) model established by the Air Resources Laboratory of the National Oceanic and Atmospheric Administration (NOAA) is used in this paper to analyze sources and transport trajectories of air pollutants [29,30]. This model gives highly accurate and temporally continuous simulation results and has been widely used in research on the transport and diffusion of various pollutants in different areas. The HYSPLIT model is divided into two parts: the backward transport model and the forward diffusion model, which solve problems concerning sources and sinks, respectively. In this paper, the backward transport model of the HYSPLIT model is employed to simulate 72-h backward airflow transport trajectories near the ground surface in Weifang during the period of interest. The characteristics of the airflow movements in the study area are thereby reflected.

(2) Trajectory clustering analysis

Trajectory clustering analysis, a multivariate statistical analyses technique, was used to divide the trajectory data into several classes or clusters. Data in the same class or cluster share a higher degree of similarity, whereas those in different classes or clusters vary more significantly [31]. This paper uses TrajStat, a plugin of MeteoInfo; this plugin can view, query, and cluster trajectories and includes two clustering methods: Euclidean distance and angle distance. Because this paper aims to determine the direction from which the air masses that reach the site have originated, the angle distance clustering method is utilized to cluster airflow trajectories. The angle distance is often used to define the mean angle between the two trajectories, which varies between 0 and π. The details of the angle distance clustering method can be found in the work of Sirois and Bottenheim (1995) [32].

(3) Potential source contribution function (PSCF)

The PSCF model is a simple method that links residence time in upwind areas with high concentrations through a conditional probability field [33]. This method can identify pollutant sources by analyzing airflow trajectories and a given threshold [34,35]. It can be calculated as follows:

$$PSCF_{ij} = \frac{n_{ij}}{N_{ij}} \tag{1}$$

where N_{ij} is the total number of airflow trajectories' endpoints that fall in the ijth grid and n_{ij} is the total number of airflow trajectories' endpoints for which the measured $PM_{2.5}$ concentration exceeds a given threshold in the same grid. In this study, the 24-h average Grade II standard $PM_{2.5}$ concentration (75 $\mu g/m^3$) in ambient air quality standards of China (GB3095-2012) was selected as the threshold value [26]. The trajectories were calculated hourly. Studies have demonstrated that great uncertainty exists in the calculation result when N_{ij} is extremely small. To eliminate this uncertainty, an arbitrary weight function, W_{ij}, was applied when the number of the endpoints in a particular cell was less than three times the average number of endpoints for each cell [36,37].

$$WPSCF_{ij} = W_{ij} \times PSCF_{ij} \tag{2}$$

$$W_{ij} = \begin{cases} 1.00 & N_{ij} > 80 \\ 0.70 & 20 < N_{ij} \leq 80 \\ 0.42 & 10 < N_{ij} \leq 20 \\ 0.05 & 0 < N_{ij} \leq 10 \end{cases} \tag{3}$$

(4) Concentration weighted trajectory (CWT)

The CWT method first computes the weighted concentrations of trajectories and then obtains the weighted concentrations of grids [38]. The calculation formula of the CWT method is given as follows:

$$CWT_{ij} = \frac{\sum_{k=1}^{N} C_k \tau_{ijk}}{\sum_{k=1}^{N} \tau_{ijk}} \tag{4}$$

where CWT_{ij} is the weighted average concentration of grid ij; N is the total number of trajectories; k denotes a trajectory; C_k is the $PM_{2.5}$ concentration of trajectory k when it passes through grid ij, which can be calculated by the HYSPLIT model; and τ_{ijk} is the duration in which trajectory k stays in grid ij [39,40]. In addition, the CWT method gives rise to great uncertainties, thus the weight coefficient W_{ij} is needed to reduce these uncertainties. Similarly, W_{ij} is determined using Equation (3), and the introduction of the coefficient is as follows:

$$WCWT_{ij} = W_{ij} \times CWT_{ij} \tag{5}$$

For PSCF and CWT methods, the input data and applicable resolution of grid are the same. The difference between CWT and PSCF is that PSCF usually uses a concentration threshold to evaluate the potential sources of $PM_{2.5}$. It means that it may have the same PSCF value when sample concentrations are either only slightly higher or much higher than the criterion. As a result, it may not distinguish moderate sources from strong ones. For CWT method, the limitation of PSCF can be overcome by assigning a weighted concentration by averaging the sample concentrations that have associated trajectories that cross the grid cell.

4. Results and Analyses

4.1. $PM_{2.5}$ Pollution Characteristics

Hourly $PM_{2.5}$ concentration data for Weifang from 2015 to 2016 are examined to analyze the annual, seasonal, and monthly characteristics of the $PM_{2.5}$ concentration. The annual average $PM_{2.5}$ concentration in Weifang is 73.03 $\mu g/m^3$, which is more than twice the national second-level standard (35 $\mu g/m^3$) and nearly five-fold the national first-level standard (15 $\mu g/m^3$). Figure 3 presents the seasonal and monthly variations in the $PM_{2.5}$ concentration. In general, the monthly average $PM_{2.5}$ concentration presents a U-shaped curve. The $PM_{2.5}$ concentration is higher in winter (December, January, and February), approximately 101.64 $\mu g/m^3$, which significantly exceeds the national 24-h atmospheric quality second-level standard (75 $\mu g/m^3$). The concentrations in autumn (September to November) and spring (March to May) are 74.81 and 66.53 $\mu g/m^3$, respectively. The lowest concentration occurs in summer (June to August), with only 49.12 $\mu g/m^3$. The $PM_{2.5}$ concentrations in the four seasons all exceed the national 24- h atmospheric quality first-level standard (35 $\mu g/m^3$). The highest concentration occurs in January (122.86 $\mu g/m^3$) and the lowest in July (48.02 $\mu g/m^3$). The monthly average considerably decreases in March but rapidly increases in October.

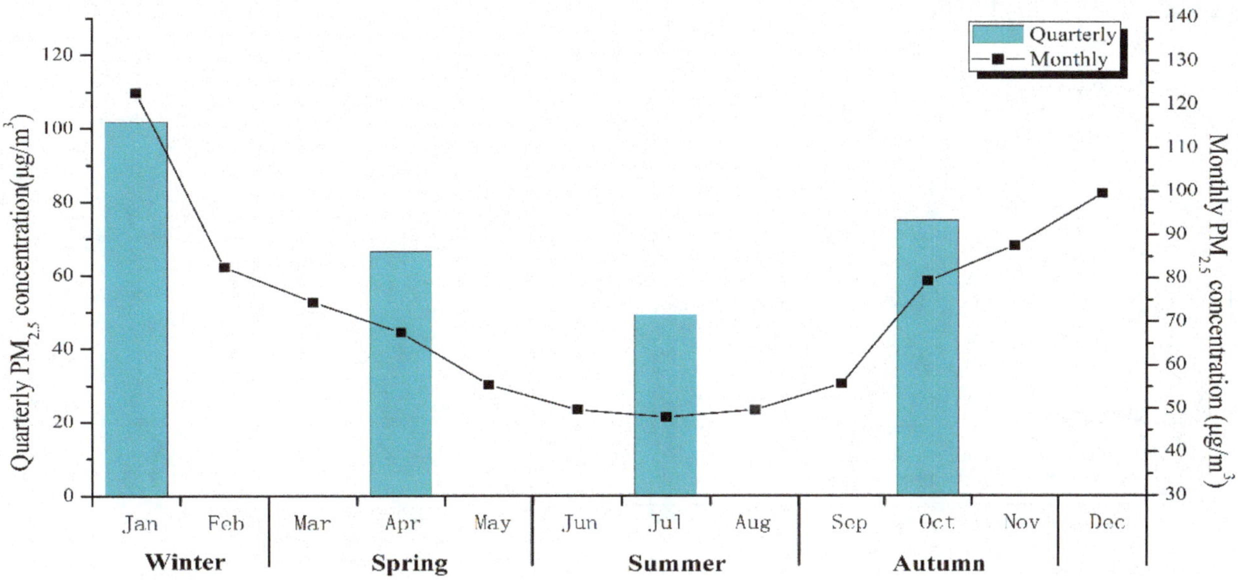

Figure 3. Quarterly and monthly characteristics of $PM_{2.5}$ concentration in Weifang.

4.2. Backward Trajectory Clustering

TrajStat was used to process airflow data in Weifang in different seasons to obtain transport trajectories in different seasons. However, from these airflow trajectories, it is not possible to determine the exact number of trajectories from different directions. Therefore, according to the consistency in the spatial distributions of various airflow trajectory types, trajectories in spring, summer, autumn, and winter are integrated into four, five, two, and two clusters, respectively (Figure 4).

Figure 4 shows that, in spring, the airflow trajectories originating from northern inland areas (Trajectories 1–3) dominate. More specifically, airflows originating from Russia, Mongolia, central Inner Mongolia, and Hebei account for 36.91% of the total trajectories, and those from the Yellow Sea and central Shandong account for 29.38%. In summer, the airflow trajectories are shorter and follow a star-shaped distribution. Influenced by warm and humid airflows from the ocean, these trajectories are dominated by southerly and southeasterly winds and account for 67.1% of the total trajectories. In autumn, as cold air masses move southward, the airflow trajectories from the southeast weaken. As a result, the airflow trajectories originating from southeastern Mongolia, central Inner Mongolia, and northeastern Hebei account for 59.07%. The airflow trajectories from northeastern Hebei are shorter, accounting for 40.30%. They pass over the waters of the Bohai Sea and finally return to Weifang via the inland areas of Shandong. In winter, under the influence of the Siberian cold current, the airflow trajectories originating from the northwest dominate (68.91%). These trajectories are longer, and the air masses move more quickly. Moreover, in winter, the airflow trajectories from southwestern Shandong account for 31.06% and are shorter and slower.

4.3. Transport Trajectories in Different Seasons

Based on the airflow backward trajectory clustering results in different seasons, combined with the hourly $PM_{2.5}$ concentration data, this paper analyzes the pollution characteristics of different transport trajectories. According to the second-level $PM_{2.5}$ concentration (75 μg/m^3) in Ambient Air Quality Standards (GB 3905-2012), back-trajectories are classified into "low polluted" trajectories (<75 μg/m^3) and "polluted" trajectories (75 μg/m^3). The statistics for different types of trajectories are described in Table 1.

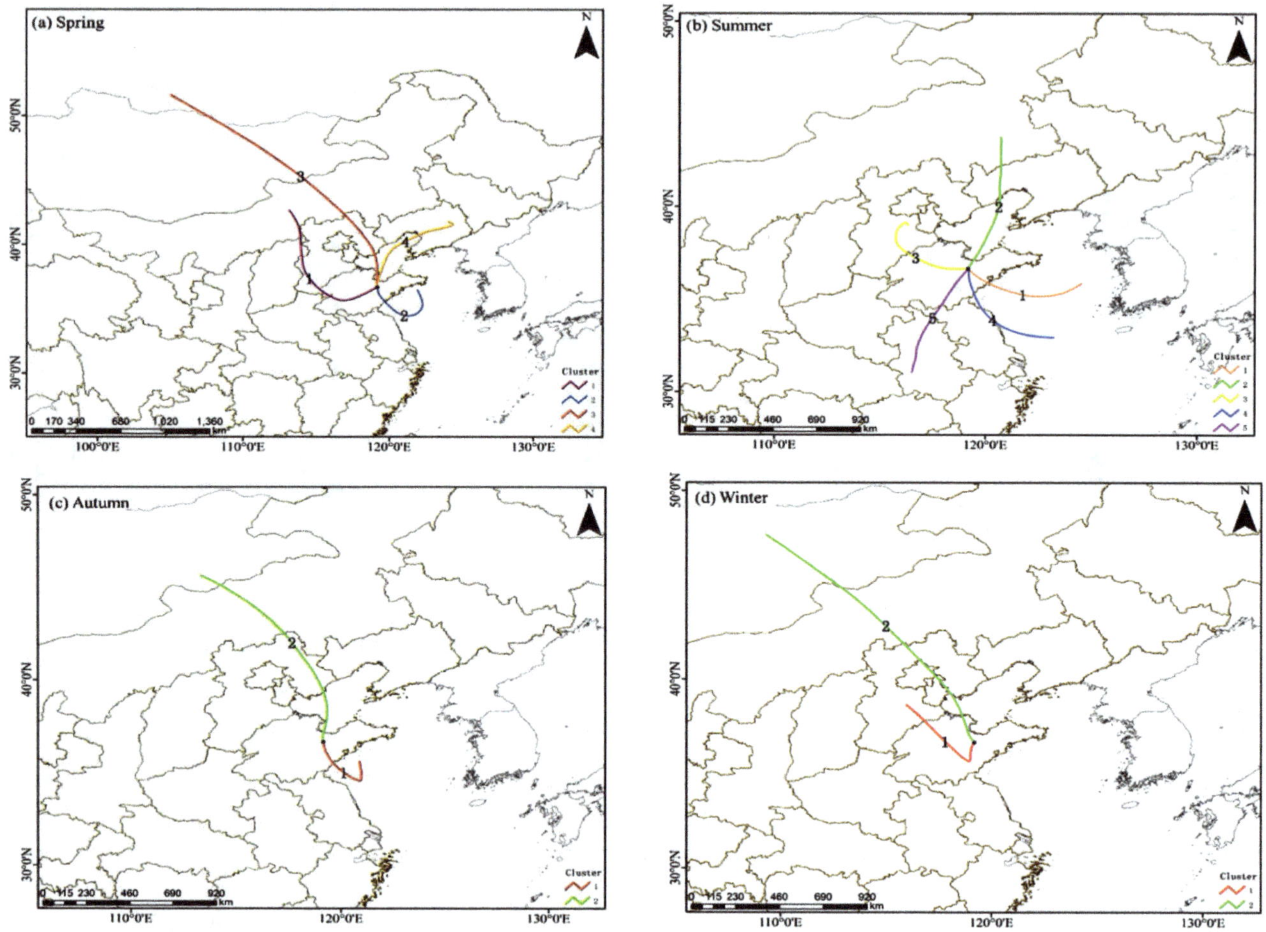

Figure 4. Cluster mean back-trajectories in different seasons in Weifang from March 2015 to February 2016.

Table 1. Trajectory proportions and $PM_{2.5}$ concentrations based on all trajectories and pollution trajectories.

Season	Trajectory Cluster	All Trajectories		Polluted Trajectories	
		Proportion (%)	Average (μg/m^3)	Proportion of Seasonal Total Polluted Trajectories (%)	Average (μg/m^3)
Spring	1	12.08	83.35	29.43	109.71
	2	45.36	69.63	31.55	99.46
	3	27.15	53.72	18.83	98.21
	4	15.41	76.64	20.20	111.62
	All		70.83		104.87
Summer	1	34.06	41.50	12.04	89.41
	2	22.71	42.93	4.45	88.29
	3	6.63	62.95	20.68	99.94
	4	21.16	68.16	43.19	94.13
	5	15.44	47.59	19.63	96.24
	All		52.63		94.92
Autumn	1	25.80	94.52	52.60	124.22
	2	74.20	66.53	47.40	116.84
	All		80.53		120.72
Winter	1	37.11	148.08	55.72	158.08
	2	62.89	82.99	44.28	114.11
	All		115.54		138.61

(1) Pollution characteristics in spring

The average $PM_{2.5}$ concentration of spring trajectories is 70.83 μg/m^3. The trajectory clusters in descending order are: 2 > 3 > 4> 1. The airflow trajectories originating from the north (Inner Mongolia) have the highest average $PM_{2.5}$ concentrations, reaching 83.35 μg/m^3. This may be due to two reasons: one is that these airflows mainly come from the western arid regions of China, and, in spring, northwesterly winds easily blow loose topsoil and fine sand from dry surfaces to form sandstorms; the other reason is that these trajectories pass over the industrial areas in southern Hebei and easily transport local pollutants to Weifang City to form accumulation. Type 3 trajectories from the northwest are longer and have the lowest $PM_{2.5}$ concentration of 53.72 μg/m^3. This may due to two reasons: (1) these trajectories travel large distances at high wind speeds; and (2) they subsequently pass over regions with high vegetation cover and relatively low population density [41], such as Xilingol league in Inner Mongolia, Chengde City in northern Hebei, and waters of the Bohai Sea. Both the abovementioned cases may lead to low pollution transport and effective adsorption, diffusion, and dilution of pollutants.

The average $PM_{2.5}$ concentration of the polluted trajectories is 104.87 μg/m^3. In particular, Type 1 and 2 trajectories account for 29.43% and 31.55% of the total polluted trajectories, respectively. These types together account for 60.98% of the seasonal total polluted trajectories, suggesting that $PM_{2.5}$ is mainly transported from the north and southeast directions to Weifang in spring.

(2) Pollution characteristics in summer

The average $PM_{2.5}$ concentration of summer trajectories is the lowest, only 49.40 μg/m^3. In summer, many plants flourish, and the total leaf area of surface vegetation significantly increases, which is conducive to the adsorption of atmospheric particulates. Moreover, precipitation in summer is more concentrated, and the increased precipitation greatly facilitates wet deposition and dilution of atmospheric pollutants. However, under the control of the subtropical high and typhoon, the trajectories are diverse. The trajectory clusters in descending order of average $PM_{2.5}$ concentration are: 4 > 3 > 5 > 2 > 1. The airflow trajectories from southeast have the highest value (68.16 μg/m^3), followed by those from northwest (62.95 μg/m^3). This pattern indicates that external pollutant emissions from southeast and northwest have a substantial effect on $PM_{2.5}$ pollution in Weifang. Although Type 4 originated from the Yellow Sea, affected by subtropical auxiliary high pressure and wheat straw burning in south China, pollutants and suspended particles from eastern Zhejiang and southern Shandong are easily carried by the trajectory and accumulate Weifang. Type 1 trajectories originating from the Bohai Sea have the lowest average $PM_{2.5}$ concentration (41.50 μg/m^3) and make up the largest proportion of the total trajectories. These trajectories mainly pass over waters and coastal areas in eastern Shandong. These areas are characterized by strong air masses, cleanliness, and intensive atmospheric wet deposition, which facilitate wet deposition and dilution of pollutants [1,6,42].

The average $PM_{2.5}$ concentration of polluted trajectories is 94.92 μg/m^3. Types 3–5 trajectories account for 20.68%, 43.19%, and 19.63% of the total polluted summer trajectories, respectively. They together account for 83.50% of the seasonal total polluted trajectories, indicating that $PM_{2.5}$ is mainly transported in easterly directions to Weifang in summer.

(3) Pollution characteristics in autumn

Autumn trajectories have an average $PM_{2.5}$ concentration of 80.53 μg/m^3. The autumn trajectory clusters in descending order of $PM_{2.5}$ concentration are: 1 > 2. In particular, Type 2 trajectories are longer and have lower $PM_{2.5}$ concentrations. These trajectories originate mostly from northwestern Mongolia, central Inner Mongolia, and northern Hebei and travel over extensive areas at high wind speeds. In addition, they pass over the waters of the Bohai Sea and enter Weifang directly. The air masses are relatively clean, which is conducive to the diffusion of pollutants. In contrast, the $PM_{2.5}$ concentrations of Type 1 trajectories are higher, with an average of 94.52 μg/m^3. This is because short and slow Type 1 trajectories do not favor the diffusion of pollutants. Furthermore, these airflow trajectories pass over the northeastern part of Shandong before they reach Weifang and often carry pollutants from Shandong, leading to higher $PM_{2.5}$ concentrations.

The polluted trajectories have an average $PM_{2.5}$ concentration of 120.72 μg/m^3. Type 1 trajectories account for 52.60% of the total polluted trajectories. It follows that the northeasterly direction is the major $PM_{2.5}$ transport direction to Weifang in autumn.

(4) Pollution characteristics in winter

Compared to the other three seasons, the average $PM_{2.5}$ concentration of winter trajectories is the highest, reaching 115.54 μg/m^3. More specifically, Type 2 trajectories have the highest $PM_{2.5}$ concentrations, with an average of 148.08 μg/m^3. This is mostly because these airflow trajectories originate from south-central Hebei and pass over inland cities in northern Shandong. In winter, these areas are in the heating period, leading to increases in anthropogenic pollutant emissions, such as from coal burning [8,43]. As air masses pass over these areas, they usually carry soil, dust, and pollutants from the inland areas to Weifang. In addition, these airflow trajectories are relatively short and travel slowly, making pollutants not able to diffuse easily. Hence, the average $PM_{2.5}$ concentration of these trajectories is relatively high. In contrast, the average concentration of Type 2 trajectories is the lowest, but they are also significantly higher than those in the other seasons, at 82.99 μg/m^3. These trajectories originate from Siberia and arrive in Weifang through the Mongolian Plateau, Inner Mongolia, Hebei, Beijing, Tianjin, and the Bohai Sea and are relatively long and travel at higher wind speeds. These trajectories pass over the waters of the Bohai Sea, which may slightly facilitate the diffusion and elimination of pollutants.

The average $PM_{2.5}$ concentration of polluted winter trajectories is 138.61 μg/m^3. The average $PM_{2.5}$ concentration of Type 1 trajectories reaches 158.08 μg/m^3, which includes not only the external source from southern Hebei Province but also local emissions and is the main pollution source of Weifang City in winter.

Overall, in autumn and winter, Weifang is mostly influenced by pollutants from inland areas of Hebei and Shandong. Airflows from these areas often carry particulates emitted from the passed areas. This reflects the knock-on effects of $PM_{2.5}$ pollution within Shandong Province. In summer, due to the influences from coastal cities in the east and clean airflows from the ocean, the concentrations of pollutants are lower. In spring, severe pollution in Weifang is closely associated with sandstorms in the arid areas in the west, where loose topsoil and fine sand are blown from dry surfaces.

4.4. Potential Source Regions

To further investigate the sources of atmospheric pollutant transport in Weifang, this paper analyzes the potential source regions of $PM_{2.5}$ pollution. First, $PM_{2.5}$ concentration data are added to airflow trajectories, and areas covering all airflow trajectories are uniformly divided into 0.5° × 0.5° grids. Then, the weighted potential source contribution function (WPSCF) value of each grid is computed. WPSCF reveals the spatial distribution of $PM_{2.5}$ potential sources obtained by combining back-trajectories and measurements of $PM_{2.5}$ concentration. A high PSCF value signifies a potential source location. The greater the WPSCF value of a grid is, the higher the contribution levels of potential source regions to $PM_{2.5}$ pollution in Weifang, given that other factors remain stable.

Figure 5 shows the PSCF results for $PM_{2.5}$ in Weifang from 2015 to 2016. The colors represent the contribution levels of potential source areas; the black color is associated with high concentrations, while green represents low $PM_{2.5}$ concentrations. Distinct seasonal variations are noted in the distribution of the potential source areas of $PM_{2.5}$. (1) In spring, high PSCF values are mainly found in north-central Jiangsu and southwestern Shandong. Additionally, areas such as Tianjin, Liaoning and western Jilin have key influences on the sources of pollution in Weifang. (2) In summer, PSCF values are generally smaller, indicating that Weifang is less affected by pollutants from the surrounding areas in summer. Compared to those in spring, the potential sources are shifted eastward and are located in Shandong and the waters of the Bohai Sea and Yellow Sea. This is may because the southeasterly monsoon carries pollutants emitted from passed areas when it moves northward. Furthermore, stubble burning is the most intensive in some southern regions in summer, creating large fumes that easily travel northward with the monsoon. (3) In autumn, higher PSCF values are increasing and are mainly found in regions such as northern Anhui, northeastern Henan, and southwestern Shandong. In addition, areas such as Hebei, Beijing, Tianjin, eastern Shanxi, and central Inner Mongolia make certain potential contributions. (4) In winter, potential source areas transit and extend northwestward. High values are found in the whole city of Shandong, northern Jiangsu, northeastern Henan, and southern Hebei; in these regions, their situations are very similar to that of Weifang, with the same climate, industrial emission, population density, and winter heating. In addition, those values in eastern Shanxi and western Inner Mongolia increase.

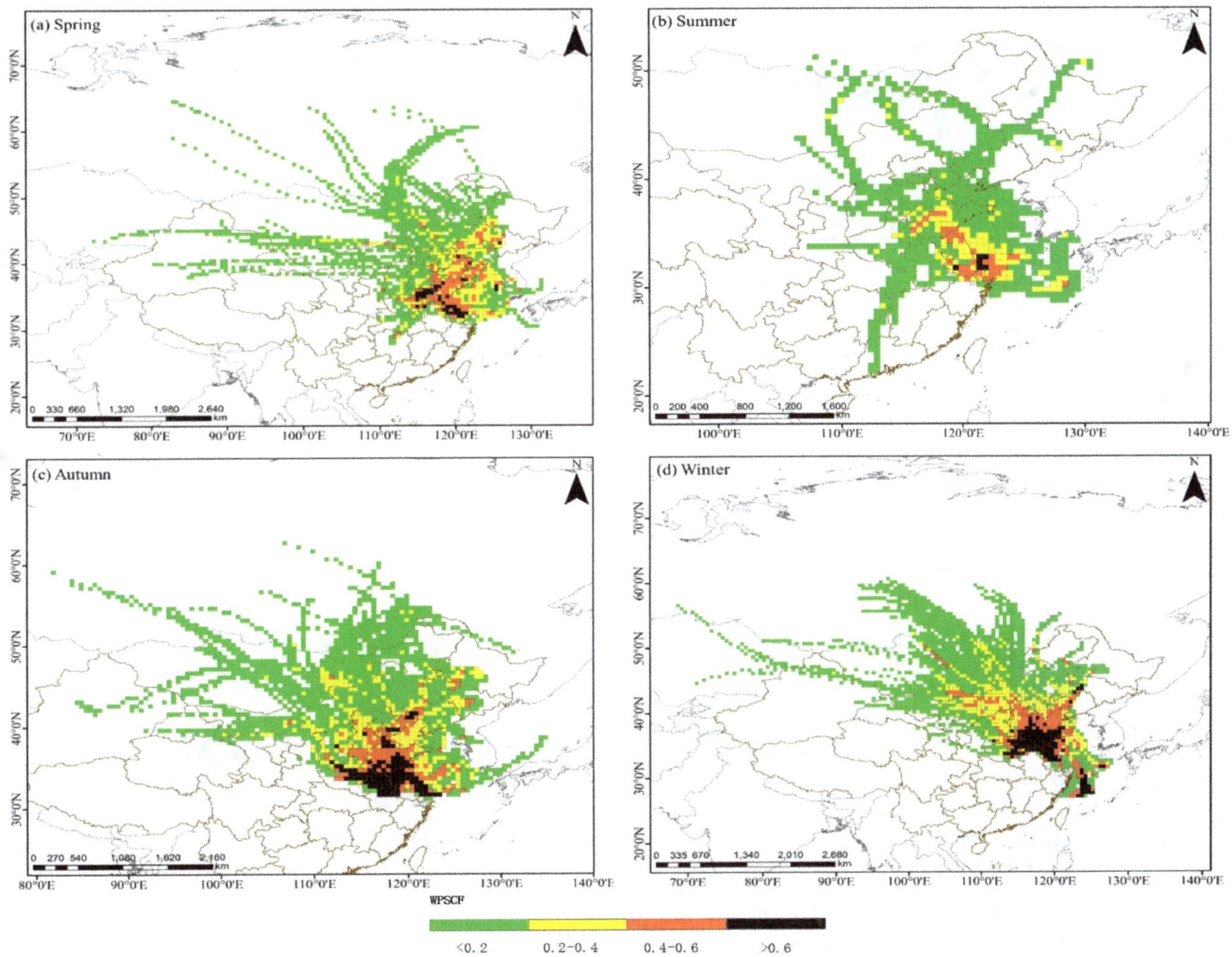

Figure 5. Spatial distribution of WPSCF values of $PM_{2.5}$ in spring, summer, autumn, and winter from March 2015 to February 2016.

In summary, in winter, spring, and autumn, Weifang is more substantially affected by nearby inland cities in Jiangsu, Henan, and Shandong, and these areas are potential source regions of pollutants in Weifang. In contrast, in summer, Weifang is less influenced by pollutants from surrounding areas. Moreover, comparisons reveal that the distributions of high PSCF values in different seasons agree well with those of the major trajectory areas shown in Table 1. This finding indicates that the potential source contribution results obtained by the PSCF method in this paper are reasonably reliable.

4.5. Potential Source Region Contributions

The CWT method is adopted to calculate the weighted concentrations of trajectories originating from various potential source regions. The CWT value represents a weighted specific concentration value by averaging sample concentrations that have associated trajectories that cross the grid cell, under the condition of other factors remaining relatively constant. The larger is the CWT value, the greater is the contribution of the grid cell to the pollutant concentration of Weifang City. The results are illustrated in Figure 6. The regions colored in black correspond to the main contributing sources associated with the highest $PM_{2.5}$ values, while the green color represents regions with low values.

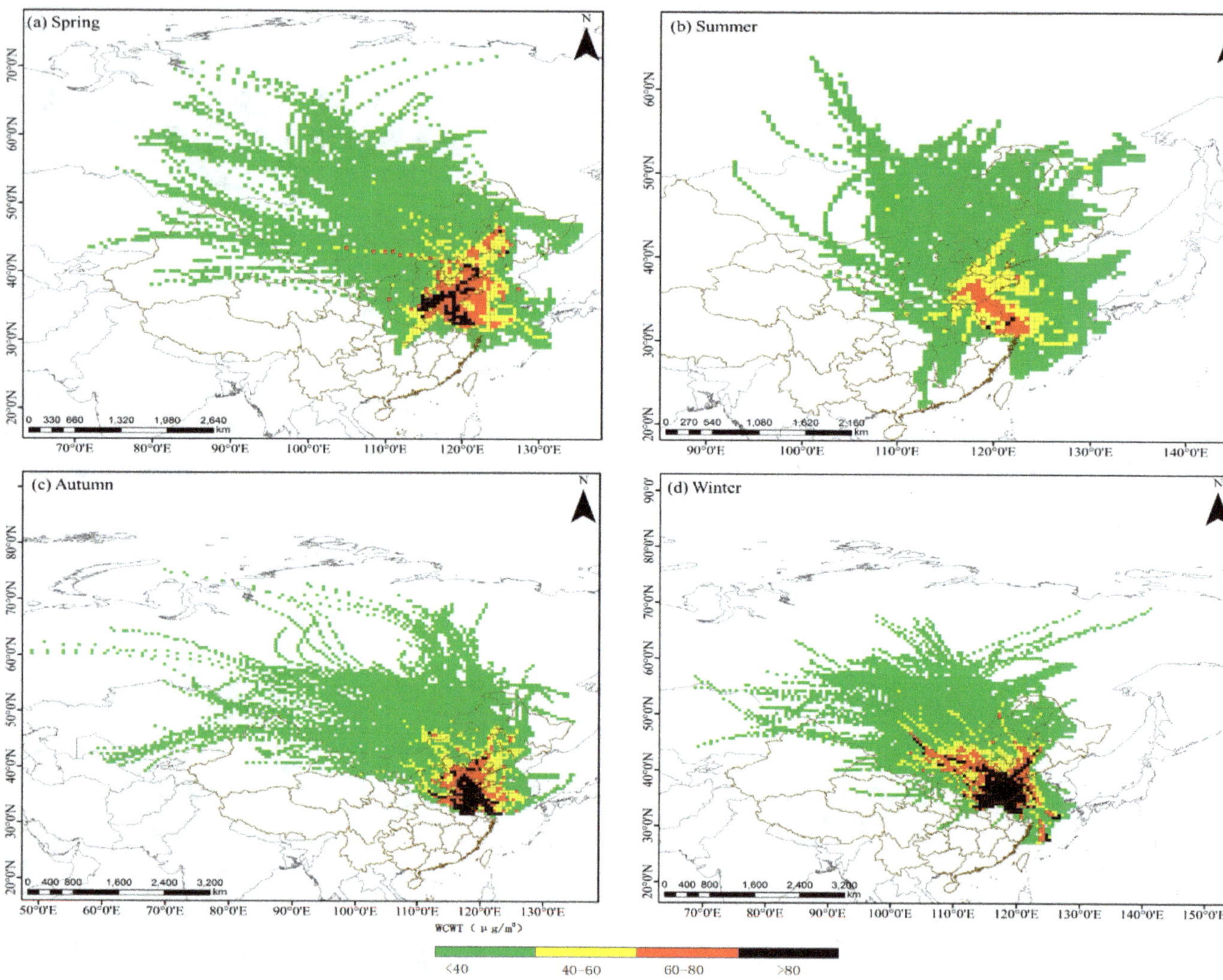

Figure 6. Spatial distribution of WCWT values of $PM_{2.5}$ in spring, summer, autumn, and winter from March 2015 to February 2016.

In spring, higher WCWT values are mostly concentrated in northeastern Henan and southwestern Shandong, which are important sources of $PM_{2.5}$ in Weifang. These areas have daily average $PM_{2.5}$ concentration contributions above 80 μg/m^3. Additionally, other regions in Shandong, western Tianjin, and the central Bohai region influence pollution in Weifang. For these areas, the daily average $PM_{2.5}$ concentration contributions are more than 60 μg/m^3.

In summer, WCWT values are generally smaller. The highest CWT values covering the map were distributed in coastal areas in the east, such as Zhejiang and Jiangsuµwith values of 60–80 μg/m^3. This is possibly because southeasterly wind dominates in Weifang in summer. Sea-salt aerosols from the south are easily transported to the Weifang region via the monsoon traveling northward.

Compared to those in summer, higher WCWT values in autumn are found in areas further south and northeast. These values are mainly observed at the intersection of Anhui, Jiangsu, and Shandong, with daily average $PM_{2.5}$ concentration contributions of above 80 μg/m^3. Furthermore, relatively substantial contributions are observed in certain localized areas such as eastern Hebei, southern Beijing, and Tianjin, and the daily average $PM_{2.5}$ concentration contributions of these areas exceed 60 μg/m^3.

The WCWT values in winter are greater than those in other seasons. The greatest contributions to the $PM_{2.5}$ concentration in Weifang are throughout Shandong as well as certain nearby areas such as northeastern Henan, northern Jiangsu, and southwestern Hebei. The daily average contributions are more than 80 μg/m^3. This is probably because many pollutants are emitted when coal is burned for heating in the north in winter.

When CWT results are compared to the WPSCF results, they have basically consistent results for areas contributing to the $PM_{2.5}$ concentration in Weifang, which demonstrates the credibility and accuracy of the analysis results. Nevertheless, the potential source regions obtained via the CWT method for spring, summer, and autumn cover larger areas than those simulated by the PSCF method, while the results for winter are similar. These results are consistent with the findings reported by Yan et al. (2018) in Yinchuan [44]. These observations may arise because the CWT method takes into account all the concentrations rather than a subset of high concentrations in the PSCF method. The findings demonstrate that the seasonal distributions of $PM_{2.5}$ source areas obtained by these two methods share generally similar characteristics, suggesting the credibility and accuracy of the analysis results.

5. Discussion and Conclusions

Based on $PM_{2.5}$ monitoring data from 2015 to 2016 from 38 air quality monitoring stations, this paper describes an in-depth study of the seasonal variations in internal and external potential sources in Weifang by using different trajectory analysis methods.

(1) Seasonal differences in the contributions of potential source regions of $PM_{2.5}$ in Weifang. In winter, spring, and autumn, airflows are mostly from the northwesterly and northerly directions and significantly influence the $PM_{2.5}$ concentration in Weifang. However, in summer, airflow trajectories are scattered, and warm and humid airflows from the ocean in the southeastern direction dominate. In winter, spring, and autumn, Weifang is more greatly affected by pollutant transport from nearby inland cities in Shandong and Henan. These transport pathways are short in general, and the wind speeds are low, leading to accumulation of the carried pollutants in Weifang. In contrast, in summer, Weifang is less influenced by pollutants from the surrounding areas, and the potential source regions are mainly located in coastal areas in the east, such as Jiangsu, the Bohai Sea, and the Yellow Sea. It should be noted that the region covering central Inner Mongolia and southern Liaoning is also a potential source area for Weifang.

(2) Policy implications for $PM_{2.5}$ pollution in Weifang. The results indicate that, in formulating relevant pollution control and prevention measures, the government should focus on the control of pollutant sources and take the migration of regional pollution caused by these sources into account.

For example, according to the transport patterns of pollutants in Weifang, the surrounding pollutant source regions can be divided and classified (such as key control zones). This finding suggests that different control and management policies can be implemented, and region-specific pollution control and prevention measures can be formulated. Because Weifang is more significantly affected by short-distance pollutant transport from the surrounding cities and provinces as well as inland areas in Shandong Province, the government should pay more attention to short-distance transport from these regions. For instance, more intensive greening measures can be introduced to reduce the short-distance transport of $PM_{2.5}$. Furthermore, the interactions between the city and its surrounding areas should be considered, and joint control and cooperation between different regions should be enhanced.

(3) Evaluation of the analysis results. The results demonstrate that the seasonal distributions of $PM_{2.5}$ source areas obtained by two methods (PSCF and CWT) share generally similar characteristics, demonstrating the credibility and accuracy of the analysis results. Both methods can effectively reflect the potential source areas of pollution in the region. However, there is a limitation in validating the modeling results with the real transport values because emissions and depositions of air pollutants along the trajectories cannot be captured easily. Nevertheless, our results are able to locate the source direction and areas and identify source contributions to a certain extent. To support this point, we analyzed the notice issued by Shandong Provincial Environmental Protection Office, which indicates that the primary source of external dust transport from Inner Mongolia is the main reason that leads to the rapid increase in pollutants in Weifang City in April [45]. Thus, some references indicated that in northern China (such as Weifang), the main pollution sources come from anthropogenic emissions related to coal burning and their transportation [46]. These cases both support our findings regarding spring and winter. In addition, PSCF and CWT are appropriate tools that help identify source contributions to the concentration variations at the destination, assuming that other elements/factors remain the same, such as the period of the kite festival (in April every year). In the future, additional work that combines emission sources and external monitored $PM_{2.5}$ concentration data is needed to improve the prediction of $PM_{2.5}$ source regions and validate the analysis results quantitatively.

This paper analyzes the external and internal potential source regions of $PM_{2.5}$ pollution and their influences on Weifang City. Our findings can provide scientific support for the design of region-specific measures for atmospheric pollution prevention and the development of a chemical transport model combined with meteorology. However, this work also has limitations.

For instance, the resolution of the study grid based on the backward trajectory model is not very high and may not be applied to small-scale regions. Furthermore, the estimation of the sources of $PM_{2.5}$ is not perfect because the contributing sources were calculated only based on meteorological data without information such as the production and deposition of dust.

In future research, based on the results in this paper, auxiliary parameters such as the local emission sources and external $PM_{2.5}$ monitoring value will be combined to improve the simulation analysis accuracy and evaluate the accuracy of potential sources contributing to the $PM_{2.5}$ in Weifang quantitatively. Furthermore, analyses for various years will be further conducted to assess the inter-annual variability.

Author Contributions: Conceptualization, C.L.; methodology, C.L. and Z.D.; software, Z.D. and X.L.; validation, Z.D., X.L. and P.W.; formal analysis, Z.D., X.L. and P.W.; investigation, Z.D. and X.L.; resources, C.L.; writing—original draft preparation, Z.D.; writing—review and editing, C.L. and Z.D.; supervision, C.L.; and funding acquisition, C.L. and Z.D. All authors have read and agreed to the published version of the manuscript.

Appendix A

Table A1. Information of Air Monitoring Stations.

Order	Station Name	Latitude	Longitude	Type	Availability
1	Weifang arbitration committee	36.702	119.122	National monitoring stations	Available
2	Weifang environmental protection bureau	36.702	119.144		
3	Hanting Station	36.774	119.191		
4	Weifang No. 7 High School	36.687	119.017		
5	Weifang Fangzi post	36.652	119.164		
6	Weifang College	36.715	119.176	Provincial monitoring stations	
7	Weifang government	36.728	119.018		
8	Jincheng Middle School	36.772	119.098		
9	Xinhui group	36.637	119.107		
10	Shouguang monitoring station	36.869	118.735	City monitoring stations	
11	Zhucheng Safety Supervision Bureau	36.004	119.406		
12	Changle Sports Bureaus	36.730	118.834		
13	Changyi No. 7 High School	36.859	119.431		
14	East side in Binhai district	37.020	119.145		
15	Qingzhou Guangtong group	36.742	118.494		
16	Gaomi college town	36.343	119.748		
17	Anqiu Qingyunhu village	36.479	119.223		
18	Changyi Xiaying school	37.051	119.479		
19	Zhucheng Technology School	36.045	119.404		
20	Xiashan water works	36.503	119.411		
21	Gaomi Ruiguang electronic	36.411	119.807		
22	Changle Wutu street	36.684	118.887		
23	Fangzi Luneng school	36.614	119.124		
24	Linqu qushan	36.503	118.543		
25	Experimental school in Gaoxin district	36.687	119.198		
26	West side in Binhai district	37.116	118.999		
27	Hanting foreign language school	36.758	119.205		
28	Shouguang business district	36.859	118.787		
29	Shouguang Hou village	37.047	119.075		
30	Qingzhou monitoring station	36.681	118.491		
31	Shouguang Yangkou village	37.240	118.879		
32	Qingzhou Shuangbeistadium	36.657	118.456		
33	Normal university of Special education	36.734	119.078		
34	Linqu water works	36.500	118.520		
35	Gaomi Sports Bureaus	36.359	119.802		
36	Changle Zhuliu street	36.715	118.875		
37	Changyi highway bureau	36.842	119.390		
38	Anqiu Qingyunshan scenic spot	36.437	119.240		

References

1. He, J.; Ding, S.; Liu, D.F. Exploring the spatiotemporal pattern of PM2.5 distribution and its determinants in Chinese cities based on a multilevel analysis approach. *Sci. Total Environ.* **2019**, *659*, 1513–1525. [CrossRef] [PubMed]
2. Xie, R.; Zhao, G.; Zhu, B.Z.; Chevallier, J. Examining the Factors Affecting Air Pollution Emission Growth in China. *Environ. Modeling Assess.* **2018**, *23*, 389–400. [CrossRef]
3. Liu, J.; Kiesewetter, G.; Klimont, Z.; Cofala, J.; Heyes, C.; Schöpp, W.; Zhu, T.; Cao, G.; Gómez-Sanabria, A.; Sander, R.; et al. Mitigation pathways of air pollution from residential emissions in the Beijing-Tianjin-Hebei region in China. *Environ. Int.* **2019**, *125*, 236–244. [CrossRef] [PubMed]
4. Burnett, R.; Chen, H.; Szyszkowicz, M.; Fann, N.; Hubbell, B.; Pope, C.A.; Apte, J.S.; Brauer, M.; Cohen, A.; Weichenthal, S.; et al. Global estimates of mortality associated with long-term exposure to outdoor fine particulate matter. *Proc. Natl. Acad. Sci. USA* **2018**, *115*, 9592–9597. [CrossRef] [PubMed]

5. Fang, D.; Wang, Q.G.; Li, H.M.; Yu, Y.; Lu, Y.; Qian, X. Mortality effects assessment of ambient PM2.5 pollution in the 74 leading cities of China. *Sci. Total Environ.* **2016**, *569–570*, 1545–1552. [CrossRef] [PubMed]
6. Li, Y.; Dai, Z.; Liu, X. Analysis of Spatial-Temporal Characteristics of the PM2.5 Concentrations in Weifang City, China. *Sustainability* **2018**, *10*, 2960. [CrossRef]
7. Cheng, Z.H.; Li, L.S.; Liu, J. Identifying the spatial effects and driving factors of urban PM2.5 pollution in China. *Ecol. Indic.* **2017**, *82*, 61–75. [CrossRef]
8. Liu, H.; Fang, C.; Zhang, X.; Wang, Z.; Bao, C.; Li, F. The effect of natural and anthropogenic factors on haze pollution in Chinese cities: A spatial econometrics approach. *J. Clean. Prod.* **2017**, *165*, 323–333. [CrossRef]
9. Ma, Y.; Ji, Q.; Fan, Y. Spatial linkage analysis of the impact of regional economic activities on PM2.5 pollution in China. *J. Clean. Prod.* **2016**, *139*, 1157–1167. [CrossRef]
10. Wei, X.; Liu, M.; Yang, J.; Du, W.-N.; Sun, X.; Huang, Y.-P.; Zhang, X.; Khalil, S.K.; Luo, D.-M.; Zhou, Y.-D. Characterization of PM2.5-bound PAHs and carbonaceous aerosols during three-month severe haze episode in Shanghai, China: Chemical composition, source apportionment and long-range transportation. *Atmos. Environ.* **2019**, *203*, 1–9. [CrossRef]
11. Yan, D.; Lei, Y.; Shi, Y.; Zhu, Q.; Li, L.; Zhang, Z. Evolution of the spatiotemporal pattern of PM2.5 concentrations in China–A case study from the Beijing-Tianjin-Hebei region. *Atmos. Environ.* **2018**, *183*, 225–233. [CrossRef]
12. Zhang, Z.; Wang, W.; Cheng, M.; Liu, S.; Xu, J.; He, Y.; Meng, F. The contribution of residential coal combustion to PM2.5 pollution over China's Beijing-Tianjin-Hebei region in winter. *Atmos. Environ.* **2017**, *159*, 147–161. [CrossRef]
13. Zhang, Y.; Chen, J.; Yang, H.; Li, R.; Yu, Q. Seasonal variation and potential source regions of PM2.5-bound PAHs in the megacity Beijing, China: Impact of regional transport. *Environ. Pollut.* **2017**, *231*, 329–338. [CrossRef]
14. Zhang, Y.; Zhang, H.; Deng, J.; Du, W.; Hong, Y.; Xu, L.; Qiu, Y.; Hong, Z.; Wu, X.; Ma, Q.; et al. Source regions and transport pathways of PM2.5 at a regional background site in East China. *Atmos. Environ.* **2017**, *167*, 202–211. [CrossRef]
15. Li, C.; Dai, Z.; Yang, L.; Ma, Z. Spatiotemporal Characteristics of Air Quality across Weifang from 2014–2018. *Int. J. Environ. Res. Public Health* **2019**, *16*, 3122. [CrossRef]
16. Ghosh, S.; Biswas, J.; Guttikunda, S.; Roychowdhury, S.; Nayak, M. An investigation of potential regional and local source regions affecting fine particulate matter concentrations in Delhi, India. *J. Air Waste Manag. Assoc.* **2015**, *65*, 218–231. [CrossRef] [PubMed]
17. Gogikar, P.; Tyagi, B. Assessment of particulate matter variation during 2011–2015 over a tropical station Agra, India. *Atmos. Environ.* **2016**, *147*, 11–21. [CrossRef]
18. Hao, T.; Cai, Z.; Chen, S.; Han, S.; Yao, Q.; Fan, W. Transport Pathways and Potential Source Regions of PM2.5 on the West Coast of Bohai Bay during 2009–2018. *Atmosphere* **2019**, *10*, 345. [CrossRef]
19. Yan, R.; Yu, S.; Zhang, Q.; Li, P.; Wang, S.; Chen, B.; Liu, W. A heavy haze episode in Beijing in February of 2014: Characteristics, origins and implications. *Atmos. Pollut. Res.* **2015**, *6*, 867–876. [CrossRef]
20. Donnelly, A.; Broderick, B.; Misstear, B. The effect of long-range air mass transport pathways on PM10 and NO_2 concentrations at urban and rural background sites in Ireland: Quantification using clustering techniques. *J. Environ. Sci. Health Part A* **2015**, *50*, 647–658. [CrossRef] [PubMed]
21. Lee, S.; Ho, C.-H.; Choi, Y.-S. High-PM10 concentration episodes in Seoul, Korea: Background sources and related meteorological conditions. *Atmos. Environ.* **2011**, *45*, 7240–7247. [CrossRef]
22. Tiwari, S.; Dumka, U.; Gautam, A.; Kaskaoutis, D.; Srivastava, A.; Bisht, D.; Chakrabarty, R.; Sumlin, B.; Solmon, F. Assessment of PM2.5 and PM10 over Guwahati in Brahmaputra River Valley: Temporal evolution, source apportionment and meteorological dependence. *Atmos. Pollut. Res.* **2017**, *8*, 13–28. [CrossRef]
23. Li, D.; Liu, J.; Zhang, J.; Gui, H.; Du, P.; Yu, T.; Wang, J.; Lu, Y.; Liu, W.; Cheng, Y. Identification of long-range transport pathways and potential sources of PM2.5 and PM10 in Beijing from 2014 to 2015. *J. Environ. Sci.* **2017**, *56*, 214–229. [CrossRef] [PubMed]
24. Li, M.; Huang, X.; Zhu, L.; Li, J.; Song, Y.; Cai, X.; Xie, S. Analysis of the transport pathways and potential sources of PM10 in Shanghai based on three methods. *Sci. Total Environ.* **2014**, *414*, 525–534. [CrossRef] [PubMed]
25. Yang, W.; Wang, G.; Bi, C. Analysis of Long-Range Transport Effects on PM2.5 during a Short Severe Haze in Beijing, China. *Aerosol Air Qual. Res.* **2017**, *17*, 1610–1622. [CrossRef]

26. Liao, T.; Wang, S.; Ai, J. Gui, K.; Duan, B.; Zhao, Q.; Zhang, X.; Jiang, W.; Sun, Y. Heavy pollution episodes, transport pathways and potential sources of PM2.5 during the winter of 2013 in Chengdu (China). *Sci. Total Environ.* **2017**, *584–585*, 1056–1065. [CrossRef]
27. Xin, Y.; Wang, G.; Chen, L. Identification of Long-Range Transport Pathways and Potential Sources of PM10 in Tibetan Plateau Uplift Area: Case Study of Xining, China. *Environ. Pollut.* **2014**, *224*, 44–53. [CrossRef]
28. Lv, B.; Liu, Y.; Yu, P.; Zhang, B.; Bai, Y. Characterizations of PM2.5 Pollution Pathways and Sources Analysis in Four Large Cities in China. *Aerosol Air Qual. Res.* **2015**, *15*, 1836–1843. [CrossRef]
29. Draxler, R.; Hess, G. An overview of HYSPLIT-4 modeling system for trajectories dispersion and deposition. *Aust. Met. Mag.* **1998**, *47*, 295–308, ISSN: 0004-9743.
30. Sahu, S.; Zhang, H.; Guo, H.; Hu, J.; Ying, Q.; Kota, S. Health risk associated with potential source regions of PM2.5 in Indian cities. *Air Qual. Atmos. Health* **2019**, *12*, 327–340. [CrossRef]
31. Yang, K.; Li, Q.; Yuan, M.; Guo, M.; Wang, Y.; Li, S.; Tian, C.; Tang, J.; Sun, J.; Li, J.; et al. Temporal variations and potential sources of organophosphate esters in PM2.5 in Xinxiang, North China. *Chemosphere* **2019**, *215*, 500–506. [CrossRef] [PubMed]
32. Sirois, A.; Bottenheim, J. Use of backward trajectories to interpret the 5-year record of pan and O_3 ambient air concentrations at Kejimkujik National Park, Nova Scotia. *J. Geophys. Res.* **1995**, *100*, 2867–2881. [CrossRef]
33. Ashbaugh, L.; Malm, W.; Sadeh, W. A residence time probability analysis of sulfur concentrations at Grand Canyon National Park. *Atmos. Environ.* **1985**, *19*, 1263–1270. [CrossRef]
34. Wang, Y.; Zhang, X.; Draxler, R. TrajStat: GIS-based software that uses various trajectory statistical analysis methods to identify potential sources from long-term air pollution measurement data. *Environ. Model. Softw.* **2009**, *24*, 938–939. [CrossRef]
35. Dimitriou, K.; Kassomenos, P. Combining AOT, Angstrom Exponent and PM concentration data, with PSCF model, to distinguish fine and coarse aerosol intrusions in Southern France. *Atmos. Res.* **2016**, *172–173*, 74–82. [CrossRef]
36. Polissar, A.; Hopke, P.; Poirot, R. Atmospheric aerosol over Vermont: Chemical composition and sources. *Environ. Sci. Technol.* **2001**, *35*, 4604–4621. [CrossRef]
37. Wang, L.; Liu, Z.; Sun, Y.; Ji, D.; Wang, Y. Long-range transport and regional sources of PM2.5 in Beijing based on long-term observations from 2005 to 2010. *Atmos. Res.* **2015**, *157*, 37–48. [CrossRef]
38. Zhang, K.; Shang, X.; Herrmann, H.; Meng, F.; Mo, Z.; Chen, J.; Lv, W. Approaches for identifying PM2.5 source types and source areas at a remote background site of South China in spring. *Sci. Total Environ.* **2019**, *691*, 1320–1327. [CrossRef] [PubMed]
39. Stohl, A. Trajectory statistics-a new method to establish source-receptor relationships of air pollutants and its application to the transport of particulate sulfate in Europe. *Atmos. Environ.* **1996**, *30*, 579–587. [CrossRef]
40. Stohl, H.; Krom-kolb, H. Origin of ozone in Vienna and surroundings, Austria. *Atmos. Environ.* **1994**, *28*, 1255–1266. [CrossRef]
41. Liu, H.; Fang, C.; Huang, J.; Zhu, X.D.; Zhou, Y.; Wang, Z.B.; Zhang, Q. The spatial-temporal characteristics and influencing factors of air pollution in Beijing-Tianjin-Hebei urban agglomeration. *J. Geogr. Sci.* **2018**, *73*, 177–191. [CrossRef]
42. Yang, Q.; Yuan, Q.; Li, T.; Shen, H.; Zhang, L. The Relationships between PM2.5 and Meteorological Factors in China: Seasonal and Regional Variations. *Int. J. Environ. Res. Public Health* **2017**, *14*, 1510. [CrossRef]
43. Zhan, D.; Kwan, M.; Zhang, W.; Yu, X.; Meng, B.; Liu, Q. The driving factors of air quality index in China. *J. Clean. Prod.* **2018**, *197*, 1342–1351. [CrossRef]
44. Yan, X.; Gou, X.; Wu, W.; Huang, F.; Yang, J.; Liu, Y. Analysis of atmospheric particulate transport path and potential source area in Yinchuan. *Acta Sci. Circumstantiae* **2018**, *38*, 1727–1738. [CrossRef]
45. Shandong's Outer Dust in April Became the Main Force of PM10 Pollution. Available online: http://weifang.iqilu.com/rdjj/2016/0512/2785473.shtml (accessed on 12 May 2016).
46. Wang, S.; Zhou, C.; Wang, Z.; Feng, K.; Hubacek, K. The characteristics and drivers of fine particulate matter (PM2.5) distribution in China. *J. Clean. Prod.* **2017**, *142*, 1800–1809. [CrossRef]

4

Evaluation of Analysis by Cross-Validation, Part I: Using Verification Metrics

Richard Ménard *and Martin Deshaies-Jacques

Air Quality Research Division, Environment and Climate Change Canada, 2121 Transcanada Highway, Dorval, QC H9P 1J3, Canada; martin.deshaies-jacques@canada.ca

* Correspondence: richard.menard@canada.ca.

Abstract: We examine how passive and active observations are useful to evaluate an air quality analysis. By leaving out observations from the analysis, we form passive observations, and the observations used in the analysis are called active observations. We evaluated the surface air quality analysis of O_3 and $PM_{2.5}$ against passive and active observations using standard model verification metrics such as bias, fractional bias, fraction of correct within a factor of 2, correlation and variance. The results show that verification of analyses against active observations always give an overestimation of the correlation and an underestimation of the variance. Evaluation against passive or any independent observations display a minimum of variance and maximum of correlation as we vary the observation weight, thus providing a mean to obtain the optimal observation weight. For the time and dates considered, the correlation between (independent) observations and the model is 0.55 for O_3 and 0.3 for $PM_{2.5}$ and for the analysis, with optimal observation weight, increases to 0.74 for O_3 and 0.54 for $PM_{2.5}$. We show that bias can be a misleading measure of evaluation and recommend the use of a fractional bias such as the modified normalized mean bias (MNMB). An evaluation of the model bias and variance as a function of model values also show a clear linear dependence with the model values for both O_3 and $PM_{2.5}$.

Keywords: chemical data assimilation; air quality model diagnostics; cross-validation

---

1. Introduction

Since 2003, Environment and Climate Change Canada (ECCC) has been producing hourly surface analyses of pollutants covering North America [1,2] which became operational products in February 2013 [3]. The analyses are produced using an optimum interpolation scheme that combines the operational air quality forecast model GEM-MACH output [4] (CHRONOS model output was used prior to 2010 [5]) with real-time hourly observations of O_3, $PM_{2.5}$, PM_{10}, NO_2, and SO_2 from the AirNow gateway with additional observations from Canada. As those surface analyses are not used to initialize an air quality model, it raises the issue on how to evaluate them. We conduct routine evaluations using the same set of observations as those used to produce the analysis. Once in a while, when there is a change in the system, a more thorough evaluation is conducted where we leave out a certain fraction of the observations and use them as independent observations, a process known as cross-validation. Observations used in producing the analysis are called *active observations* while those not used for evaluation are *passive observations*. Cross-validation is used to validate any model that depends on data. In air quality applications it has been used, for example, for mapping and exposure models [6–8]. The purpose of this two-part paper is to examine the relative merit of using active or passive observations (or independent observations in general) viewed from different evaluation metrics, but also to develop, in the second part, a mathematical framework to estimate the analysis error, and in doing so, to improve the analysis.

The evaluation of an analysis is important, even in the case where it is used to initialize an air quality forecast model, since the evaluation of the resulting air quality forecast may not be a good measure of the quality of the analysis. In air quality forecasting, the forecast error growth is small, depicts little sensitivity to initial conditions and is in fact more sensitive to numerous modeling errors such as: photochemistry, clouds, meteorology, boundary conditions and emissions just to name a few [9–13]. Furthermore, chemical species that are observed are incomplete compared to species needed to initialize an air quality model, incomplete in terms of the number of species observed as well as in their kind [9,11,13,14]. Only a fraction of the observed species (either of secondary or primary pollutants) are usable for data assimilation; important chemical mechanisms are left completely unobserved and for aerosols, information on size distribution is quite limited and almost nonexistent when it comes to speciation [9,13]. In addition, the observational coverage is limited to the surface or to total column measurements which, up until now, were available at one or two local times per day. There are thus many assumptions to be made from an analysis to a proper 3D initial chemical condition and surface emission correction and its subsequent impact on the air quality forecast. These considerations warrant an independent evaluation of the quality of the analysis on its own [15].

Evaluating an analysis with observations is quite different from evaluating a model with observations, since analyses are created from observations. From a statistical point of view, the observation and analysis cannot be considered independent. However, let us assume that observation errors are spatially uncorrelated. Then, since the passive and active observation sites are never collocated, then the errors from passive observations are uncorrelated with errors of active observations (i.e., observations that are used for the analysis). Furthermore, since the modelling errors are usually assumed to be uncorrelated with observation errors, then it is also uncorrelated with the analysis errors. Cross-validation thus offers a means to evaluate analyses with statistically independent (passive) observations [16].

In part one of this paper, we evaluate the relative merit of passive and active observations in the evaluation of analyses using standard metrics used for model evaluation. We show how and when the use of active observations can be misleading and that passive observations can provide a means to identify optimal analyses. Our examples show that optimal analyses, at the independent observation sites, have much smaller biases than the model biases and increase the correlation coefficient by nearly a factor of 2.

The paper is thus organized as follows. First we present the analysis scheme we will be using, as well as the cross-validation design, the evaluation metrics and the configuration of the experiments. Then in Section 3, we assess the quality of the analyses in both active and passive observation spaces using standard air quality evaluation metrics, identify some pitfalls of some metrics and advocate using active observations. Conclusions are presented in Section 4.

2. Experimental Design

2.1. Design of the Objective Analysis Solver

In optimum interpolation there is no use of an explicit interpolation observation operator. The correlation between a pair of locations, either from two observation sites or from an observation site to a model grid point, is computed as a function of distance using a prescribed correlation function. The observation operator is in effect a delta function applied over a continuous spatial domain [17].

In this study we interpolate the gridded analysis field to observation locations, using bilinear interpolation, to compute residuals such as observation-minus-analysis. Thus there can be a discrepancy between the observation operator used to generate the analysis, i.e., delta functions, and the observation operator used to interpolate the analysis field at the observation location, i.e., bilinear interpolation. To eliminate this discrepancy in observation operators we have revised the optimum interpolation scheme to use explicitly the same bilinear interpolation in handling the error covariance. We will give details below.

As in the operational optimum interpolation, the inversion of the innovation covariance matrix for the analysis solver is done using Choleski decomposition on the full matrix. The number of observations to be processed per analysis being of the order of a thousand or less, there was no need for computational simplification for large number of observations by using either data selection [18] or compact support correlation functions [17,19]. Thus, the analysis scheme used in this study computes explicitly the gain matrix $\widetilde{\mathbf{K}}$ as,

$$\widetilde{\mathbf{K}} = \widetilde{\mathbf{B}}\mathbf{H}^T(\mathbf{H}\widetilde{\mathbf{B}}\mathbf{H}^T + \widetilde{\mathbf{R}})^{-1} \quad (1)$$

where $\mathbf{H}$ is a bilinear interpolation operator, $\widetilde{\mathbf{B}}$ is the prescribed background error covariance and $\widetilde{\mathbf{R}}$ is the prescribed observation error covariance. The tilde $(\widetilde{\cdot})$ emphasizes that these are prescribed, potentially suboptimal, quantities.

The computational demand of the Kalman gain was kept low by computing the background error correlation function only at model grid points needed for the bilinear interpolation. For example, to calculate the correlation between a pair of observations requires the computation of correlation between four points surrounding observation 1 (needed for the bilinear interpolation) and the other four points surrounding observation 2, thus forming a 4×4 correlation matrix $\mathbf{C}$ between the target model grid points. Then we calculate $\mathbf{HCH}^T$ which gives the correlation between two observation sites. This procedure is generalized for the N observations needed for the analysis. Equation (1) also involves the computation of $\widetilde{\mathbf{B}}\mathbf{H}^T$ that we compute as a set of N representers (i.e., columns of $\widetilde{\mathbf{B}}\mathbf{H}^T$), each being a 2D field that maps the background error covariance in model space with a single observation location, using again the bilinear interpolation approach to get a single interpolated representor for each observation location. By doing so we keep the consistency between the observation operators used for interpolation of a field and the observation operator used to manipulate matrices.

2.2. Cross-Validation

Cross-validation is a technique to evaluate an analysis (or in general any model that depends on observations) by partitioning the original observation data set into a training set, used to create the analysis, and an independent (or passive) set, used to evaluate the analysis. The most common cross-validation designs are: the *k*-fold cross-validation, where the original observation data set is partitioned into *k* equal size subsamples and the leave-one-out cross-validation, where *N* subsamples are created, each with one different observation set aside for the evaluation while the other $N - 1$ observations are used in producing the analysis. The cross-validation is then repeated with all the different sets until all observations have been used for evaluation. Clearly, there are *k* analyses computed in the *k*-fold cross-validation and *N* in the leave-one-out cross-validation, which is being computationally demanding when *N* is large. The main disadvantage of the *k*-fold cross-validation is that the analyses being evaluated uses a smaller number of observations (actually $(k-1)N/k$) than the original observation data set, whereas the leave-one-out cross-validation evaluates analysis that uses nearly the same number of observations (actually $N - 1$) as the original observation data set. This actually matters with the *k*-fold cross-validation if we need an estimate of the analysis error variance (or any other second moments) as the analysis error variance depends on the number of observations used.

Let $\mathbf{O}_j$ be a vector that contains the *j*th set of observations used for evaluation, and let $\mathbf{A}_{(j)}$ be a vector of analysis value interpolated at the verification observation locations of $\mathbf{O}_j$ and where the analysis used all observations except those in $\mathbf{O}_j$ (the index in parenthesis, i.e., (j), indicates all sets except the set *j*). It is customary in cross-validation literature (e.g., [20]) to construct a mean square error cost function, often denoted by CV,

$$\mathrm{CV} = \sum_j (\mathbf{O}_j - \mathbf{A}_{(j)})^T(\mathbf{O}_j - \mathbf{A}_{(j)}) \quad (2)$$

that represents a misfit quadratic error of the model **A**- in our case the analysis. Different model **A** can be compared and selected from which the CV value is smallest. Likewise, a tunable parameter in **A** can be obtained by minimizing the cost function CV with respect to that parameter. As we shall discuss later in this paper, in Section 4 and onwards, the bias of $(\mathbf{O}_j - \mathbf{A}_{(j)})$ needs to be removed from the cost function in order to estimate the input error covariance parameters.

In applications and thus in all experiments that follow, the analyses and verification against passive observations are made only with a set of observations that have passed a quality control. The quality control is nearly identical to the quality control used for the operational implementation of the analysis of surface pollutants at ECCC (see supplementary material in Robichaud et al. [3]). It consists in discarding observations that report a negative value, or whose value exceeds a certain unrealistic threshold set to 300 ppbv for ozone (300 $\mu g/m^3$ for $PM_{2.5}$). Observations are also discarded based on innovations (or observed-minus-background values) when, for ozone, they exceed 50 ppbv (100 $\mu g/m^3$ for $PM_{2.5}$) in absolute value. The quality-controlled observations are then separated into three sets of observations of equal numbers, i.e., a 3-fold cross-validation procedure, as illustrated in Figure 1.

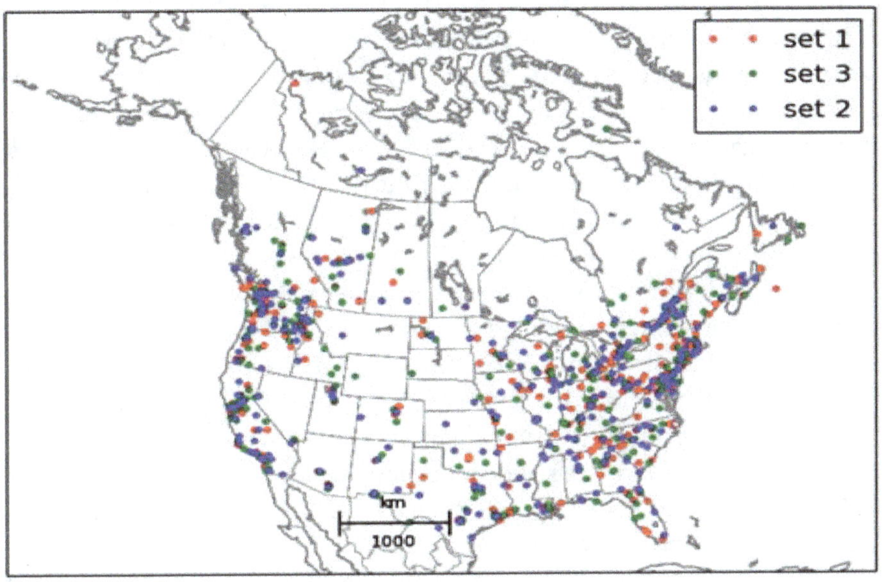

Figure 1. Spatial distribution of the three subsets of $PM_{2.5}$ observations used for cross-validation. The selection algorithm is based on regular picking of station by ID number.

The selection into three sets is made by station ID number, selecting on a regular basis each fourth station, starting with station 1 for the first set, station 2 for the second set and station 3 for the third set, and resulting in locally spatially random distribution of each sets of stations. The cross-validation is then made by leaving one set out of the three sets, and using the remaining two sets to produce the analysis.

2.3. Verification Metrics

We will evaluate the analyses against passive and active observations with the following standard evaluation metrics used for air quality models [21–24]; the bias, the modified normalized mean bias (MNMB), the fraction of correct within a factor of 2 (FC2), the variance ($\text{var}(O - A)$) and the correlation coefficient ($cor(O, A)$), where the statistics is computed over time t for each station, and then the resulting metric is averaged over all the verifying station i,

$$\text{bias} = \frac{1}{N_i}\sum_i \left\{ \frac{1}{N_k}\sum_k (O_i(t_k) - A_i(t_k)) \right\} \tag{3}$$

$$\text{MNMB} = \frac{1}{N_i}\sum_i \left\{ \frac{2}{N_k}\sum_k \left(\frac{O_i(t_k) - A_i(t_k)}{O_i(t_k) + A_i(t_k)} \right) \right\} \tag{4}$$

$$\mathrm{FC2} = \frac{1}{N_i}\sum_i \left\{ \frac{1}{N_k} count\left\{ 0.5 \leq \frac{A_i(t_k)}{O_i(t_k)} \leq 2 \right\} \right\} \tag{5}$$

$$\mathrm{var}(O - A) = \frac{1}{N_i}\sum_i \left\{ \frac{1}{N_k - 1}\sum_k \left[(O_i(t_k) - A_i(t_k)) - \overline{(O_i - A_i)} \right]^2 \right\} \tag{6}$$

$$\mathrm{cor}(O, A) = \frac{1}{N_i}\sum_i \left\{ \frac{1}{N_k - 1} \frac{\sum_k (O_i(t_k) - \overline{O}_i)(A_i(t_k) - \overline{A}_i)}{\sqrt{\sum_k (O_i(t_k) - \overline{O}_i)^2 \sum_k (A_i(t_k) - \overline{A}_i)^2}} \right\} \tag{7}$$

where $O_i(t_k)$ is the observed value at time t_k at the station i, $A_i(t_k)$ is the analysis at time t_k interpolated at the location of the station i, N_k is the total number of time sample per station, N_s is the total number of stations (in the sample or over the domain), and the overbar $\overline{(\ )}$ denotes the time average. The bias and the MNMB are metrics of the first moment that have distinctive properties. The bias gives a representative measure of the systematic discrepancy between analyzed and observed values over the whole set of observations used for verification. However, since atmospheric constituents exhibit a range of values that can vary in time and space, and different constituents have different range of values and may as well be expressed with different units, a relative error measure such as the MNMB is often preferred [24]. The MNMB is a dimensionless quantity that falls in the range $[-2, +2]$. The factor of 2 is introduced so to give a % error interpretation to the MNMB. This metric has the additional advantage of treating over- and under-estimation in a symmetric way [24]. However, the MNMB is relatively insensitive to relatively large discrepancies between analysis (or model) values and observed values, that is when its values are close to +2 (200%) or -2 (-200%) [23].

The fraction of correction within a factor of 2 (FC2) is a measure of reliability. It is based on counts and has the distinctive advantage that it is insensitive to outliers. It is worth mentioning that it accounts both high values outliers and also low values outliers that is a unique property of this metric [22]. The FC2 metric is also symmetric with respect to permutation of A and O, it is also dimensionless and its values must lie between 0 and 1. Our experience with this metric indicates that it is relatively insentive for relatively good agreement between analysis and observed values.

The variance, $\mathrm{var}(O - A)$, and the correlation coefficient $\mathrm{cor}(O, A)$ are metrics that depend on the spread of the discrepancy between analysis and observed values. The variance is not a dimensionless metric. It gives a representative measure of the spread of the discrepancy between analyses and observations and is not sensitive to systematic errors. As we will show in Section 4 and also shown in Marseille et al. [16], $\mathrm{var}(O - A)$ with passive observations has the distinct advantage of providing a measure of the true analysis error variance (i.e., the error with respect to the truth) and $\mathrm{var}(O - A)$ can be considered as a cross-validation cost function CV, Equation (2), with debiased $(O - A)$ increments. As for any second moment metric, $\mathrm{var}(O - A)$ is sensitive to outliers; they must be removed, and this is done by gross check of the $(O - B)$, as explained in the previous subsection Section 2.2. Finally, the correlation coefficient is a dimensionless quantity that lies in the range $[-1, +1]$. It is also invariant to shifts in the mean (i.e., not sensitive to systematic errors), and multiplicative rescaling of either analysis or observations. The correlation is also relatively insensitive to improvement when the correlation is close to 1 or -1.

2.4. Description of the Ensemble of Analyses and Their Verification Statistics

A series of hourly analyses of O_3 and $PM_{2.5}$ at 21 UTC for a period of 60 days (14 June to 12 August 2014) were performed with given input error statistics using the operational model GEM-MACH and the real-time AirNow observations as described in the introduction and with quality controlled observations (see Section 2.2 above). In all experiments, the observation and background error variances, σ_o^2 and σ_b^2, used in the analysis are uniform. The prescribed observation error and background error covariances are given as $\widetilde{\mathbf{R}} = \sigma_o^2 \mathbf{I}$, $\widetilde{\mathbf{B}} = \sigma_b^2 \mathbf{C}$, where the correlation model $\mathbf{C}$ is a homogeneous isotropic second-order autoregressive model with a correlation length obtained by maximum likelihood, as in Ménard et al. [17]. Note that aside from quality control, that ends up rejecting some observations, the analysis uses the observation values and model realizations as is, with no bias correction.

We repeat the series of 60 day analyses for different observation and background error variances chosen in such a way that their sum $\sigma_o^2 + \sigma_b^2$ is equal to $\mathrm{var}(O - B)$ but with different ratios of error variances $\gamma = \sigma_o^2/\sigma_b^2$. We perform the series of analyses over a wide range of γ ratios in the interval $[10^{-2}, 10^2]$, thus creating on one end analyses with very large observation weights, i.e., $\gamma \ll 1$, such that the analysis interpolated at the active observation sites tend to match the observed value, and on the other end, with $\gamma \gg 1$, creating analyses with very small observation weight producing analyses that are very close to the background (model) state.

The condition $\sigma_o^2 + \sigma_b^2 = \mathrm{var}(O - B)$, called the *innovation variance consistency*, is an important constraint that is useful for the estimation of the *true* error statistics [25]. Indeed, the stronger condition for the full covariance matrices, the *innovation covariance consistency* criterion, takes the form: $< (O - B)(O - B)^T >= \widetilde{\mathbf{R}} + \mathbf{H}\widetilde{\mathbf{B}}\mathbf{H}^T$, where $<>$ represents the mean over an ensemble of realizations, $\mathbf{H}$ is the interpolation from model grid to the observation location (or observation operator). It is one of the two necessary and sufficient condition to obtain the *true* error covariance statistics (in observation space) [25,26].

As explained in Section 2.3 above, the verification metrics are first calculated over 60 days for a given hour and for each station. Then the metric is averaged over all the verifying stations, resulting in one value of the metric for each hour of the day. Here, however, we computed the metrics for 21 UTC only. If N_s is the total number of stations, the statistics over one of the 3-fold subset then involves an average of the metric over $N_s/3$ passive stations. Doing this for all three subsets, and taking the average of the subsets' results, is equivalent to taking the average of the metric over all stations. In the results that will be presented in the following sections, we always present the average metric over the three passive subsets so that, in the end, the sample size of the passive observation experiments and of the active observation experiments are equal and thus can be presented side by side on the same graphic.

3. Verification against Passive and Active Observations

In this series of experiments, analyses of O_3 and $PM_{2.5}$ were produced using a fixed homogeneous isotropic correlation function, where the correlation length was obtained by maximum likelihood using a second-order auto-regressive model and error variances computed using a local Hollingsworth-Lönnberg fit [17]. A correlation length of 124 km was obtained for O_3 and of 196 km for $PM_{2.5}$. Our correlation length is defined from the curvature at the origin as in Daley [27] and is different from the length-scale parameter of the correlation model (see Ménard et al. [17] for a discussion of these issues). We did a series of 60-days analyses for different values of σ_o^2 and σ_b^2 but such that their sum respects the innovation variance consistency, $\sigma_o^2 + \sigma_b^2 = \mathrm{var}(O - B)$, an important condition for an optimal analysis [25], as explained in Section 2.4. This is the experimental procedure that has been used to generate the Figures 2–7. The results are shown for a wide range of variance ratios $\gamma = \sigma_o^2/\sigma_b^2$ from 10^{-2} to 10^2 in Figures 2–5 and 7 in particular. Note that $\gamma \ll 1$ corresponds to a very large observation weight while $\gamma \gg 1$ correspond to very small observation weight.

The $\mathrm{var}(O - A)$ using passive observations (red curve with circles) and active observations (black curve with squares) is presented in Figure 2 for O_3 (left panel) and $PM_{2.5}$ (right panel). The solid blue line represents $\mathrm{var}(O - B)$, the variance of observation-minus-model, i.e., prior to an analysis. As mentioned in Section 2.4, in the cross-validation experiments we averaged the verification metric over the 3-fold subsets so that, in effect, the total number of observations that end up being used for verification is N_s, the total number of stations. We thus argue that the verification sampling error for the cross-validation experiments (red curve) is the same as for the active observations using the full analysis (i.e., analysis using the total number of stations; black curve). In addition, note that from Figure 1 the station sampling strategy gives rise to spatially random selection of stations, so that the individual metric on each set should be comparable. Furthermore, there is roughly 1300 quality controlled O_3 observations over the domain and 750 $PM_{2.5}$ quality controlled observations, each with 60 time samples or less. To give some qualitative idea of the sampling error, the different metric values for the individual 3-fold sets are presented in the Supplementary Material Figures S1 and S2, where we can see that for $\mathrm{var}(O-A)$ and $\mathrm{cor}(O,A)$ the metric values for the individual sets are nearly indistinguishable from the means of the 3-subset.

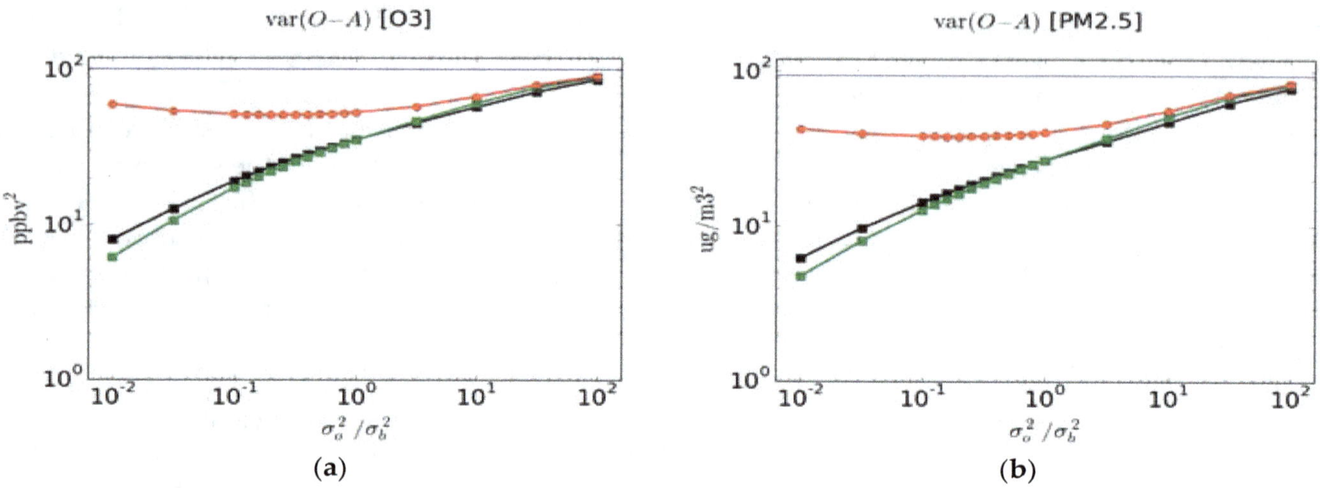

Figure 2. Variance of observation-minus-analysis residuals of O_3 and $PM_{2.5}$ for both active and cross-validation passive observations as a function of $\gamma = \sigma_o^2/\sigma_b^2$. (**a**) is for O_3 with ordinates in ppbv^2 units, and (**b**) is for $PM_{2.5}$ with ordinates in $(\mu\mathrm{g/m}^3)^2$. Red curve results from the evaluation at the passive observation sites (average of the 3-fold subsets). Black curve results from evaluation at the active observation sites with analyses using all observations. Green curve results from the evaluation at the active observation sites in the cross-validation experiment (i.e., using 2/3 of the observations; average of the three subsets). Blue curve is the variance of observation-minus-model.

The difference between the verification against passive observations in cross-validation analyses (red curve) and the verification against active observations using full analyses (black curve) can be attributed to two effects: (1) the analysis used in the cross-validation uses 2/3rd of the total number of observations and thus the analysis error has larger variance than analyses using all observations, (2) since the analysis error variance has typically a local minimum at the individual active observation sites and increases away from it (see for example Figure 4a in [28] i.e., Part II of this paper), an evaluation of the analysis error at passive sites (i.e., away from the active sites) has larger error variance than those evaluated at the active sites [16]. We may call this the distance effect of passive observation sites. In order to separate these two effects, we also display the 3-fold average of the metric verifying against active observation for the cross-validation analyses as a green curve with squares. Thus in summary we display a;

- red curve: using analysis with $2N_s/3$ observations with an evaluation at passive sites
- green curve: using analysis with $2N_s/3$ observations with an evaluation at active sites
- black curve: using analysis with N_s observations with an evaluation at active sites.

The difference between the red and green curves show the influence of distance between passive and active observation sites, whereas the difference between the green and black curves show the influence of having different number of observations in creating the analysis for verification.

Let us first examine the results of verifying against active observations. As the observation weights get smaller (i.e., $\gamma \gg 1$), the analysis draws closer to the background, so that $\mathrm{var}(O-A)$ increases toward $\mathrm{var}(O-B)$. On the other end, when $\gamma \ll 1$, the $\mathrm{var}(O-A)$ continuously decreases as γ diminishes to ultimately reach zero. This is in effect an expected result from the inner working of an analysis scheme that the analysis error variance goes when the observation error variance goes to zero. This effect does not depend on the observed values or the model values. For this reason, the var$(O - A)$ using active observations cannot provide a true measure of the quality of an analysis.

Now let us examine the results of verifying against passive observations with cross-validation analyses. As the observation weights get smaller (i.e., $\gamma \gg 1$), as for active observations the analysis draws closer to the background, so that var$(O - A)$ increases toward var$(O - B)$. On the other end when $\gamma \ll 1$, the var$(O - A)$ using passive observation increases as γ diminishes, whereas the var$(O - A)$ evaluated at the active observation sites (green and black curves) decreases, indicating that the analysis tries to overfit active observations which results in a spatially noisy analysis between the active observation sites. Somewhere in between these two extreme values of γ lies a minimum of var$(O - A)$ where there is neither an overfitting nor an underfitting to the active observations. This "optimal" ratio, that is found by inspection, actually corresponds the optimal analysis. It is also where the analysis error variance with respect to the truth is minimum, but to show this last statement requires an extensive analysis of the problem that we will discuss in part two of this study.

We also computed the verification of the subset of active observations used in the cross-validation experiments with green curves. The difference between the black and green curves indicate the effect of having more observations in the analysis. One would expect that having a larger number of observations in the full analysis active var$(O - A)$ compared to the active var$(O - A)$ for the cross-validation analyses would result in slightly smaller var$(O - A)$. This is indeed observed between the black and green curves when the observation weight is small (i.e., $\gamma \gg 1$). However, surprisingly, when the observation weight is large, $\gamma \ll 1$, we observe the opposite. This intriguing behavior may indicate an inconsistency between the assumption of uniform error variances for σ_o^2 and σ_b^2 (assumed in the input error statistics) and the real spatial distribution of error variances. This discrepancy being simply amplified when the observation weight is large and when there are less observations to produce the analysis.

The difference between var$(O - A)$ at passive sites and active sites (with the same number of observations to construct the analyses) is substantial. For O_3 and for an optimal ratio, the var$(O - A)$ at passive sites is 51.02 ppbv^2 (red curve) while at active sites is 22.77 ppbv^2 (green curve). For PM2.5 and for an optimal ratio, the var$(O - A)$ at passive sites is 38.09 $(\mu g/m^3)^2$(red curve) while at active sites is 15.41 $(\mu g/m^3)^2$(green curve). For both species, the error variance at active sites gives a significant overestimation of the error variance by more than a factor of 2.

In Figure 3, we present the correlation metric between the observations and the analysis using, as in Figure 2, the verification against passive observations in cross-validation analyses (red curve), the verification against active observations using full analyses (black curve) and the verification against active observations in the cross-validation analyses (green curve). The blue curve depicts the correlation between the model and the observations, that is the prior correlation.

The evaluation against passive observations with cross-validation analyses (red curve) shows a maximum at the same values of $\gamma = \sigma_o^2/\sigma_b^2$ than for the var$(O - A)$. We argue that the same arguments of underfitting and overfitting are responsible for this maximum. The correlation between the active observations and the analysis (black and green curves) increases as the observation weight increases (γ decreases), theoretically reaching a value 1 for $\sigma_o^2 = 0$, which is again unrealistic and simply shows the impact of ill-prescribed error statistics in an analysis scheme. The gain in correlation between independent observations and analysis is significant. For O_3, it increases from a value of 0.55 with respect to the model to a value of 0.74 with respect to an optimal analysis (when $\gamma = \sigma_o^2/\sigma_b^2$ is optimal). For $PM_{2.5}$, the correlation against the model has a value of 0.3 which basically has no skill, to a value of 0.54 for optimal analysis, which represent a modest but useable skill. The correlation evaluated at the active sites for an optimal ratio, is 0.85 for O_3 (green curve) and 0.74 for $PM_{2.5}$ (green curve), being a substantial overestimation with respect to values obtained at passive sites.

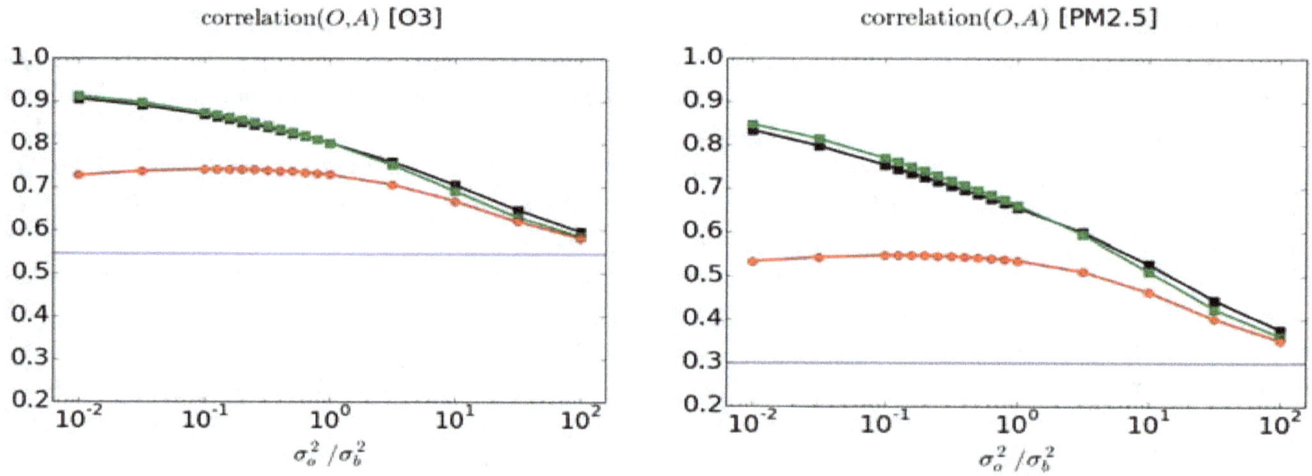

(a) (b)

Figure 3. Correlation between observations and analysis for (**a**) O_3 and (**b**) $PM_{2.5}$ for both active and cross-validation passive observations as a function of $\gamma = \sigma_o^2/\sigma_b^2$. The red, black and green curves are as in Figure 2.

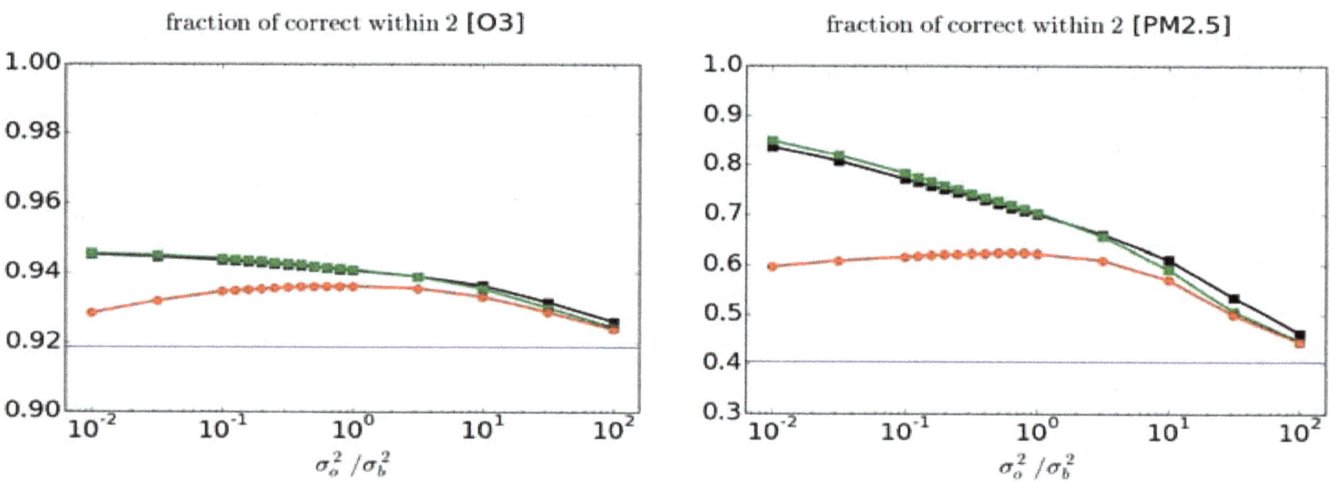

(a) (b)

Figure 4. FC2 for (**a**) O_3 and (**b**) $PM_{2.5}$ for both active and cross-validation passive observations as a function of $\gamma = \sigma_o^2/\sigma_b^2$. The red, black and green curves are as in Figure 2.

Another metric that we have considered is the FC2, Equation (5) [3]. The evaluation of this metric against passive and active observations is presented in Figure 4 for O_3 (left panel) and $PM_{2.5}$ (right panel). Note that the scale in the ordinate is quite different between the left and right panels. Although the results bear similarity with the correlation between O and A presented in Figure 3, the maximum with passive observations is reached at larger γ values than those obtained for $\text{var}(O - A)$ or $\text{cor}(O, A)$, which are identical. Individual fold results are presented in the supplementary materials Figure S3.

The interpretation of this metric is, however, not clear. Although the ratio $z = A/O$ is a dimensionless quantity the spread of z is generally not independent of the variance of A or O and there are cases where it is. So to count the number of occurrence of z between the dimensionless values 0.5 and 2 is confusing. As a simplified illustration, suppose that A is normally distributed as $N(0, \sigma_a^2)$ and similarly with $O \sim N(0, \sigma_o^2)$. The ratio of these two random variables is then a Cauchy distribution whose probability density function (pdf) is $\sigma_o\sigma_a/[\pi(\sigma_o^2 z^2 + \sigma_a^2)]$. The mean, variance and higher moments of Cauchy probability distributions are not defined since the integral of the pdf is not bounded; only the mode is defined. Cauchy distributions also have a spread parameter, which in this case is equal to

σ_a/σ_o. If the variance of A and O are equal, then the number count between the dimensionless bounds 0.5 and 2 depends only on the shape of the probability distribution function, not on the variance. If the variance of A and O are different, then it also depends on the ratio of variances. Furthermore, in principle this metric also depends on the bias (which is not the case here for these analyses). It may be a difficult metric to interpret but if used as a quality control, the FC2 have the unique ability of rejecting too low as well as too high values of z.

In Figure 5 we present the bias between observations and analyses, and where the verification is made against passive and active observations as done with the other metrics. Bias is not a dimensionless quantity; note that the range and scale presented for O_3 and $PM_{2.5}$ in Figure 5 are different. The blue curve is the mean $(O - B)$ and thus indicates that for O_3 in average over all observation stations (for the time and dates considered) the model overpredicts, and that for $PM_{2.5}$ the model underpredicts.

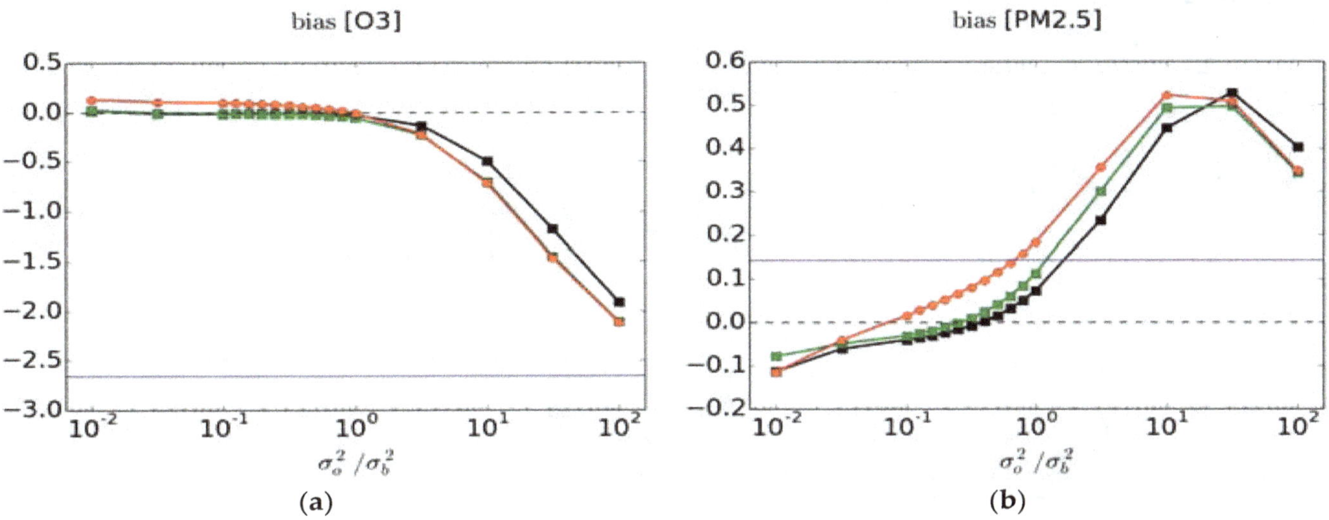

Figure 5. Bias between observation and analysis for (**a**) O_3 and (**b**) $PM_{2.5}$ for both active and cross-validation passive observations as a function of $\gamma = \sigma_o^2/\sigma_b^2$. The red, black and green curves are as in Figure 2.

Contrary to all metric results seen so far, the 3-fold variability of the bias is substantial: it is of the order of ± 0.5 ppbv (in average) for O_3 at passive sites and of the order of ± 0.1 $\mu g/m^3$ (in average) for the $PM_{2.5}$ at passive sites (results shown in the supplementary material Figure S4). The distinction between the red, black and green curves may not be statistically significant for both O_3 and $PM_{2.5}$. However, the difference between the analysis bias and model bias is large and statistically significant (see supplementary material). For O_3, the model bias is eliminated at the passive observation sites (red curve) as long as the observation weight $\gamma \leq 1$. The situation is not so clear for $PM_{2.5}$. In fact, when the observation weight is small, we get the intruiging result that the bias of the analysis is larger bias than the model. How can that be when the observation weight is small (i.e., $\gamma > 1$); should the analysis not be close to the model values? This apparent contradiction reveals a more complex issue underlying the bias metric.

To explore the possible causes, we have calculated the bias per bin of model values, displayed in Figure 6. In order to have a decent sample size per bin, we collect all the $(O - A)$ and $(O - B)$ over time and observation sites, create bins of model values and calculate the statistic per bin (and not per station as before). The result shows that the model bias is nearly linearly dependent on the model values (black boxes in the bias panel). Both O_3 and $PM_{2.5}$ show an underprediction for low model values and an overprediction for large model values. The origin of this bias is not known but one would argue that it is not directly related to chemistry as such since both constituents, O_3 and $PM_{2.5}$, present the same feature. Possible explanations could be related to the model boundary layer, the emissions being too low for low polluted areas and too large for polluted areas, insufficient transport away from polluted areas to unpolluted areas, species destruction/scavenging could be too low in low polluted areas and too high in polluted areas. The lower panels of Figure 6a,b show the count of stations per model bin size. We observe that the majority of stations have O_3 model values in the range of 40 to 55 ppbv, where the bias is negative. Over all the stations, this gives rise to a negative mean $(O - B)$, and this is how we make the claim that the model overpredicts. However, for $PM_{2.5}$ the situation is different: the majority of stations lie in the low model value range, and there are gradually less stations for increasingly larger model values. Although the $(O - B)$ have large negative values in the high model value bin while small model value bins have positives $(O - B)$'s, the effect over all stations is to yield a modestly positive mean $(O - B)$ and thus the model underestimates the $PM_{2.5}$. The results of the analysis evaluated at the passive observation sites are presented with the yellow and grey histogram boxes. In yellow, near optimal analyses with optimal observation weight, as determined by the minimum of $\mathrm{var}(O - A)$ are used, and in grey non-optimal analyses with $\gamma = 10$.

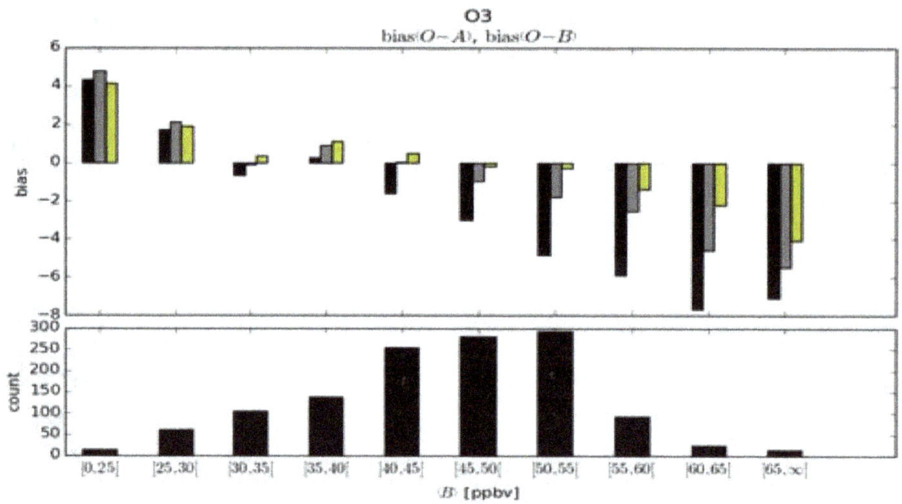

(a)

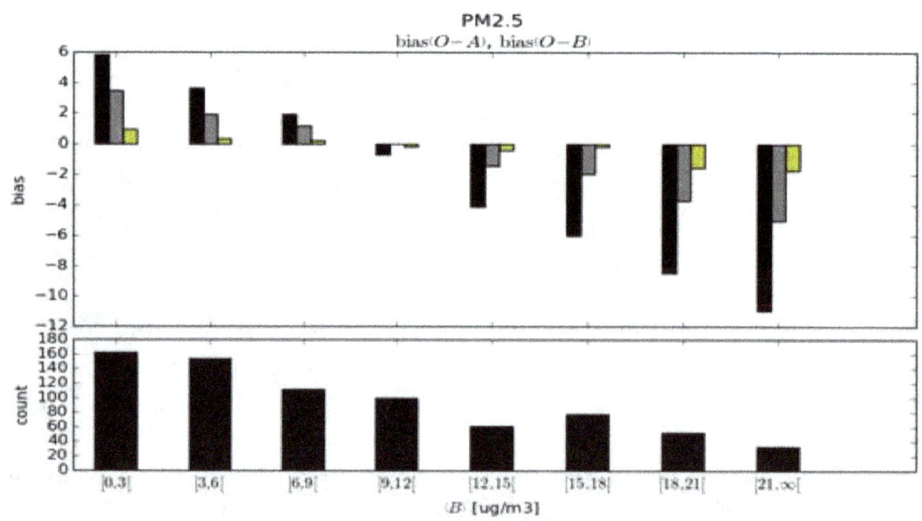

(b)

Figure 6. Biases per bin of model values. Figure (**a**), presents the statistics for O_3 and in (**b**), for $PM_{2.5}$. In the upper portion, (**a**,**b**) are the residual statistics per bin; in black, the $(O - B)$, in grey, the $(O - A)$ at passive observation sites (mean of the 3-fold subsets) for a non-optimal analysis with $\gamma = 10$, and in yellow, the $(O - A)$ at passive observation sites (mean of the 3-fold subsets) using the optimal observation weight. In the lower portion, (**a**,**b**) are the station number count per model values.

We observe that the effect of the optimal analysis is nearly insentive to model bin values, where near zero biases are obtained in most of the range except for very small and very large model values. The fact that we are not able to capture the full benefit of analysis on all model values may be an artefact of the assumption that we are using uniform observation and background error variances whereas the model values varies considerably. In grey, we used the non-optimal analyses with a small observation weight were we set $\gamma = 10$. In the non-optimal case, the state-dependent bias is still present but appears to be nearly perfectly anti-symmetric, positive in the low model value bins and nearly the exact opposite in high model value bins. Since for O_3 the majority of observations lie in the range 40 to 55 ppbv, $(O - A)$ for the optimal analyses at passive observation sites is nearly zero. However, for the non-optimal analysis with $\gamma = 10$, the $(O - A)$ at passive sites is negative, i.e., the analysis is overpredicting, as shown in Figure 5.

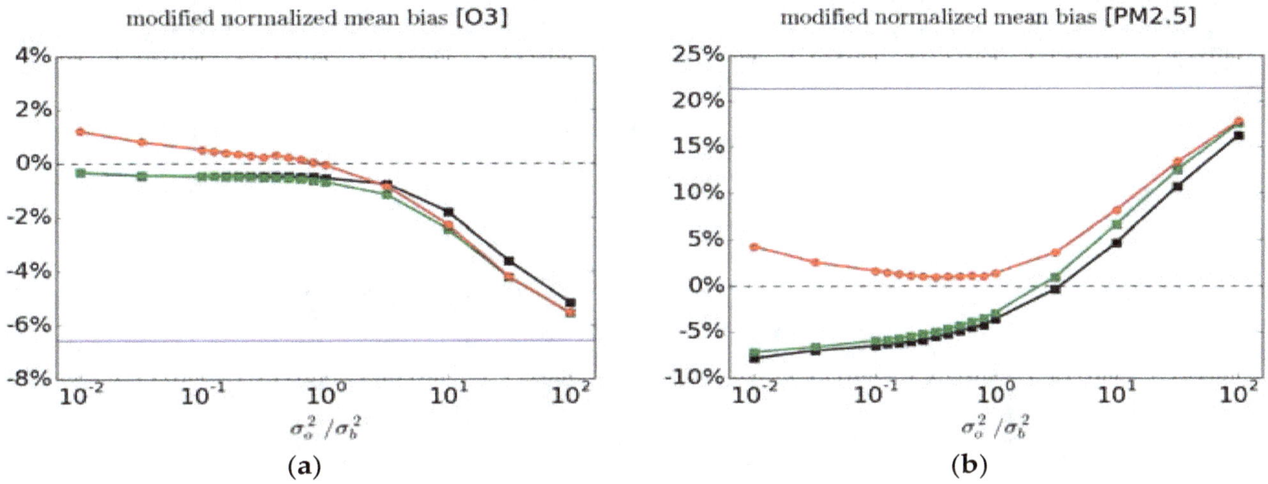

Figure 7. Modified normalized mean bias (MNMB) between observation and analysis for (**a**) O_3 and (**b**) $PM_{2.5}$ for both active and cross-validation passive observations as a function of $\gamma = \sigma_o^2 / \sigma_b^2$. The red, black and green curves are as in Figure 2.

For $PM_{2.5}$, the weighted sum of the $(O - A)$ bins is such that over all stations the bias for an optimal analysis is nearly zero. In the case of the non-optimal analysis with $\gamma = 10$, the weighted sum of the nearly anti-symmetric $(O - A)$ bias per bin gives more weight to the positive bias at smaller model values, so that overall there is a positive $(O - A)$, as in Figure 5.

To circumvent the state-dependency of the $(O - A)$ biases it is useful to consider instead a fractional bias metric, such as the modified normalized mean bias, MNMB Equation (4). The MNMB metric is a dimensionless measure and as defined with a factor of 2, Equation (4), represents a % error. The MNMB metric is a relative measure with respect to the mean observed-analysis value and is thus less sensitive to spatially varying distribution of the concentrations, revealing instead the intrinsic difference between the fields. The MNMB for O_3 and $PM_{2.5}$ for passive and active observations are displayed in Figure 7 using the same color as in Figure 2. We note immediately that the MNMB analysis bias does not exceed the MNMB model bias as we observed for the bias metric of $PM_{2.5}$ (Figure 5 right panel). The MNMB bias also varies smoothly as a function of γ (at variance with the bias metric for $PM_{2.5}$—Figure 5).

Furthermore, examining the 3-fold variability of the cross-validation analysis MNMB at the passive sites and the variability of the MNMB at the active sites (see Figure S5 in supplementary materials), we infer that for $PM_{2.5}$, where we can actually deduce that the difference between the cross-validation and the validation against active observations is statistically significant when $\gamma < 1$. There is also another important point to make; although analyses are designed to reduce the error variance, it so happens that for a near optimal analysis the fractional bias MNMB is very small, around 1% for O_3 and about 1–2% for $PM_{2.5}$. We argue that it results from an optimal use of observations.

There is also some information to gain from the variance of observed-minus-analysis per bin size, as illustrated in Figure 8, using the same color histograms as in Figure 6. We note that for O_3, the model error variance against observations increases gradually with larger model values. However, the fraction of analysis variance vs. model variance is roughly uniform across all bins. This can be explained by the fact that the observation and background error variances are uniform, and thus the reduction of variance across all bins is uniform as well. However, the situation is different for $PM_{2.5}$. We note a relatively poor performance of the model at low model values, with standard deviation of 7 $\mu g/m^3$. For slightly larger model values (3–6 $\mu g/m^3$), the error variance is smaller to 5.5 $\mu g/m^3$ and then increases almost linearly with model values. The fraction of analysis variance vs. model variance decreases steadily with larger model values. These results thus indicate that the assumption that observation and background error variances are uniform and independent of the model value may have to be revisited.

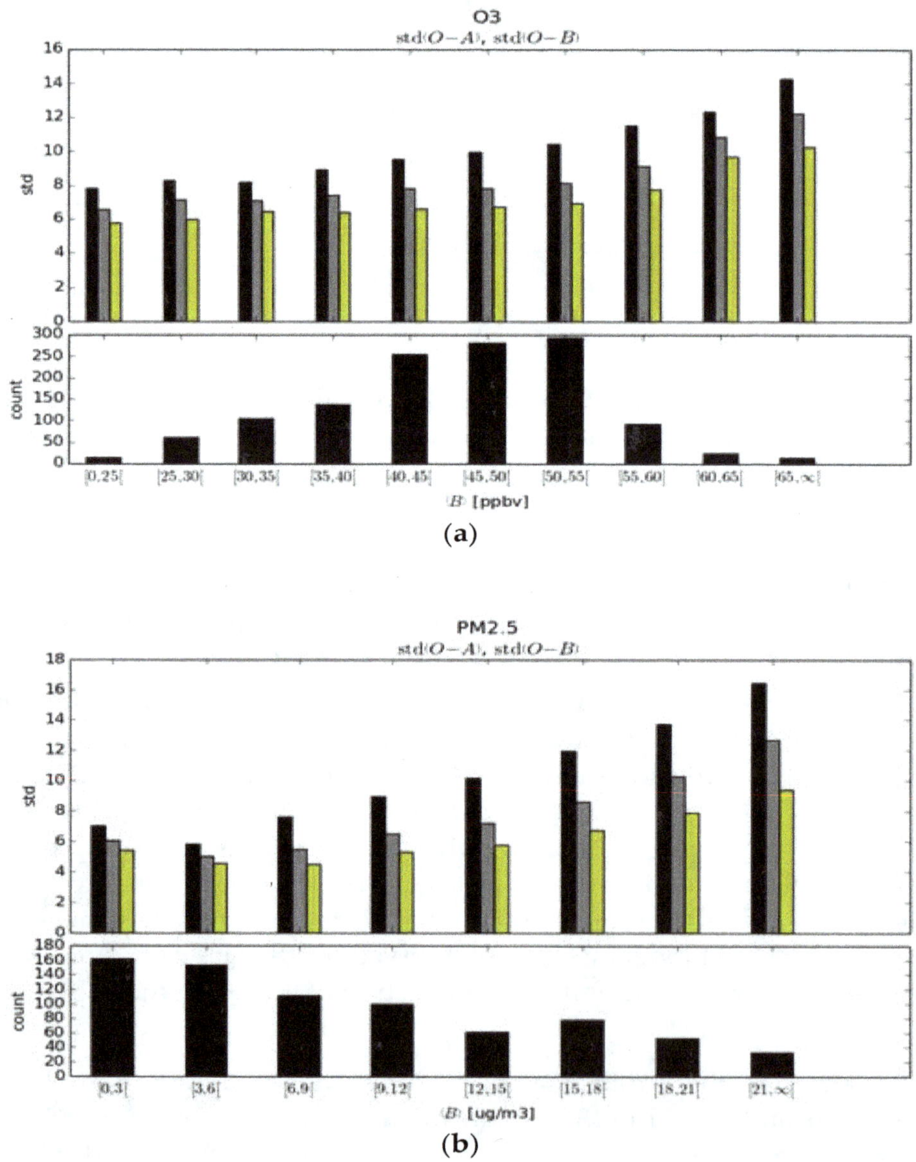

Figure 8. Same as Figure 6 except that we display the variance of analysis-minus-passive observations per bin of model values.

4. Conclusions

We have developed an approach by which analyses can be evaluated and optimized without using a model forecast but rather by partitioning the original observation data set into a training set, to create the analysis, and an independent (or passive) set, used to evaluate the analysis. This kind of evaluation by partitioning is called cross-validation.

The need for such a technique came about from our desire to evaluate our operational surface air quality analyses that are created off-line with no assimilation cycling. Evaluating a surface air quality analysis based on its chemical forecast does in fact require additional information or assumptions, such as vertical correlation, aerosol speciation and bin distribution (while surface measurement is primarily about mass) or unobserved chemical variables correlations, and so on. So that the quality of the chemical forecast is not solely dependent on the quality of the analysis and, if there are compensating errors, can actually be a misleading assessment of the quality of the analysis.

We have applied this cross-validation procedure to the operational analyses of surface O_3 and $PM_{2.5}$ over North America for a period of 60 days and present an evaluation using different metrics; bias, modified normalized mean bias, variance of observation-minus-analysis residuals, correlation between observation and analysis, and fraction of correction within a factor of 2.

Our results show that, in terms of variance and correlation, the verification of analyses against active observations always yield an overestimation of the accuracy of the analysis. This overestimation also increases as the observation weight increases. On the other hand for biases, the distinction between the verification against active observations and passive observations is unclear and drowned in the sample variability. However, using a fractional bias metric, in particular the MNMB, shows that the verification against passive observations can be close to one percent for an optimal analysis while the verification against active observations is much larger.

Results also show the importance of having an optimal analysis for verification. The variance of the analysis with respect to independent observations is minimum and the correlation between the analysis and independent observations is maximum for an optimal analysis. By being a compromise between an overfit to the active observations (which produce noisy analysis field) and an underfit, the optimal analysis offers the best use of observations throughout. At optimality, the analysis fractional bias (MNMB) at the passive observation sites has only one or two percent error whereas the fractional bias of the model is 6.5% for O_3 and 21% for $PM_{2.5}$. The correlation between the analysis and independent observations is also significantly improved with an optimal analysis: the correlation between the model and independent observations is 0.55 for O_3 and increases to 0.74 with the analysis, while for $PM_{2.5}$ the correlation between the model and independent observations is only 0.3 (which is basically no skill) but rises to 0.54 for the analysis.

We also argue that the fraction of correct within a factor of 2, is a metric whose interpretation is unclear as it mixes information about bias, variance and probability distribution in a non-uniform way and does not seem to add anything new to other metrics. The bias is also very sensitive to sample variability and can lead to wrong conclusions. For example, we have seen that the mean analysis bias can be larger than the mean model bias, whether verifying against active or passive observations. However, since an analysis is always closer to the truth than its prior (i.e., the model), it results in an apparent contradiction. This implies that the bias metric cannot be used to faithfully compare model states accurately. Such wrongful conclusions do not arise, however, with the MNMB. We thus recommend avoiding using bias as a measure of truthfulness, and use instead a fractional bias measure such as the MNMB.

We also found that errors in the GEM-MACH model grow almost linearly with the model value. This is particularly evident for the bias where the model underestimates at small model values and overestimates at large model values. Furthermore, this occurs in equal ways for O_3 and $PM_{2.5}$, thus indicating that the source of this bias is not related to chemistry. The fact that, over the entire domain, the model overestimates O_3, and underestimates $PM_{2.5}$ is simply a result of the concentrations. We have not conducted a systematic study of model error for other times of the day and other periods of the year, but it would be very interesting to look at this, to see whether or not changes of biases are due primarily to changes in the distribution of values rather than a fundamental change in the bias per model value bin.

Finally, we have also examined the variance against independent observations per model value bin, and concluded that the error variance is not quite uniform with model values but increases slowly with model values for O_3 and in a more pronounced way for $PM_{2.5}$.

In part two, we will focus on the estimation of the analysis error variance and develop a mathematical formalism that permits the comparison of different diagnostics of variance under different assumptions, optimizes the analysis parameters and gains confidence on the estimate of analysis error as we obtain coherent estimated values across different diagnostics.

Acknowledgments: We are grateful to the US/EPA for the use of the AIRNow database for surface pollutants and to all provincial governments and territories of Canada for kindly transmitting their data to the Canadian Meteorological Centre to produce the surface analysis of atmospheric pollutants. We are also thankful for the proof read by Kerill Semeniuk, and for three anonymous reviewers for their comments and help in improving the manuscript.

Author Contributions: This research was conducted as a joint effort by both authors. R.M. contributed to the theoretical development and wrote the paper, and M.D.-J. conducted all experiments design and execution, proof reading and introduced a new diagnostic of optimal analysis error that was furher extended to passive observations space.

References

1. Ménard, R.; Robichaud, A. The chemistry-forecast system at the Meteorological Service of Canada. In Proceedings of the ECMWF Seminar Proceedings on Global Earth-System Monitoring, Reading, UK, 5–9 September 2005; pp. 297–308.
2. Robichaud, A.; Ménard, R. Multi-year objective analysis of warm season ground-level ozone and $PM_{2.5}$ over North-America using real-time observations and Canadian operational air quality models. *Atmos. Chem. Phys.* **2014**, *14*, 1769–1800. [CrossRef]
3. Robichaud, A.; Ménard, R.; Zaïtseva, Y.; Anselmo, D. Multi-pollutant surface objective analyses and mapping of air quality health index over North America. *Air Qual. Atmos. Health* **2016**, *9*, 743–759. [CrossRef] [PubMed]
4. Moran, M.D.; Ménard, S.; Pavlovic, R.; Anselmo, D.; Antonopoulus, S.; Robichaud, A.; Gravel, S.; Makar, P.A.; Gong, W.; Stroud, C.; et al. *Recent Advances in Canada's National Operational Air Quality Forecasting System*, 32nd ed.; Springer: Dordrecht, The Netherlands, 2014.
5. Pudykiewicz, J.A.; Kallaur, A.; Smolarkiewcz, P.K. Semi-lagrangian modelling of tropospheric ozone. *Tellus B* **1997**, *49*, 231–248. [CrossRef]
6. Cressie, N.; Wikle, C.K. *Statistics for Spatio-Temporal Data*; Wiley: Hoboken, NJ, USA, 2011.
7. Schneider, P.; Castell, N.; Vogt, M.; Dauge, F.R.; Lahoz, W.A.; Bartonova, A. Mapping urban air quality in near real-time using observations from low-cost sensors and model information. *Environ. Int.* **2017**, *106*, 234–247. [CrossRef] [PubMed]
8. Lindström, J.; Szpiro, A.A.; Oron, P.D.; Richards, M.; Larson, T.V.; Sheppard, L. A flexible spatio-temporal model for air pollution and spatio-temporal covariates. *Environ. Ecol. Stat.* **2014**, *21*, 411–433. [CrossRef] [PubMed]
9. Carmichael, G.R.; Sandu, A.; Chai, T.; Daescu, D.N.; Constantinescu, E.M.; Tang, Y. Predicting air quality: Improvements through advanced methods to integrate models and measurements. *J. Comput. Phys.* **2008**, *227*, 3540–3571. [CrossRef]
10. Dabberdt, W.F.; Carroll, M.A.; Baumgardner, D.; Carmichael, G.; Cohen, R.; Dye, T.; Ellis, J.; Grell, G.; Grimmond, S.; Hanna, S.; et al. Meteorological research needs for improved air quality forecasting: Report of the 11th prospectus development team of the US weather research program. *Bull. Am. Meteorol. Soc.* **2004**, *85*, 563–586. [CrossRef]
11. Sportisse, B. A review of current issues in air pollution modeling and simulation. *Comput. Geosci.* **2007**, *11*, 159–181. [CrossRef]

12. Elbern, H.; Strunk, A.; Nieradzik, L. Inverse modelling and combined state-source estimation for chemical weather. In *Data Assimilation*; Lahoz, W., Khattatov, B., Ménard, R., Eds.; Springer: Berlin/Heidelberg, Germany, 2010; pp. 491–513.
13. Bocquet, M.; Elbern, H.; Eskes, H.; Hirtl, M.; Žabkar, R.; Carmicheal, G.R.; Flemming, J.; Inness, A.; Pagaoski, M.; Pérez Camaño, J.L.; et al. Data assimilation in atmospheric chemistry models; current status and future prospects for coupled chemistry meteorology models. *Atmos. Chem. Phys.* **2015**, *15*, 5325–5358. [CrossRef]
14. Chai, T.; Carmichael, G.R.; Sandu, A.; Tang, Y.H.; Daescu, D.N. Chemical data assimilation of transport and chemical evolution over the pacific (TRACE-P) aircraft measurements. *J. Geophys. Res.* **2006**, *111*, D02301. [CrossRef]
15. Sandu, A.; Chai, T. Chemical data assimilation—An overview. *Atmosphere* **2011**, *2*, 426–463. [CrossRef]
16. Marseille, G.J.; Barkmeijer, J.; De Haan, S.; Verkle, W. Assessment and tuning of data assimilation systems using passive observations. *Q. J. R. Meteorol. Soc.* **2016**, *142*, 3001–3014. [CrossRef]
17. Ménard, R.; Deshaies-Jacques, M.; Gasset, N. A comparison of correlation-length estimation methods for the objective analysis of surface pollutants at Environment and Climate Change Canada. *J. Air Waste Manag. Assoc.* **2016**, *66*, 874–895. [CrossRef] [PubMed]
18. Cohn, S.E.; Da Silva, A.; Guo, J.; Sienkiewicz, M.; Lamich, D. Assessing the effects of data selection with the DAO physical-space statistical analysis system. *Mon. Weather Rev.* **1998**, *126*, 2913–2926. [CrossRef]
19. Houtekamer, P.L.; Mitchell, H.L. A sequential ensemble Kalman filter for atmospheric data assimilation. *Mon. Weather Rev.* **2001**, *129*, 123–137. [CrossRef]
20. Efron, B.; Tibshirani, R.J. *An Introduction to Boostrap*; Chapman & Hall: New York, NY, USA, 1993.
21. Seigneur, C.; Pun, B.; Pai, P.; Louis, J.F.; Solomon, P.; Emery, C.; Morris, R.; Zahniser, M.; Worsnop, D.; Koutrakis, P.; et al. Guidance for the performance evaluation of three-dimensional air quality modeling systems for particulate matter and visibility. *J. Air Waste Manag. Assoc.* **2000**, *50*, 588–599. [CrossRef] [PubMed]
22. Chang, J.C.; Hanna, S.R. Air quality model performance evaluation. *Meteorol. Atmos. Phys.* **2004**, *87*, 167–196. [CrossRef]
23. Savage, N.H.; Agnew, P.; Davis, L.S.; Ordóñez, C.; Thorpe, R.; Johnson, C.E.; O'Connor, F.M.; Dalvi, M. Air quality modelling using the Met Office Unified Model (AQUM OS24-26): Model description and initial evaluation. *Geosci. Model Dev.* **2013**, *6*, 353–372. [CrossRef]
24. Katragkou, E.; Zanis, P.; Tsikerdekis, A.; Kapsomenakis, J.; Melas, D.; Eskes, H.; Flemming, J.; Huijnen, V.; Inness, A.; Schultz, M.G.; et al. Evaluation of near surface ozone over Europe from the MACC reanalysis. *Geosci. Model Dev.* **2015**, *8*, 2299–2314. [CrossRef]
25. Ménard, R. Error covariance estimation methods based on analysis residuals: Theoretical foundation and convergence properties derived from simplified observation networks. *Q. J. R. Meteorol. Soc.* **2016**, *142*, 257–273. [CrossRef]
26. Desroziers, G.; Berre, L.; Chapnik, B.; Poli, P. Diagnosis of observation, background, and analysis-error statistics in observation space. *Q. J. R. Meteorol. Soc.* **2005**, *131*, 3385–3396. [CrossRef]
27. Daley, R. *Atmospheric Data Analysis*; Cambridge University Press: New York, NY, USA, 1991; p. 457.
28. Ménard, R.; Deshaies-Jacques, M. Evaluation of analysis by cross-validation, Part II: Diagnostic and optimization of analysis error covariance. *Atmosphere* **2018**, *9*, 70. [CrossRef]

5

Evaluation of Analysis by Cross-Validation, Part II: Diagnostic and Optimization of Analysis Error Covariance

Richard Ménard *and Martin Deshaies-Jacques

Air Quality Research Division, Environment and Climate Change Canada, 2121 Transcanada Highway, Dorval, QC H9P 1J3, Canada; martin.deshaies-jacques@canada.ca

* Correspondence: richard.menard@canada.ca.

Abstract: We present a general theory of estimation of analysis error covariances based on cross-validation as well as a geometric interpretation of the method. In particular, we use the variance of passive observation-minus-analysis residuals and show that the true analysis error variance can be estimated, without relying on the optimality assumption. This approach is used to obtain near optimal analyses that are then used to evaluate the air quality analysis error using several different methods at active and passive observation sites. We compare the estimates according to the method of Hollingsworth-Lönnberg, Desroziers et al., a new diagnostic we developed, and the perceived analysis error computed from the analysis scheme, to conclude that, as long as the analysis is near optimal, all estimates agree within a certain error margin.

Keywords: data assimilation; statistical diagnostics of analysis residuals; estimation of analysis error; air quality model diagnostics; Desroziers et al. method; cross-validation

---

1. Introduction

At Environment and Climate Change Canada (ECCC) we have been producing hourly surface pollutants analyses covering North America [1–3] using an optimum interpolation scheme which combines the operational air quality forecast model GEM-MACH output [4] with real-time hourly observations of O_3, $PM_{2.5}$, PM_{10}, NO_2, and SO_2 from the AirNow gateway with additional observations from Canada. These analyses are not used to initialize the air quality model and we wish to evaluate them by cross-validation, that is by leaving out a subset of observations from the analysis to use them for verification. Observations used to produce the analysis are called active observations while those used for verification are called passive observations.

In a first-part paper of this study, i.e., Ménard and Deshaies-Jacques [5], we have examined different verification metrics using either active or passive observations. As we changed the ratio of observation error to background error variances $\gamma = \sigma_o^2/\sigma_b^2$, while keeping the sum $\sigma_o^2 + \sigma_b^2$ equal to $\mathrm{var}(O - B)$, we found a minimum in $\mathrm{var}(O - A)$ in the passive observation space. In this second-part paper, we formalize this result, develop the principles of estimation of the analysis error covariance by cross-validation, and apply it to estimate and optimize the analysis error covariance of ECCC's surface analyses of O_3 and $PM_{2.5}$.

When we refer to analysis error, or analysis error covariance, it is important to distinguish the perceived analysis error with the true analysis error [6]. The perceived analysis error is the analysis error that results from the analysis algorithm itself, whereas the true analysis error is the difference between the analysis and the true state. Analysis schemes are usually derived from an optimization of some sort. In a variational analysis scheme for example, the analysis is obtained by minimizing a cost function with some given or prescribed observation error and background error covariances, $\widetilde{\mathbf{R}}$ and $\widetilde{\mathbf{B}}$

respectively. In a linear unbiased analysis scheme, the gain matrix $\widetilde{\mathbf{K}}$ is obtained by minimum variance estimation, yielding an expression of the form, $\widetilde{\mathbf{K}} = \widetilde{\mathbf{B}}\mathbf{H}(\mathbf{H}\widetilde{\mathbf{B}}\mathbf{H} + \widetilde{\mathbf{R}})^{-1}$, where $\mathbf{H}$ is the observation operator. The perceived analysis error covariance is then derived as $\widetilde{\mathbf{A}} = (\mathbf{I} - \widetilde{\mathbf{K}}\mathbf{H})\widetilde{\mathbf{B}}$. In order to derive an expression for the perceived analysis error covariance we in fact assume that given error covariances, $\widetilde{\mathbf{R}}$ and $\widetilde{\mathbf{B}}$, are error covariances with respect to the true state, i.e., the true error covariances. We also assume that the observation operator is not an approximation with some error, but is the true error-free observation operator. Of course, in real applications neither of $\widetilde{\mathbf{R}}$ and $\widetilde{\mathbf{B}}$ are covariance measures with respect to the true state, but only a more or less accurate estimate of those. Daley [6] have argued that in principle for an arbitrary gain matrix $\widetilde{\mathbf{K}}$, the true analysis error covariance $\mathbf{A}$ can be computed as $\mathbf{A} = (\mathbf{I} - \widetilde{\mathbf{K}}\mathbf{H})\mathbf{B}(\mathbf{I} - \widetilde{\mathbf{K}}\mathbf{H})^T + \widetilde{\mathbf{K}}\mathbf{R}\widetilde{\mathbf{K}}^T$ provided that we know the true observation and background error covariances, $\mathbf{R}$ and $\mathbf{B}$. This expression is a quadratic matrix equation, and has the property that the true analysis error variance, $tr(\mathbf{A})$ is minimum when $\widetilde{\mathbf{K}} = \mathbf{B}\mathbf{H}(\mathbf{H}\mathbf{B}\mathbf{H} + \mathbf{R})^{-1} = \mathbf{K}$. In that sense, the analysis is truly optimal. The optimal gain matrix $\mathbf{K}$ is called the Kalman gain. It thus illustrates that although an analysis is obtained through some minimization principle, the resulting analysis error is not necessarily the true analysis error.

One of the main sources of information to obtain the true $\mathbf{R}$ and $\mathbf{B}$ is from the $\mathrm{var}(O - B)$ statistic. However, it has always been argued that this is not possible without making some assumptions [7–9], the most useful one being that background errors are spatially correlated while the observation errors are spatially uncorrelated, or at least on a much shorter length-scale. Even under those assumptions, different estimation methods such as the Hollingsworth-Lönnberg method [10], the maximum likelihood gives different error variances and different correlation lengths [11]. Other methods use $\mathrm{var}(O - B)$ for rescaling but assume that the observation error is known. The assumption that the observation error is known is also debated as they contain representativeness errors [12] that include observation operator errors. How to obtain an optimal analysis is thus unclear.

The evaluation of the true or perceived analysis error covariance using its own active observations is also a misleading problem unless the analysis is already optimal. Hollingsworth and Lönnberg [13] addressed this issue for the first time where they noted that in the case of an optimal gain (i.e., optimal analysis), the statistics of observation-minus-analysis residuals $O - \hat{A}$ are related to the analysis error by $\mathrm{E}[(O - \hat{A})(O - \hat{A})^T] = \mathbf{R} - \mathbf{H}\hat{\mathbf{A}}\mathbf{H}^T$, where $\hat{\mathbf{A}}$ is the optimal analysis error covariance and $\mathbf{H}$ and $\mathbf{R}$ are the observation operator and observation error covariance respectively. The caret (ˆ) over $\mathbf{A}$ indicates that the analysis uses an optimal gain. In the context of spatially uncorrelated observation errors, the off-diagonal elements of $\mathrm{E}[(O - \hat{A})(O - \hat{A})^T]$ would then give the analysis error covariance in observation space. Hollingsworth and Lönnberg [13] argued that for most practical purposes, the negative intercept of $\mathrm{E}[(O - \hat{A})(O - \hat{A})^T]$ at zero distance and the prescribed observation weight should be nearly equal, and thus could be used as an assessment of optimality of an analysis. However, in case where such agreement does not exist, an estimate of the actual analysis error is not possible. Another method, proposed by Desroziers et al. [14], argued that the diagnostic $\mathrm{E}[(O - \hat{A})(\hat{A} - B)^T]$ should be equal to the analysis error covariance in observation space but, again, only if the gain is optimal and the innovation covariance consistency is respected [15].

The impasse of the estimation of the true analysis error seems to be tied with using active observations, i.e., using the same observations as those used to create the analysis. A robust approach that does not require an optimal analysis is to use observations whose errors are uncorrelated with the analysis error. For example, if we assume that observation errors are temporarily (serially) uncorrelated, an estimation of the analysis error can be made with the help of a forecast model initialized by the analysis by verifying the forecast against these observations. This is the essential assumption used traditionally in meteorological data assimilation to assess indirectly the analysis error by comparing the resulting forecast with observations valid at the forecast time. As forecast error grows with time, the observation-minus-forecast can be used to assess whether an analysis is better than another. In a somewhat different method but making the same assumption, Daley [6] used the temporal (serial) correlation of the innovations to diagnose the optimality of the gain matrix. This property was first established in the context of Kalman filter estimation theory by Kailath [16]. However, both the traditional meteorological forecast approach and the Daley method [6] are subject to limitations: they assume that the model forecast has no bias and the analysis corrections are made correctly on all the variables needed to initialize the model. In practice, improper initialization of unobserved meteorological variables gives rise to spin-up problems or imbalances. Furthermore, with the traditional meteorological approach, compensation due to model error can occur, so that an optimal analysis does not necessarily yield an optimal forecast [7].

An alternative approach introduced by Marseille et al. [17], which we will follow here, is to use independent observation or passive observations to assess the analysis error. The essential assumption of this method is that the observations have spatially uncorrelated errors, so that the observations used for verification, i.e., the passive observations, have uncorrelated errors with the analysis. The advantage of this approach is that it does not involve any model to propagate the analysis information to a later time. Marseille et al. [17] then showed that by multiplying the Kalman gain with an appropriate scalar value, one can reduce the analysis error. In this paper, we go further by using principles of error covariance estimation to obtain a near optimal Kalman gain. In addition we impose the innovation covariance consistency [15] and show that all diagnostics of analysis error variance nearly agree with one another. These include the Hollingsworth and Lönnberg [13], the Desroziers et al. [14] and new diagnostics that we will introduce.

The paper is organized as follows. First we present in Section 2 the theory and diagnostics of analysis error covariance in both passive and active observation spaces, as well as a geometrical representation. This leads us to a method to minimize the true analysis error variance. In Section 3, we present the experimental setup on how we obtain near optimal analyses and presents the results of several diagnostics in active and passive observation spaces, and compare with the analysis error variance obtained from the optimum interpolation scheme itself. In Section 4, we discuss the statistical assumptions being used, how and if they can be extended and how this formalism can be used in other applications such as the estimation of correlated observation errors with satellite observations. Finally, we draw some conclusions in Section 5.

2. Theoretical Framework

This section is composed of mainly three parts. In Sections 2.1 and 2.2, we first describe the diagnostics to obtain the true error covariances using passive observations whether the analysis is optimal or not. We then give a geometric interpretation in Section 2.3 that indicate the way to obtain the optimal analysis, that is, one that minimizes the true analysis error. Lastly, in Sections 2.4 and 2.5, we formulate the diagnostics of analysis error for optimal analyses.

2.1. Diagnostic of Analysis Error Covariance in Passive Observation Space

Let us decompose the observation space in two disjoint sets; the active observation set or training set $\{\mathbf{y}\}$ used to create the analysis, and the independent or passive observation set $\{\mathbf{y}_c\}$ used to evaluate the analysis. An analysis built from prescribed background and observation error covariances, $\widetilde{\mathbf{B}}$ and $\widetilde{\mathbf{R}}$ respectively, is given by

$$\mathbf{x}^a = \mathbf{x}^f + \widetilde{\mathbf{B}}\mathbf{H}^T(\mathbf{H}\widetilde{\mathbf{B}}\mathbf{H}^T + \widetilde{\mathbf{R}})^{-1}\mathbf{d} = \mathbf{x}^f + \widetilde{\mathbf{K}}\mathbf{d} \tag{1}$$

where $\widetilde{\mathbf{K}}$ is the gain matrix built from the prescribed error covariances, $\mathbf{H}$ is the observation operator for the active observation set, $\mathbf{d}$ is the active innovation vector, $\mathbf{d} = \mathbf{y} - \mathbf{H}\mathbf{x}^f = \boldsymbol{\varepsilon}^o - \mathbf{H}\boldsymbol{\varepsilon}^f$, $\mathbf{x}^f$ is the background state or model forecast, $\boldsymbol{\varepsilon}^o$ is the active observation error and $\boldsymbol{\varepsilon}^f$ the background error. The observation-minus-analysis residual $(O - A)$ for the active set is given by [14,15,18].

$$\begin{aligned}(O-A) &= \mathbf{y} - \mathbf{H}\mathbf{x}^a = \boldsymbol{\varepsilon}^o - \mathbf{H}\boldsymbol{\varepsilon}^a \\ &= \mathbf{d} - \mathbf{H}\widetilde{\mathbf{B}}\mathbf{H}^T(\mathbf{H}\widetilde{\mathbf{B}}\mathbf{H}^T + \widetilde{\mathbf{R}})^{-1}\mathbf{d} \\ &= \widetilde{\mathbf{R}}(\mathbf{H}\widetilde{\mathbf{B}}\mathbf{H}^T + \widetilde{\mathbf{R}})^{-1}\mathbf{d}\end{aligned} \quad (2)$$

where $\boldsymbol{\varepsilon}^a$ is the analysis error. The analysis interpolated at the passive observation sites can be denoted by $\mathbf{H}_c\mathbf{x}^a$, where $\mathbf{H}_c$ is the observation operator at passive observation sites. The observation-minus-analysis residual at the passive observation sites $(O - A)_c$ is then given by

$$\begin{aligned}(O-A)_c &= \mathbf{y}_c - \mathbf{H}_c\mathbf{x}^a = \boldsymbol{\varepsilon}_c^o - \mathbf{H}_c\boldsymbol{\varepsilon}^a \\ &= \mathbf{d}_c - \mathbf{H}_c\widetilde{\mathbf{B}}\mathbf{H}^T(\mathbf{H}\widetilde{\mathbf{B}}\mathbf{H}^T + \widetilde{\mathbf{R}})^{-1}\mathbf{d}\end{aligned}, \quad (3)$$

where $\mathbf{d}_c = \mathbf{y}_c - \mathbf{H}_c\mathbf{x}^f = \boldsymbol{\varepsilon}_c^o - \mathbf{H}_c\boldsymbol{\varepsilon}^f$ is the innovation at the passive observation sites. Note that the formalism introduced here is general, and can be used with any set of independent observations such as different instruments or observation networks as long as the $\mathbf{H}_c$ operator is properly defined. Consequently, for generality, we distinguish the passive observation errors or independent observation error, $\boldsymbol{\varepsilon}_c^o$, from the active observation error $\boldsymbol{\varepsilon}^o$.

There are two important statistical assumptions from which we derive cross-validation diagnostics. Assuming that the observation errors are spatially uncorrelated, it follows that

$$\mathrm{E}[\boldsymbol{\varepsilon}^o(\boldsymbol{\varepsilon}_c^o)^T] = \mathbf{0} \quad (4)$$

where E[ ] is the mathematical expectation that represents the mean over an ensemble of realizations. It has been argued by Marseille et al. [17] that representativeness error can violate this assumption for a close pair of active-passive observations, but we will neglect this effect. Also, assuming that observation errors are uncorrelated with background error, we have

$$\mathrm{E}[\boldsymbol{\varepsilon}^o(\mathbf{H}\boldsymbol{\varepsilon}^f)^T] = \mathbf{0},\ \mathrm{E}[\boldsymbol{\varepsilon}_c^o(\mathbf{H}_c\boldsymbol{\varepsilon}^f)^T] = \mathbf{0} \quad (5)$$

We come now to the most important property: since the analysis is a linear combination of the active observations and the background state, the analysis error is then uncorrelated with the passive observation errors,

$$\mathrm{E}[({}_c\boldsymbol{\varepsilon}^a)(\boldsymbol{\varepsilon}_c^o)^T] = \mathbf{0} \quad (6)$$

and thus we get the following cross-validation diagnostic in passive observation space,

$$\mathrm{E}[(O-A)_c(O-A)_c^T] = \mathbf{R}_c + \mathbf{H}_c\mathbf{A}\mathbf{H}_c^T \quad (7)$$

similarly to Marseille et al. [17]. A very important point to note is that $\mathbf{A}$ is the true analysis error covariance—it does not assume that the gain is optimal. The matrices in Equation (7) are of the dimension of the passive observation space and $\mathbf{R}_c$ is the observation error covariance matrix for the passive or independent observations.

2.2. A Complete Set of Diagnostics of Error Covariances in Passive Observation Space

It is also possible to define a set of diagnostics that would determine, in principle, the true error covariances, $\mathbf{R}, \mathbf{B}$ and $\mathbf{A}$. From Equations (4) and (5) we get another cross-validation diagnostic,

$$\mathrm{E}[(O-B)_c(O-B)^T] = \mathbf{H}_c\mathbf{B}\mathbf{H}^T \quad (8)$$

This diagnostic is related to the Hollingsworth-Lönnberg [10] estimation of the spatially correlated part of the innovation, and in practice can be dominated by sampling error. Note that it is not a square matrix and an estimation of parameters of **B** may not be trivial.

We can also obtain the innovation covariance matrix in passive observation space (identical in fact to the one in active observations space) as

$$\mathrm{E}[(O-B)_c(O-B)_c^T] = \mathbf{R}_c + \mathbf{H}_c\mathbf{B}\mathbf{H}_c^T \tag{9}$$

The system of Equations (7)–(9) gives a complete set of equations to determine the true **R**, **B** and **A** at the passive observation sites provided that by interpolation/extrapolation we can obtain $\mathbf{H}_c\mathbf{B}\mathbf{H}_c^T$ from $\mathbf{H}_c\mathbf{B}\mathbf{H}^T$.

Nevertheless, and for sake of completeness, we also investigated the meaning of the statistics $\mathrm{E}[(O-A)_c(O-B)^T]$ and came to the conclusion that it can interpreted as a misfit to the Desroziers et al. estimate of **B** [14] with the true value **B**. We recall that the first iterate of the Desroziers et al. estimate for **B** in active observation space is given by $\mathbf{H}\mathbf{B}^D\mathbf{H}^T = \mathrm{E}[(\mathbf{H}\mathbf{x}^a - \mathbf{H}\mathbf{x}^f)(\mathbf{y}-\mathbf{H}\mathbf{x}^f)^T]$ [15], where we used the superscript D indicate the Desroziers et al. first iterate estimate. We can actually generalize this diagnostic to be a cross-covariance between state space and active observation space as

$$\mathbf{B}^D\mathbf{H}^T = \mathrm{E}[(\mathbf{x}^a - \mathbf{x}^f)(\mathbf{y}-\mathbf{H}\mathbf{x}^f)^T] \tag{10}$$

By applying $\mathbf{H}_c$ to Equation (10) we can then introduce a generalized Desroziers et al. estimate of the background error covariance **B** between passive and active observation spaces as,

$$\mathrm{E}[(A-B)_c(O-B)^T] = \mathbf{H}_c\mathbf{B}^D\mathbf{H}^T \tag{11}$$

Since $(O-A)_c = (O-B)_c - (A-B)_c$, we get with Equations (8) and (10),

$$\mathrm{E}[(O-A)_c(O-B)^T] = \mathbf{H}_c(\mathbf{B}-\mathbf{B}^D)\mathbf{H}^T \tag{12}$$

That is the difference between the true **B** and the Desroziers et al. first estimate $\mathbf{B}^D$ in the cross passive-active observation spaces. Similarly to Equation (8) this diagnostics requires spatial interpolation of error covariances from active sites to passive sites of basically noisy statistics. The estimation of **B** from this diagnostics is further complicated by the fact that what is being interpolated, that is $\mathbf{B}-\mathbf{B}^D$, may not even be positive definite, but the augmented matrix,

$$\mathrm{cov}\begin{pmatrix} (O-B) \\ (O-A)_c \end{pmatrix} = \begin{bmatrix} \mathbf{R}+\mathbf{H}\mathbf{B}\mathbf{H}^T & \mathbf{H}(\mathbf{B}-\mathbf{B}^D)\mathbf{H}_c^T \\ \mathbf{H}_c(\mathbf{B}-\mathbf{B}^D)\mathbf{H}^T & \mathbf{R}_c+\mathbf{H}_c\mathbf{A}\mathbf{H}_c^T \end{bmatrix} \tag{13}$$

is positive definite. Except for the diagonal of Equation (13), we have not attempted in this study to conduct this complete estimation of **R**, **B** and **A**, but rather focused on getting a reliable estimate of the analysis error covariance **A**.

2.3. Geometrical Interpretation

A geometrical illustration of some of the relationships obtained above can be made by using a Hilbert space representation of random variables in observation space. A 2D representation for the analysis of a scalar quantity was used in Desroziers et al. [14] to illustrate their a posteriori diagnostics. We will generalize this approach to include passive observations by considering a 3D representation.

As in Desroziers et al. [14] let's consider the analysis of a scalar quantity. Several variables are to be considered in this observation space: y^o the active observation (or measurement) of the scalar quantity, y^b the background (or prior) value equivalent in observation space (i.e., $y^b = H\,x^b$), y^a the analysis in observation space (i.e., $y^a = H\,x^a$), and for verification y^c an independent observation (or passive observation) that is not used to compute the analysis. Each of these quantities is a random variable as they each contain random errors, and any linear combination of random variables in observation space also belong to observation space. For example, $y^o - y^b$ is the innovation (commonly denoted by O-B) and that belongs to observation space, $y^a - y^b$ is the analysis increment in observation space (commonly denoted by A-B), and $y^o - y^a$ is the analysis residual in observation space (commonly denoted by O-A). We can also define an inner product of any random variables in observation space. y_1, y_2, as

$$\langle y_1, y_2 \rangle := \mathrm{E}[(y_1 - \mathrm{E}(y_1))(y_2 - \mathrm{E}(y_2))] \tag{14}$$

The squared norm then represents the variance,

$$\|y\|^2 := \langle y, y \rangle = \sigma_y^2 \tag{15}$$

so the inner product has the following geometric interpretation

$$\langle y_1, y_2 \rangle = \|y_1\| \|y_2\| \cos\theta \tag{16}$$

where $\cos\theta$ is the correlation coefficient. Uncorrelated random variables are thus statistically orthogonal. With this inner product, the observation space forms a Hilbert space of random variables.

Figure 1 illustrates the statistical relationship in observation space between: the active observation y^o (illustrated as O in the figure), the prior or background y^b (i.e., B), the analysis y^a (i.e., A), and the independent observation y^c (i.e., O_c). The origin T corresponds to the truth of the scalar quantity, and also corresponds to the zero of the central moment of each random variables, e.g., $y - \mathrm{E}[y]$, since each variables are assumed to be unbiased. We also assume that the background, active and passive observations errors are uncorrelated to one another, so the three axes; ε^o for the active observation error, ε^b for the background error, and ε_c^o for the passive observation error are orthogonal. The plane defined by ε^o and ε^b axes is the space where the analysis takes place, and is called the analysis plane. However, since we define the analysis to be linear and unbiased, only linear combinations of the form $y^a = ky^o + (1-k)y^b$ where k is a constant are allowed. The analysis A then lies on the line (B, O). The thick lines in Figure 1 represent the norm of the associated error. For example, the thick line along the ε^o axis depict the (active) observation standard deviation σ_o, and similarly for the other axes and other random variables. Since the active observation error is uncorrelated with the background error, the triangle ΔOTB is a right triangle, and by Pythagoras theorem we have, $\overline{(y^o - y^b)^2} := \langle (O-B), (O-B) \rangle = \sigma_o^2 + \sigma_b^2$. This is the usual statement that the innovation variance is the sum of background and observation error variances. The analysis is optimum when the analysis error $||\varepsilon^a||_2 = \sigma_a^2$ is minimum, in which case the line (T, A) is perpendicular to line (O, B).

Now let's consider the passive observation O_c . The passive observation error is perpendicular to the analysis plane, thus the triangle $\Delta O_c TA$ is a right triangle,

$$\overline{(y^c - y^a)^2} := \langle (O-A)_c, (O-A)_c \rangle = \sigma_c^2 + \sigma_a^2 \tag{17}$$

where σ_c^2 is the passive observation error variance. The most important fact to stress here is that the orthogonality expressed in Equation (17) is true whether or not the analysis is optimal. Furthermore, as the distance $\overline{(y^c - y^a)^2}$ varies with the position of A along the line (O, B), the distance $\overline{(y^c - y^a)^2}$ reaches a minimum value when σ_a^2 is minimum that is when the analysis is optimal. We thus also argue from this representation that there is always a minimum, and the minimum is unique. Finally, we note that ΔBTO_c is also a right triangle so that $\overline{(y^c - y^b)^2} := \langle (O-B)_c, (O-B)_c \rangle = \sigma_c^2 + \sigma_b^2$, which is the scalar version of Equation (9).

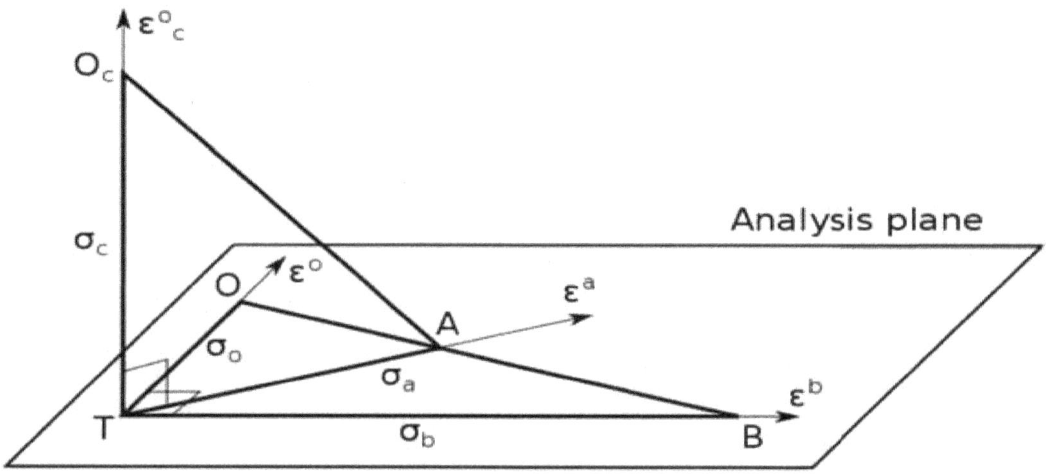

Figure 1. Hilbert space representation of a scalar analysis and cross-validation problem. The arrows indicate the directions of variability of the random variables, and the plane defined by the background and observation errors ε^b, ε^o defines the analysis plane. The thick lines represent the norm associated with the different random variables. T indicate the truth, O the active observation, B the background, A the analysis and O_c the passive observation.

To extend this formalism to random vectors requires to define a proper matricial inner product as briefly described in Section 1.2 of Caines [19]. It is important here to distinguish the stochastic metric space from the observation vector space. In the stochastic metric space a "scalar" needs not to be a number, but only a non-random quantity invariant with respect to E[ ]. Hence, we can define a Hilbert space with the following (stochastic scalar) matricial inner product $\langle \mathbf{y}, \mathbf{w} \rangle = \mathrm{E}[(\mathbf{y} - \mathrm{E}(\mathbf{y}))(\mathbf{w} - \mathrm{E}(\mathbf{w})^T)] = \mathrm{cov}(\mathbf{y}, \mathbf{w})$. The matrix nature of $\mathrm{cov}(\mathbf{y}, \mathbf{w})$ pertains to the observation vector space, but it remains a scalar with respect to the stochastic Hilbert space herein defined. In order to obtain a true scalar ($\in$ R), one would need to define a metric on the observation space matrices as well, such as the trace. Also to be able to compare active and passive observations, projectors need to be introduced on the observation space, implying yet another metric structure on the observation space. We do not carry out this formalism here as it would represent a rather lengthy development that would distract us from the main purpose of this paper; this will be considered in a future manuscript. Finally, we remark that Hilbert space representation of random variables in infinite dimensional space (i.e., continuous space) can also be defined, see Appendix 1 of Cohn [20].

2.4. Error Covariance Diagnostics in Active Observation Space for Optimal Analysis

An analysis is optimal if the analysis error $\mathrm{E}[(\varepsilon^a)^T \varepsilon^a]$ is minimum. This implies that the gain matrix using the prescribed error covariances, $\widetilde{\mathbf{K}}$ in Equation (1), must be identical to the gain using the true error covariances, i.e., $\widetilde{\mathbf{K}} = \mathbf{B}\mathbf{H}^T(\mathbf{H}\mathbf{B}\mathbf{H}^T + \mathbf{R})^{-1}$ [14,15]. It is important to mention that necessary and sufficient conditions to obtain the true error covariances $\mathbf{H}\mathbf{B}\mathbf{H}^T$ and $\mathbf{R}$ in observation space, are: 1—the Kalman gain condition, $\mathbf{H}\widetilde{\mathbf{K}} = \mathbf{H}\mathbf{K}^{true}$ and 2—the innovation covariance consistency, $\mathrm{E}[(O - B)(O - B)^T] = \mathbf{H}\widetilde{\mathbf{B}}\mathbf{H}^T + \widetilde{\mathbf{R}}$. For a proof see the Theorem on error covariance estimates in Ménard [15].

From the optimality of the analysis (or Kalman gain) alone, we derive that $\mathrm{E}[(\hat{A} - T)(O - B)^T] = \mathbf{0}$ or $\mathrm{E}[(\hat{A} - T)(O - \hat{A})^T] = \mathbf{0}$. Indeed, from Equation (2), we get $(O - \hat{A}) = \mathbf{R}(\mathbf{H}\mathbf{B}\mathbf{H} + \mathbf{R})^{-1}\mathbf{d}$, and for the analysis error in observation space we get, $(\hat{A} - T) = \mathbf{R}(\mathbf{H}\mathbf{B}\mathbf{H}^T + \mathbf{R})^{-1}\mathbf{H}\varepsilon^f + \mathbf{H}\mathbf{B}\mathbf{H}^T(\mathbf{H}\mathbf{B}\mathbf{H}^T + \mathbf{R})^{-1}\varepsilon^o$, from which we derive the expectations above. Using the geometrical representation in Section 2.3 the distance between A and T is minimum, when ΔTAO and ΔTAB (Figure 1) are right triangles. We should also note that for the scalar problem, the Kalman gain depends only on the ratio of observation to background error variances and thus the scalar Kalman gain is optimal if the ratio of the prescribed error variances is equal to the ratio of the true error variances.

If in addition to the optimality of the analysis or Kalman gain we add the innovation covariance consistency then we get three different statistical diagnostics of the (optimal) analysis error covariance. Hollingsworth and Lönnberg [13] was the first to introduce a statistical diagnostic of analysis error in the active observation space, as:

$$\mathrm{E}[(O-\hat{A})(O-\hat{A})^T] = \mathbf{R} - \mathbf{H}\hat{\mathbf{A}}_{HL}\mathbf{H}^T \tag{18}$$

Here we use a subscript, *HL* to indicate that this is the Hollingsworth-Lönnberg estimate. Equation (18) can obtained from the covariance of $(O-\hat{A})$ and that, for an optimal gain matrix, $\mathbf{H}\hat{\mathbf{A}}\mathbf{H}^T = \mathbf{R} - \mathbf{R}(\mathbf{HBH}+\mathbf{R})^{-1}\mathbf{R}$ which derives from the usual formula, $\hat{\mathbf{A}} = \mathbf{B} - \mathbf{BH}^T(\mathbf{HBH}+\mathbf{R})^{-1}\mathbf{HB}$. Geometrically it derives from the fact that the triangle ΔTÂO is a right triangle, and from the innovation covariance consistency that implies that the triangle ΔOTB is a right triangle. Using data to construct $\mathrm{E}[(O-\hat{A})(O-\hat{A})^T]$, an estimated analysis error covariance obtained from Equation (18) is symmetric but could be non-positive definite as it is obtained by subtracting two positive definite matrices. The effect of misspecification in the prescribed error covariances resulting from a lack of innovation covariance consistency will be discussed in the result Section 3 and in Appendix B.

Inspired from the geometrical interpretation that ΔTAB is also be a right triangle we derived the following diagnostic,

$$\mathrm{E}[(\hat{A}-B)(\hat{A}-B)^T] = \mathbf{HBH}^T - \mathbf{H}\hat{\mathbf{A}}_{MDJ}\mathbf{H}^T = \mathbf{H}(\mathbf{B}-\hat{\mathbf{A}}_{MDJ})\mathbf{H}^T \tag{19}$$

where *MDJ* stands for Ménard-Deshaies-Jacques. This relationship is obtained by using the expression $(\hat{A}-B) = \mathbf{HBH}^T(\mathbf{HBH}+\mathbf{R})^{-1}\mathbf{d}$, the innovation covariance consistency and the formula for the optimal analysis error covariance $\hat{\mathbf{A}} = \mathbf{B} - \mathbf{BH}^T(\mathbf{HBH}+\mathbf{R})^{-1}\mathbf{HB}$. As for the *HL* diagnostics, the estimated error covariance obtained from this diagnostic is symmetric by construction but may not be positive definite. Another way of looking at Equation (19) is that it expresses, in observation space, the error reduction due to the use of observations.

Another diagnostic of analysis error covariance was proposed by Desroziers et al. [14]. By combining $(O-\hat{A})$ and $(\hat{A}-B)$ we get

$$\mathrm{E}[(O-\hat{A})(\hat{A}-B)^T] = \mathbf{H}\hat{\mathbf{A}}_D\mathbf{H}^T \tag{20}$$

where the subscript *D* denotes the Desroziers et al. estimate. By construction, the estimated analysis error covariance is not necessarily symmetric. A geometrical derivation is provided in Appendix A. We also provide in Appendix B a sensitivity analysis on the departure from innovation covariance consistency for each diagnostics in both active and passive observation spaces.

2.5. Error Covariance Diagnostics in Passive Observation Space for Optimal Analysis

We can also derive optimal analysis diagnostics in the passive observation space. Considering the 3D geometric interpretation, and in particular the tetrahedron $(\mathrm{O_c}, \mathrm{T}, \mathrm{A}, \mathrm{B})$, we notice that since ΔTAB is a right triangle, so is $\Delta\mathrm{O_c}\mathrm{AB}$, which is a projection of the triangle ΔTAB on the plane passing through $\mathrm{O_c}$, O and B. We thus have $\mathrm{E}[(\hat{A}-B)_c(\hat{A}-B)_c^T] + \mathrm{E}[(O-\hat{A})_c(O-\hat{A})_c^T] = \mathrm{E}[(O-B)_c(O-B)_c^T]$. Combining this result with Equation (7) and using, $\mathrm{E}[(O-B)_c(O-B)_c^T] = \mathbf{H}_c\mathbf{BH}_c^T + \mathbf{R}_c$, we then get

$$\mathrm{E}[(\hat{A}-B)_c(\hat{A}-B)_c^T] = \mathbf{H}_c\mathbf{BH}_c^T - \mathbf{H}_c\hat{\mathbf{A}}_{MDJ}\mathbf{H}_c^T \tag{21}$$

Note that our analysis diagnostic is the only diagnostic that is valid in both active and passive observation spaces, i.e., Equation (21) is similar to Equation (19).

The other, less direct, diagnostic for optimal analysis is simply based on Equation (7) that is

$$\underset{\gamma,\, L_c}{\operatorname{argmin}}\{\mathrm{E}[(O-A)_c(O-A)_c^T]\} = \mathbf{R}_c + \mathbf{H}_c\hat{\mathbf{A}}(\gamma, L_c)\mathbf{H}_c^T \tag{22}$$

These five diagnostics will be used later in the results section.

3. Results with Near Optimal Analyses

3.1. Experimental Setup

We will just give here a short summary of the experimental setup we are using in this study. More details can be found in the Part I paper (Ménard and Deshaies-Jacques [5]). A series of hourly analyses of O_3 and $PM_{2.5}$ at 21 UTC for a period of 60 days (14 June to 12 August 2014) were performed using an optimum interpolation scheme combining the operational air quality model GEM-MACH forecast and the real-time AirNow observations (see Section 2 of [5] for further details). The analyses are made off-line so they are not used to initialize the model. As input error covariances, we use uniform observation and background error variances, with $\widetilde{\mathbf{R}} = \sigma_o^2\mathbf{I}$ and $\widetilde{\mathbf{B}} = \sigma_b^2\mathbf{C}$, where $\mathbf{C}$ is a homogeneous isotropic error correlation based on a second-order autoregressive model. The correlation length is estimated by using a maximum likelihood method using at first, error variances obtained from a local Hollingworth-Lönnberg fit [11] (and only for the purpose of obtaining a first estimate of the correlation length). We then conduct a series of analyses by changing error variance ratio $\gamma = \sigma_o^2/\sigma_b^2$ while at the same time respecting the innovation variance consistency condition, $\sigma_o^2 + \sigma_b^2 = \mathrm{var}(O-B)$. This corresponds basically in searching for the minimum of the *tr* (trace) of Equation (7) while the trace of the innovation covariance consistency, $tr[\mathbf{R} + \mathbf{HBH}^T] = tr\{\mathrm{E}[(O-B)(O-B)^T]\}$, is respected.

First the observations are separated into 3 sets of observations of equal number and distributed randomly in space. By leaving out one set of observations for verification and constructing analyses with the remaining 2 other sets, we construct a cross-validation setup from which we can evaluate the diagnostic $\mathrm{var}(O-A)_c = tr\{\mathrm{E}[(O-A)_c(O-A)_c^T\}$ in passive observation space. This constitutes our first guess experiment that we will refer to as **iter 0**. No observation or model bias correction was applied, nor were the mean of the innovation at the stations were removed prior to performing the analysis. The variance statistics are first computed at the station using the 60-day members, and then averaged over the domain to give the equivalent of $tr\{\mathrm{E}[(O-A)_c(O-A)_c^T\}$. We repeat the procedure by exhausting all permutations possible, in this case 3. The mean value statistic for the three verifying subsets are then averaged. More details can be found in Section 2 and beginning of Section 3 of Part I [5].

The red curve on Figure 2 illustrates how this diagnostic varies with and exhibits a minimum. This minimum can easily be understood by referring to Figure 1: as the analysis point A changes position along the line (O, B) the distance $\|O_c - A\|$ reaches a minimum, and this is what we observe in Figure 2.

In our next step, **iter 1**, we first re-estimate the correlation length by applying a maximum likelihood method, as in Ménard [15], using the **iter 0** error variances that are consistent with the optimal ratio $\hat{\gamma}$ obtained in **iter 0** and the innovation variance consistency. Then, with this new correlation length, we estimate a new optimal ratio $\hat{\gamma}$ (**iter 1**), which turn out to be very close to the value obtained in **iter 0**. We recall that we use uniform error variances, both to keep things simple but also because the optimal ratio is obtained by minimizing a domain-averaged variance $\mathrm{var}(O_c - A)$. A summary of the error covariance parameters obtained for **iter 0** and **iter 1** are presented in Table 1.

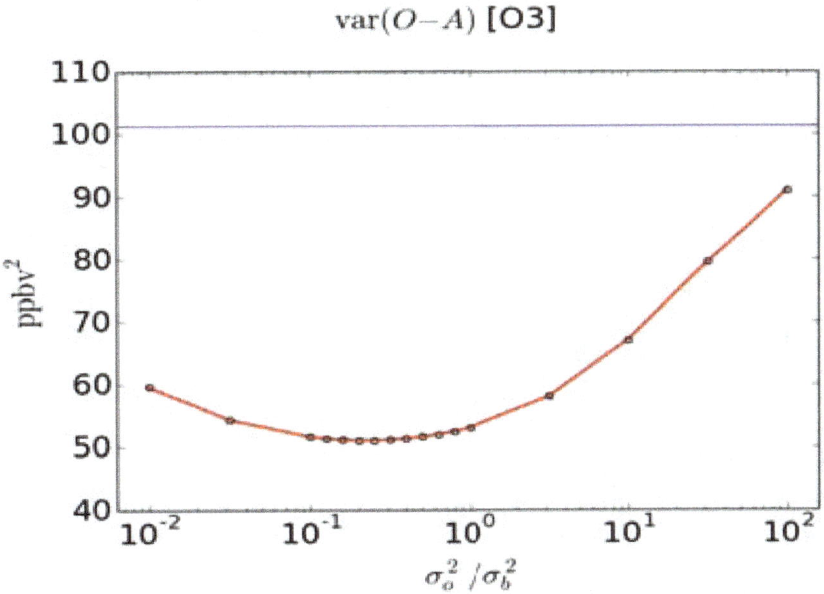

Figure 2. Red line, variance of observation-minus-analysis of O_3 in passive observation space. Blue line, variance of observation-minus-model.

Table 1. Input error statistics for the first experiment and optimized variance ratio experiment.

Experiment		L_c (km)	$\left\langle (O-B)^2 \right\rangle$	$\hat{\gamma} = \hat{\sigma}_o^2/\hat{\sigma}_b^2$	$\hat{\sigma}_o^2$	$\hat{\sigma}_b^2$	χ^2/N_s
O_3	iter 0	124	101.25	0.22	18.3	83	2.23
O_3	iter 1	45	101.25	0.25	20.2	81	1.36
$PM_{2.5}$	iter 0	196	93.93	0.17	13.6	80.3	2.04
$PM_{2.5}$	iter 1	86	93.93	0.22	16.9	77	1.25

We have also added in Table 1, the χ^2/N_s diagnostic values [15] which is the 60-day mean value of ${\chi_k}^2/N_s(k)$ where $N_s(k)$ is the total number of observations available at time t_k and ${\chi_k}^2 = \mathbf{d}_k^T\left(\mathbf{H}\widetilde{\mathbf{B}}\mathbf{H} + \widetilde{\mathbf{R}}\right)^{-1}\mathbf{d}_k$. The χ^2/N_s diagnostics [15] is closely related to the $J_{\min}$ diagnostic used in variational methods, and should have a value of 1 in the case where the innovations are consistent with the prescribed error statistics. When there is innovation covariance consistency then $\chi^2/N_s = 1$ but the reverse is not true. We observe that there was a significant improvement from **iter 0** to **iter 1** in terms of χ^2/N_s but is still not equal to one. We thus refer the analysis of **iter 1** as near optimal.

The repeated application of this sequence of estimation methods, i.e., find the correlation length by maximum likelihood and optimize the variance ratio, converges really fast and in practice there is no need to go beyond **iter 1**. Figure 3 displays iterates 0 to 4 with our estimation procedure for O_3. With one iteration update we nearly converge. A similar procedure was used in Ménard [15], where the variance and correlation length (estimated by maximum likelihood) were estimated in sequence, which taught us that a slow and fictitious drift in estimated variances and correlation length can occur when the correlation model is not the true correlation. So in regard of similar considerations that may occur here, we do not extend our iteration procedure beyond the first iterate.

3.2. Statistical Diagnostics of Analysis Error Variance

For each of these experiments, statistics related diagnostics for analysis error variance, discussed in Section 2, are computed and the results are presented in Table 2 for the verification made against active observations, and in Table 3 to the verification made against the passive observations. If the analysis was truly optimal the different diagnostics would all agree.

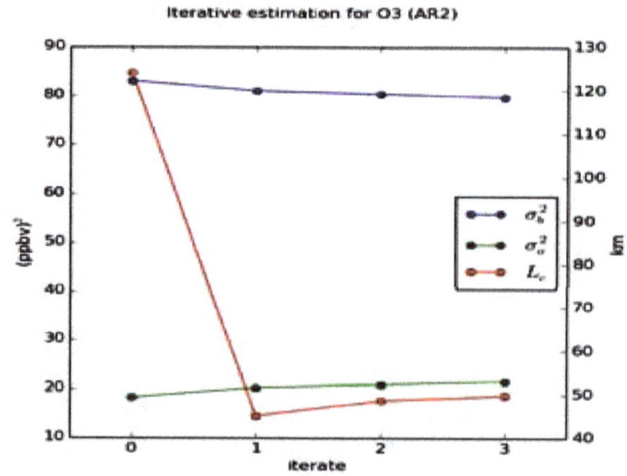

Figure 3. Optimal estimates of σ_o^2, σ_b^2 and maximum likelihood estimate of correlation length L_c for the first four iterates. Blue, is the optimal background error variance, green, the optimal observation error variance and in red the correlation length (in km, with labels on the right side of the figure).

Table 2. Analysis statistics against active observations.

Experiment		Active $\mathrm{var}(\hat{A}-B)$	Active $tr(\mathbf{H}\hat{\mathbf{A}}_{MDJ}\mathbf{H}^T)/N_s$	Active $tr(\mathbf{H}\hat{\mathbf{A}}_{D}\mathbf{H}^T)/N_s$	Active $\mathrm{var}(O-\hat{A})$	Active $tr(\mathbf{H}\hat{\mathbf{A}}_{HL}\mathbf{H}^T)/N_s$
O_3	**iter 0**	60.29	22.69	9.61	24.33	−6.03
O_3	**iter 1**	67.66	13.32	13.68	11.26	8.94
$PM_{2.5}$	**iter 0**	62.29	17.98	7.71	16.78	−3.18
$PM_{2.5}$	**iter 1**	66.3	10.68	9.51	9.57	7.33

In the second, third and last column of Table 2 are tabulated estimates of the analysis error variance at the active location sites, i.e., $tr(\mathbf{HAH}^T)/N_s$, obtained by three different methods. The second column is an estimate given with our method $\sigma_b^2 - \mathrm{var}(\hat{A}-B) = tr(\mathbf{H}\hat{\mathbf{A}}_{MDJ}\mathbf{H}^T)/N_s$. The third column is the Desroziers et al. estimate of analysis error [14], Equation (20), and the last column is the estimate using the method proposed by Hollingsworth and Lönnberg [13], Equation (18). We note that the analysis error variance estimate provided by the first two methods is fairly consistent for an updated correlation length estimate, i.e., **iter 1** (but not **iter 0**). We also note that χ^2/N_s is closer to one for **iter 1**. These two facts indicate that the updated correlation length (**iter 1**) with uniform error variances is closer to the innovation covariance consistency. The Hollingsworth and Lönnberg [13] method however, is very sensitive and negatively biased in the lack of innovation covariance consistency.

Estimate of the analysis error variance at the passive observation locations, i.e., $tr(\mathbf{H}_c\mathbf{AH}_c^T)/N_s$, provided by two different methods are given by Equation (21) in column 3 and by Equation (22) in column 5 of Table 3. As for the estimate at the active locations (Table 2), there is a general agreement on the analysis error estimates with the updated correlation length (**iter 1**), although this distinction is not that clear for $PM_{2.5}$.

Table 3. Analysis statistics against passive observations.

Experiment		Passive $\mathrm{var}[(\hat{A}-B)_c]$	Passive $tr(\mathbf{H}_c\hat{\mathbf{A}}_{MDJ}\mathbf{H}_c^T)/N_s$	Passive $\mathrm{var}[(O-\hat{A})_c]$	Passive $\mathrm{var}[(O-\hat{A})_c]-\sigma_{oc}^2$
O_3	**iter 0**	56.95	26.03	51.02	32.72
O_3	**iter 1**	52.04	28.95	48.95	28.75
$PM_{2.5}$	**iter 0**	62.29	22.65	38.09	24.49
$PM_{2.5}$	**iter 1**	66.3	24.62	38.28	21.38

We note also that the analysis error variance at the active sites is smaller than the analysis error variance at the passive observation sites. This involves in particular the fact that since the passive observation are away from the active observation sites, the reduction of variance at the passive observation sites is smaller than at the active observation sites.

3.3. Comparison with the Perceived Analysis Error Variance

We computed the analysis error covariance **A** resulting from the analysis scheme, the so-called perceived analysis covariance [6], using the expression,

$$\mathbf{A} = \widetilde{\mathbf{B}} - \widetilde{\mathbf{B}}\mathbf{H}^T(\mathbf{H}\widetilde{\mathbf{B}}\mathbf{H} + \widetilde{\mathbf{R}})^{-1}\mathbf{H}\widetilde{\mathbf{B}} = \widetilde{\mathbf{B}} - \mathbf{G}\mathbf{G}^T \quad (23)$$

Contrary to the statistical diagnostics above, the perceived analysis error variance is obtained at each model grid point. We then compared the perceived analysis error variance at the active observation sites with the estimated active analysis error variance obtained from previous diagnostics. Since we showed that in both active and passive observations sites the different diagnostics agrees when the analysis is optimal, this would indicate that if the perceived analysis error variance agrees with the analysis error variance diagnostics, that the whole map of analysis error variance is credible.

In order to calculate the perceived analysis error covariance Equation (23) we first perform a Choleski decomposition of $\mathbf{H}\widetilde{\mathbf{B}}\mathbf{H}^T + \widetilde{\mathbf{R}} = \mathbf{L}\mathbf{L}^T$, where **L** is a lower triangular matrix. Then with a forward substitution we obtain $\mathbf{L}^{-1}$, from which we compute $\mathbf{G} = \widetilde{\mathbf{B}}\mathbf{H}^T\mathbf{L}^{-T}$. The perceived analysis error variance for the ozone optimal analysis (i.e., O_3 **iter 1**) is displayed in Figure 4 (A similar figure but for $PM_{2.5}$ is given in supplementary material). We note that although the input statistics used for the analysis are uniform (i.e., uniform background and observation error variances, and homogeneous correlation model), the computed analysis error variance at the active observation location displays large variations, which is attributed to the non-uniform spatial distribution of the active observations.

In Figure 5 we display a histogram of those variances for the ozone optimal analysis O_3 **iter 1** (panel **b**) and for the first experiment O_3 **iter 0** (panel **a**) without optimization (A similar figure is but for $PM_{2.5}$ is given in supplementary material).

Note that median or mean values of variances are significantly different between the optimal and non-optimal analysis cases. Although the observation and background errors are uniform in both optimal and non-optimal analyses, in the optimal analysis case the perceived analysis uncertainty at the observation location is distributed more equally across all values, indicating that there is a better propagation of information, measured by analysis error variance, across the different observation sites. The exception being the isolated observation sites, for which we observe a maxima on the high end of the histograms. At those sites the analysis error variance is simply obtained by the scalar equation $1/\sigma_a^2 = 1/\sigma_o^2 + 1/\sigma_b^2$. For O_3 **iter 1** the scalar analysis error variance gives 16.2, and for O_3 **iter 0** we get 15.0, thus explaining the secondary maxima on the high end of the histogram.

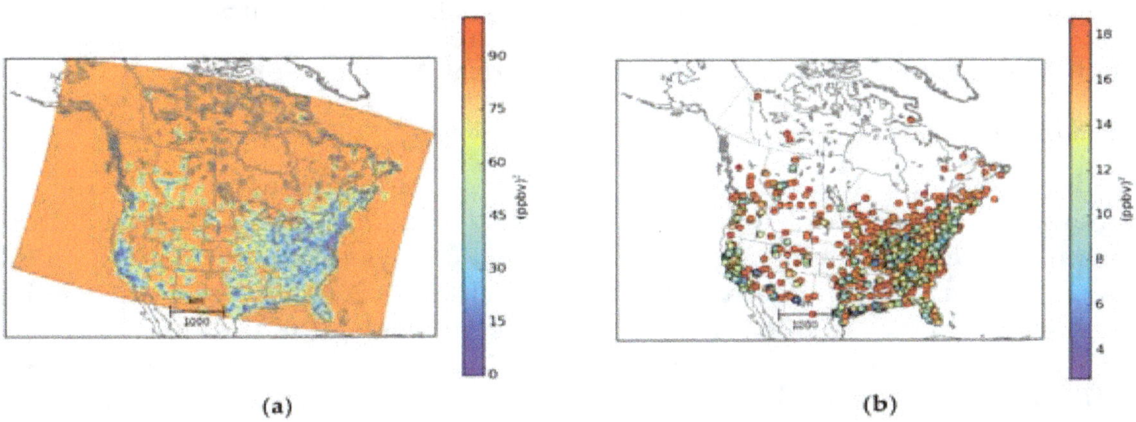

Figure 4. Analysis error variance for ozone optimal analysis case O_3 **iter 1**. (**a**) is the analysis error on the model grid and (**b**) at the active observation sites. Note that the color bar of the left and right panels are different. The maximum of the color bar for the left panel correspond to $\sigma_o^2 + \sigma_b^2$.

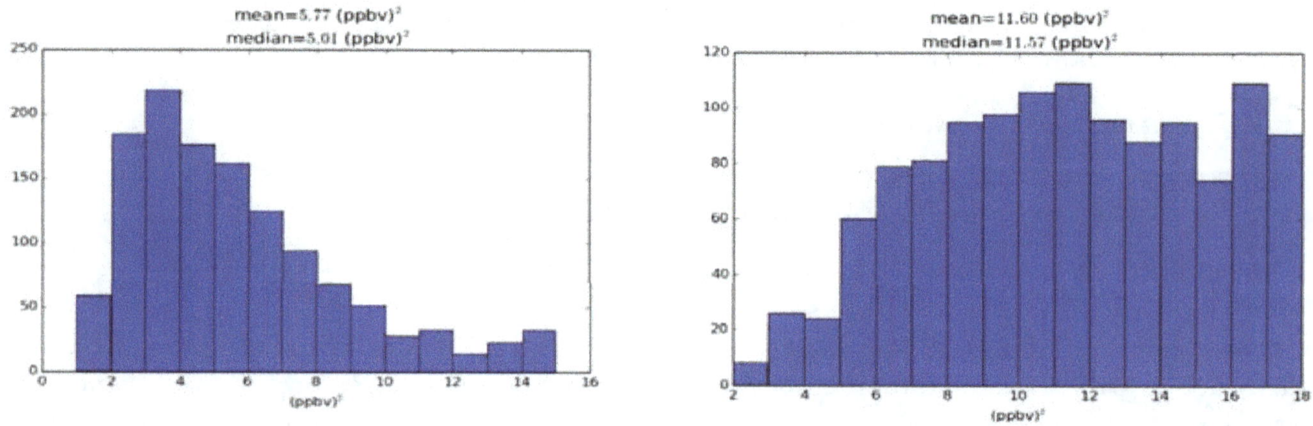

(a) (b)

Figure 5. Distribution (histogram) of the ozone analysis error variance at the active observation locations. (**a**) First analysis experiment O_3 **iter 0** (no optimization) on the left panel; (**b**) Optimal analysis case O_3 **iter 1**. Note that the scales are different between the left and right panels.

The mean perceived analysis error variance for all experiments is presented in Table 4. Comparing these values with the estimated values of analysis error variance based on diagnostics in Table 2 we note that for both optimal experiments, O_3 **iter 1** and $PM_{2.5}$ **iter 1**, the perceived analysis error variance roughly agrees with all analysis error variances estimated with diagnostics (Table 2).

Table 4. Perceived analysis error variance. Mean over active observation sites.

Experiment		Perceived $tr(\mathbf{H}\hat{\mathbf{A}}_P\mathbf{H}^T)/N_s$
O_3	**iter 0**	5.77
O_3	**iter 1**	11.60
$PM_{2.5}$	**iter 0**	4.37
$PM_{2.5}$	**iter 1**	8.21

However, for the non-optimal analyses, O_3 **iter 0** and $PM_{2.5}$ **iter 0**, there is a general disagreement between all estimated values. Looking more closely, however, we note that the agreement in the optimal case is not perfect. The perceived analysis error variance is about 20% lower than the best estimates $tr(\mathbf{H}\hat{\mathbf{A}}_{MDJ}\mathbf{H}^T)/N_s$ and $tr(\mathbf{H}\hat{\mathbf{A}}_D\mathbf{H}^T)/N_s$. The optimal χ^2/N_s values in the "optimal" cases are slightly above one, thus indicating that the innovation covariance consistency is not exact and some further tuning of the error statistics could be done. More on that matter will be presented in Section 4.5.

4. Discussion on the Statistical Assumptions and Practical Applications

4.1. Representativeness Error with In situ Observations

The statistical diagnostics presented in Section 2 derive from the assumption that the observation errors are horizontally uncorrelated and uncorrelated with the background error. Although this assumption is never entirely observed in reality, there are ways to work around it. In the case of in situ observations, and assuming that any systematic error have been removed, random errors are still present, due to the difference between the observation and the model's equivalent of the observation—called representativeness error (see Janjic et al. [12] for a review). Representativeness error is due to unresolved scales and processes in the model and interpolation or forward observation model errors. These errors are typically roughly at the scale of the model grid [21,22], so typically a few tens of kilometers for air quality models. This should not be confused with the representativeness of an observation, where, for example, remote stations are representative of large area (e.g., several hundreds of kilometers), whereas urban and suburban stations are at the scale of human activity in the cities, traffic and industries, etc. and are, depending on the chemical specie, of a few kilometers and less.

Representativeness error of in situ measurements can be discarded altogether by simply filtering any pair of observations that are in the range of a few model grid sizes, both in assimilation and estimation of error statistics [1] or in pairs of passive-active observations for cross-validation [17]. Once this filtering is done, the assumption on observation errors being spatially uncorrelated and uncorrelated with the background error then applies.

4.2. Correlated Observation-Background Errors

In any case, it is interesting to show how the different diagnostics, introduced in Section 2, depends on the statistical assumptions of the observation error. One way to get an understanding of the effect of these assumptions is to look at it from a geometrical point of view, using the representation introduced in Section 2.2. Note that the same results can be obtained analytically, but the geometrical interpretation gives a simple and appealing way of looking at the problem.

Let us consider the effect on the analysis of observation error correlated with background error. The case where the observation error is uncorrelated with background error is represented in Figure 6 on the panel **a** and when they are correlated on the panel **b**.

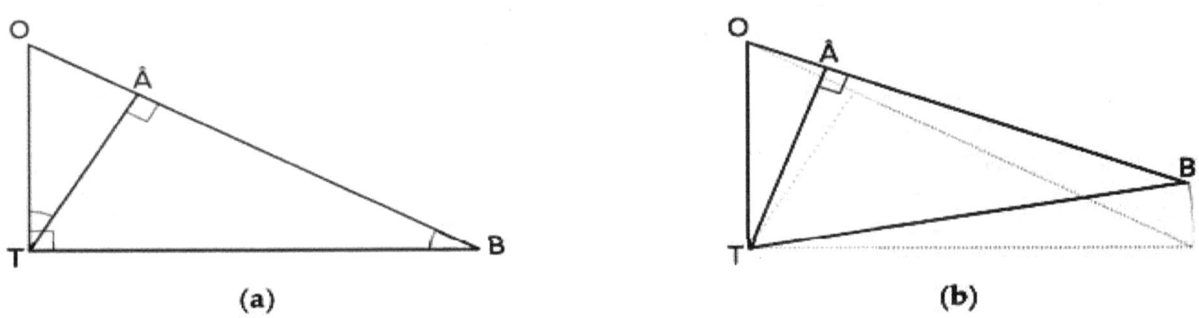

Figure 6. Geometrical representation of the analysis. (**a**) for observation errors uncorrelated with the background error. (**b**) with correlated errors. T indicate the truth, O the observation, B the background and Â the optimal analysis.

The observation and background error variances are kept unchanged, with the same (O, T) length and (B, T) length in both panels. In the case of correlated errors the angle ∠BTO is no longer a right angle. Yet, it is still possible to obtain an optimal analysis, Â, as a linear combination of the observation and the background, on the line (O, B), for which the distance Â to T (i.e., the analysis error variance) is minimum. In this case, (Â, T)⊥(O, B). Note that for strongly correlated errors and when $\sigma_b^2 > \sigma_o^2$, although Â is still on the line (O, B), it may actually lie outside the segment [O, B]. Yet, the principles and theory still hold in that case.

When the observation error is uncorrelated with the background error, (O, T)⊥(B, T), the triangles ΔOTÂ and ΔTBÂ are similar and it follows that $\langle (O - \hat{A})(\hat{A} - B) \rangle = \| (\hat{A} - T) \|^2$, which is the Desroziers et al. [14] diagnostic for analysis error variance. However, when the observation error is correlated with the background error (right panel of Figure 6), the triangles ΔOTÂ and ΔTBÂ are no longer similar triangles and the Desroziers et al. [14] diagnostics for analysis error does not hold (see derivation in Appendix A). However, the *HL*, Equation (18), and *MDJ* diagnostic, Equation (19), depend only on having right triangles ΔOTÂ and ΔTBÂ, and not on the orthogonality of (B, T) with (O, T). Therefore, the *HL* and *MDJ* diagnostics are valid with or without correlated observation-background errors.

4.3. Estimation of Satellite Observation Errors with In situ Observation Cross-Validation

One of the important problems in satellite assimilation is the estimation of the satellite observation error, which could be addressed with a simple modification of our cross-validation procedure. Let us assume that we have in situ observations that we assume to have uncorrelated errors between themselves (or use a filter with a minimum distance as discussed in Section 4.1), with the background errors and the satellite observation errors. Yet, the satellite observation errors could be correlated with the background error. Satellite observations could come from a multi-channel instrument with channel-correlated observation errors, as found with many instruments, and yet our validation procedure can still be used. Let us consider that the analyses comprise of satellite and in situ observations but, for the purpose of cross-validation, we use only 2/3rd of the in situ observations in the analysis, and keep the remaining 1/3rd as passive to carry out the cross-validation procedure.

The first thing to note is that the passive in situ observations have uncorrelated errors with the analysis error (the analysis is composed of satellite observations and 2/3rd of the in situ observations). We then use Equation (7) where the interpolation of the analysis is made only at the in situ active observations. Minimizing $\mathrm{E}[(O-A)_c^T(O-A)]$ (i.e., the trace of the left hand side of Equation (7)) results in finding the optimal in situ observation weight. Then, computing the analysis error covariance in the satellite observation space from the analysis scheme (either from a Hessian of a variational cost function, or with an explicit gain as in Equation (23)), i.e., $\mathbf{H}_{sat}\hat{\mathbf{A}}\mathbf{H}_{sat}^T$, we use the HL formulation Equation (18) to obtain the satellite observation error covariance,

$$\mathbf{H}_{sat}\hat{\mathbf{A}}\mathbf{H}_{sat}^T + \mathrm{E}[(O-A)_{sat}(O-A)_{sat}^T] = \mathbf{R}_{sat} \tag{24}$$

The Equation (24) has the important properties that the estimated observation error covariance is symmetric and positive definite by construction. Then, a new analysis could be carried out to obtain a more realistic $\mathbf{H}_{sat}\hat{\mathbf{A}}\mathbf{H}_{sat}^T$, with a resulting updated $\mathbf{R}_{sat}$, and so forth until convergence.

4.4. Remark on Cross-Validation of Satellite Retrievals

As a last remark, it appears that cross-validation of satellite retrieval observations using a k-fold approach where the observations are used as passive observations to validate the analysis can be a difficult problem. Retrievals from passive remote sensing at nadir generally involve a prior or climatology or a model assumption over different regions, and is thus likely to have spatially correlated errors and errors correlated with the background error. It does not mean, however, that nothing can be done in that case. For example, for certain sensors, such as infrared sensors, it is possible to disentangle the prior from the retrieval, so that by an appropriate transformation of the measurements, observations can be practically decorrelated from the background [23,24]. However to the authors' knowledge, such an approach have never been undertaken for visible measurements such as for NO_2 or AOD's.

4.5. Lack of Innovation Covariance Consistency and Its Relevance to the Statistical Diagnostics

The error covariance diagnostics for optimal analysis, presented in Sections 2.4 and 2.5, depends on the innovation covariance consistency, $\mathrm{E}[(O-B)(O-B)^T] = \mathbf{H}\widetilde{\mathbf{B}}\mathbf{H}^T + \widetilde{\mathbf{R}}$, and our results presented in Section 3 have shown that the different estimates for the optimal analysis error variance are close, but do not strictly agree to each other. This disagreement is related to the lack of innovation consistency as follows.

Let us introduce a departure matrix $\boldsymbol{\Delta}$ from innovation covariance consistency as,

$$\mathrm{E}[\mathbf{d}\mathbf{d}^T](\mathbf{H}\widetilde{\mathbf{B}}\mathbf{H}^T + \widetilde{\mathbf{R}})^{-1} = \mathbf{I} + \boldsymbol{\Delta} \tag{25}$$

The trace of Equation (25), which is related to χ^2, is given by

$$\mathrm{E}[\chi^2] = tr\{\mathrm{E}[\mathbf{d}\mathbf{d}^T](\mathbf{H}\widetilde{\mathbf{B}}\mathbf{H}^T + \widetilde{\mathbf{R}})^{-1}\} = N_s + tr(\boldsymbol{\Delta}) \tag{26}$$

We recall that in the experiment **iter 1** we got χ^2/N_s values of 1.36 for O_3 and 1.25 for $PM_{2.5}$ (see Table 1), indicating that the innovation covariance consistency is deficient, although less serious than with the experiment **iter 0** where values of 2 and higher have been obtained.

If we take into account the fact that there can be a difference between $\mathrm{E}\left[\mathbf{d}\mathbf{d}^T\right]$ and $(\mathbf{H}\widetilde{\mathbf{B}}\mathbf{H}^T + \widetilde{\mathbf{R}})$ and we rederive the (active) analysis error covariance for *HL*, *MDJ* and *D* schemes, we get (see Appendix B)

$$tr\{\mathbf{H}\hat{\mathbf{A}}_{HL}\mathbf{H}^T\} = tr\{\mathbf{H}\hat{\mathbf{A}}^{true}\mathbf{H}^T\} + tr\{\widetilde{\mathbf{R}}\boldsymbol{\Delta}\} - tr\{\mathbf{H}\widetilde{\mathbf{B}}\mathbf{H}^{\mathbf{T}}\boldsymbol{\Delta}\} + tr\{\mathbf{error}_{MDJ}\} \tag{27}$$

$$tr\{\mathbf{H}\hat{\mathbf{A}}_{MDJ}\mathbf{H}^T\} = tr\{\mathbf{H}\hat{\mathbf{A}}^{true}\mathbf{H}^T\} - tr\{\mathbf{error}_{MDJ}\} \tag{28}$$

$$tr\{\mathbf{H}\hat{\mathbf{A}}_{D}\mathbf{H}^T\} = tr\{\mathbf{H}\hat{\mathbf{A}}^{true}\mathbf{H}^T\} - tr\{\mathbf{error}_{D}\} \tag{29}$$

where $tr\{\mathbf{error}_{MDJ}\} = tr\{(\mathbf{H}\widetilde{\mathbf{B}}\mathbf{H}^T)(\mathbf{H}\widetilde{\mathbf{B}}\mathbf{H}^T + \widetilde{\mathbf{R}})^{-1}\boldsymbol{\Delta}(\mathbf{H}\widetilde{\mathbf{B}}\mathbf{H}^T)\}$, $tr\{\mathbf{error}_{D}\} = tr\{\widetilde{\mathbf{R}}(\mathbf{H}\widetilde{\mathbf{B}}\mathbf{H}^T + \widetilde{\mathbf{R}})^{-1}\boldsymbol{\Delta}(\mathbf{H}\widetilde{\mathbf{B}}\mathbf{H}^T)\}$. We note that although the error terms are complex expressions, they all depend linearly on $\boldsymbol{\Delta}$. Thus, the disagreement between the *HL*, *MDJ* and *D* analysis error variance estimates is due to lack of innovation covariance consistency.

5. Conclusions

We showed that analysis error variance can be estimated and optimized, without using a model forecast, by partitioning the original observation data set into a training set, to create the analysis, and an independent (or passive) set, used to evaluate the analysis. This kind of evaluation by partitioning is called cross-validation. The method derives from assuming that the observations have spatially uncorrelated errors or, minimally, that the independent (or passive) observations have uncorrelated errors with the active observation, and are uncorrelated the background error. This leads to the important property that passive observations are uncorrelated with the analysis error and can then be used to evaluate the analysis [17].

We have developed a theoretical framework and a geometric interpretation that has allowed us to derive a number of statistical estimation formulas of analysis error covariance that can be used in both passive and active observation spaces. It is shown that by minimizing the variance of observation-minus-analysis residuals in passive observation space we actually identify the optimal analysis. This has been done with respect to a single parameter, namely the ratio of observation to background error variances, to obtain a near optimal Kalman gain. The optimization is also done under the constraint of the innovation covariance consistency [14,15]. This optimization could have been done with more than one error covariance parameter but this has not been attempted here. The theory does suggest, however, that the minimum is unique.

Once an optimal analysis is identified we conduct an evaluation of the analysis error covariance using several different formulas; Desroziers et al. [14], Hollingsworth Lönnberg [13], and one that we develop in this paper which works in either active or passive observation spaces. As a way to validate the analysis error variance computed by the analysis scheme itself, the so-called perceived analysis error variance [6], we compare it with the values obtained from the different statistical diagnostics of analysis error variance.

This methodology arises from a need to assess and improve ECCC's surface air quality analyses using our operational air quality model GEM-MACH and real-time surface observations of O_3 and $PM_{2.5}$. Our method applied the theory in a simplified way. First by considering the averaged observation and background error variances and finding an optimal ratio $\gamma = \sigma_o^2/\sigma_b^2$ using as a constraint the trace of the innovation covariance consistency [15]. Second, using a single parameter correlation model, its correlation length, we used the maximum likelihood estimation [11] to obtain near optimal analyses. Also we did not attempt to account for representativeness error in the observations by, for example, filtering observations that are close. Despite all these limitations, our results show that with near optimal analyses, all estimates of analysis error variance roughly agree with each other, while disagreeing strongly when the input error statistics are not optimal. This check on estimating the analysis error variance gives us confidence that the method we propose is reliable, and provides us an objective method to evaluate different analysis components configurations, such as the type of background error correlation model, the spatial distribution of error variances and possibly the use of thinning observations to circumvent effects of representativeness errors.

The methodology introduced here for estimating analysis error variances is general and not restricted to the case of surface pollutant analysis. It would be desirable to investigate other areas of applications, such as surface analysis in meteorology and oceanography. The method could, in principle, provide guidance for any assimilation system. By considering the observation space subdomain [25], proper scaling, local averaging [26], or other methods discussed in Janjic et al. [12] it may also be possible to extend this methodology to spatially varying error statistics. Based on our verification results in Part I [5], we found that there is a dependence between model values and error variances, which we will investigate further in view of our next operational implementation of the Canadian surface air quality analysis and assimilation.

One strong limitation of the optimum interpolation scheme we are using (i.e., homogeneous isotropic error correlation and uniform error variances), which is also the case for most 3D-Var implementations, is the lack of innovation covariance consistency. Ensemble Kalman filters seem, however, much better in that regard although they have their own issues with localization and inflation. Experiments with chemical data assimilation using an ensemble Kalman filter does gives χ^2/N_s values very close to unity after simple adjustments for observation and model error variances [27]. We thus argue that ensemble methods, such as the ensemble Kalman filter, would produce analysis error variance estimates that are much more consistent between the different diagnostics.

Estimates of analysis uncertainties can also be obtained by resampling techniques, such as the jackknife method and bootstrapping [28]. In bootstrapping with replacement, the distribution of the analysis error is obtained by creating new analyses by replacing and duplicating observations from an existing set of observations [28]. This technique relies on the assumption that each member of the dataset is independent and identically distributed. For surface ozone analyses where there is persistence to next day and the statistics is spatially inhomogeneous, the assumption of statistical independence may not be adequate. The comparison of these resampling estimates of analysis uncertainties could be compared with our analysis error variance estimates to help us identify limitations and areas of improvement.

Supplementary Materials

Figure S1: Analysis error variance for ozone optimal analysis case $PM_{2.5}$ **iter 1**. Left panel is the analysis error on the model grid and on the right panel at the active observation sites. Note that the color bar of the left and right panels are different. The maximum of the color bar for the left panel correspond to $\sigma_o^2 + \sigma_b^2$, Figure S2: Distribution (histogram) of the ozone analysis error variance at the active observation locations. First analysis experiment $PM_{2.5}$ **iter 0** (no optimization) on the left panel, and optimal analysis case $PM_{2.5}$ **iter 1** on the right panel.

Acknowledgments: We are grateful to the US/EPA for the use of the AIRNow database for surface pollutants and to all provincial governments and territories of Canada for kindly transmitting their data to the Canadian Meteorological Centre to produce the surface analysis of atmospheric pollutants. We are also thankful for the proof read by Kerill Semeniuk, and for two anonymous reviewers for their comments and help in improving the manuscript.

Author Contributions: This research was conducted as a joint effort by both authors. R.M. contributed to the theoretical development and wrote the paper, and M.D.-J. conducted all experiments design and execution, proof reading and introduced a new diagnostic of optimal analysis error that was further extended to passive observations space.

Appendix A. A Geometrical Derivation of the Desroziers et al. Diagnostic

Let u and v be two random variables of a real Hilbert space as defined in Section 2.3. Two properties of Hilbert spaces are: the polarization identity

$$\langle u, v \rangle = \frac{1}{4}\left\{ \|u+v\|^2 - \|u-v\|^2 \right\} \tag{A1}$$

and the parallelogram identity [19]

$$\|u+v\|^2 + \|u-v\|^2 = 2\left(\|u\|^2 + \|v\|^2 \right) \tag{A2}$$

Combining these two equations, we get:

$$\langle u, v \rangle = \frac{1}{2}\left\{ \|u+v\|^2 - \|u\|^2 - \|v\|^2 \right\} \tag{A3}$$

Let $u = O - A$ and $v = A - B$, then $u + v = O - B$, and so we have

$$\langle (O-A), (A-B) \rangle = \frac{1}{2}\left\{ \|O-B\|^2 - \|O-A\|^2 - \|A-B\|^2 \right\} \tag{A4}$$

If we assume that the analysis is optimal, so that ∠OAT is a right triangle (see Figure 1), then

$$\|\hat{A} - T\|^2 + \|O - \hat{A}\|^2 = \|O - T\|^2 \tag{A5}$$

and similarly that ∠TAB is a right triangle,

$$\|\hat{A} - T\|^2 + \|\hat{A} - B\|^2 = \|B - T\|^2 \tag{A6}$$

and substitute these expression into Equation (A4) we get

$$\left\langle (O-\hat{A}), (\hat{A}-B) \right\rangle = \frac{1}{2}\left\{ \|O-B\|^2 - \|O-T\|^2 - \|B-T\|^2 - 2\|\hat{A}-T\|^2 \right\} \tag{A7}$$

If in addition we assume that we have uncorrelated observation-background errors, that is

$$\|O-B\|^2 = \|O-T\|^2 + \|B-T\|^2 \tag{A8}$$

we then get

$$\left\langle (O-\hat{A}), (\hat{A}-B) \right\rangle = \|\hat{A}-T\|^2 \tag{A9}$$

Note that the difference with the result obtained in Appendix A (property 6) of Ménard [15] where it is shown that the necessary and sufficient condition for the perceived analysis error covariance in active observation space to meet the Desroziers et al. diagnostics for analysis error, is to have innovation covariance consistency and uncorrelated observation-background errors. The result obtained here is different; it concerns the optimal analysis error covariance.

Appendix B. Diagnostics of Analysis error Covariance and the Innovation Covariance Consistency

Let us introduce a departure matrix $\mathbf{\Delta}$ from innovation covariance consistency as,

$$\mathrm{E}[\mathbf{d}\mathbf{d}^T](\mathbf{H}\widetilde{\mathbf{B}}\mathbf{H}^T + \widetilde{\mathbf{R}})^{-1} = \mathbf{I} + \mathbf{\Delta} \tag{A10}$$

The *HL* diagnostic, Equation (18), can be expanded as

$$\begin{aligned}\mathrm{E}[(O-\hat{A})(O-\hat{A})^T] &= \mathrm{E}[\mathbf{dd}^T] + \mathbf{H\widetilde{B}H}^T(\mathbf{H\widetilde{B}H}^T+\widetilde{\mathbf{R}})^{-1}\mathrm{E}[\mathbf{dd}^T](\mathbf{H\widetilde{B}H}^T+\widetilde{\mathbf{R}})^{-1} \\ &\quad -\mathrm{E}[\mathbf{dd}^T](\mathbf{H\widetilde{B}H}^T+\widetilde{\mathbf{R}})^{-1}\mathbf{H\widetilde{B}H}^T - \mathbf{H\widetilde{B}H}^T(\mathbf{H\widetilde{B}H}^T+\widetilde{\mathbf{R}})^{-1}\mathrm{E}[\mathbf{dd}^T] \\ &= \widetilde{\mathbf{R}} - \mathbf{H}(\widetilde{\mathbf{B}} - \widetilde{\mathbf{B}}\mathbf{H}(\mathbf{H\widetilde{B}H}^T+\widetilde{\mathbf{R}})^{-1}\mathbf{H\widetilde{B}})\mathbf{H}^T + \mathbf{error}_{HL} \\ &= \widetilde{\mathbf{R}} - \mathbf{H\hat{A}H}^T + \mathbf{error}_{HL}\end{aligned} \tag{A11}$$

where the $\mathbf{error}_{HL}$ is given as

$$\mathbf{error}_{HL} = \Delta\widetilde{\mathbf{R}} - (\mathbf{H\widetilde{B}H}^T)\Delta^T + \mathbf{H\widetilde{B}H}^T(\mathbf{H\widetilde{B}H}+\widetilde{\mathbf{R}})^{-1}\Delta(\mathbf{H\widetilde{B}H}^T) \tag{A12}$$

which includes three error terms. A scalar version of Equation (A11) is given as

$$error_{HL} = \frac{\sigma_b^2\Delta}{1+\gamma} - \sigma_b^2\Delta + \sigma_o^2\Delta \tag{A13}$$

The error analysis for the *MDJ* diagnostic, Equation (19) is

$$\begin{aligned}\mathrm{E}[(\hat{A}-B)(\hat{A}-B)] &= \mathbf{H\widetilde{K}}\mathrm{E}[\mathbf{dd}^T]\widetilde{\mathbf{K}}\mathbf{H}^T \\ &= \mathbf{H\widetilde{B}H}^T(\mathbf{H\widetilde{B}H}^T+\widetilde{\mathbf{R}})^{-1}\mathbf{H\widetilde{B}H}^T + \mathbf{error}_{MDJ}\end{aligned} \tag{A14}$$

where the $\mathbf{error}_{MDJ}$ is given as

$$\mathbf{error}_{MDJ} = \mathbf{H\widetilde{B}H}^T(\mathbf{H\widetilde{B}H}^T+\widetilde{\mathbf{R}})^{-1}\Delta(\mathbf{H\widetilde{B}H}^T) \tag{A15}$$

There is only one term, and its scalar version is given as,

$$error_{MDJ} = \frac{\sigma_b^2\Delta}{1+\gamma} \tag{A16}$$

The error analysis for the Desroziers et al. diagnostic, Equation (20) is

$$\begin{aligned}\mathrm{E}[(O-\hat{A})(\hat{A}-B)] &= \mathrm{E}[\mathbf{dd}^T](\mathbf{H\widetilde{B}H}^T+\widetilde{\mathbf{R}})^{-1}\mathbf{H\widetilde{B}H}^T - \mathbf{H\widetilde{B}H}^T(\mathbf{H\widetilde{B}H}^T+\widetilde{\mathbf{R}})\mathrm{E}[\mathbf{dd}^T](\mathbf{H\widetilde{B}H}^T+\widetilde{\mathbf{R}})^{-1}\mathbf{H\widetilde{B}H}^T \\ &= \mathbf{H\widetilde{B}H}^T - \mathbf{H\widetilde{B}H}^T(\mathbf{H\widetilde{B}H}^T+\widetilde{\mathbf{R}})^{-1}\mathbf{H\widetilde{B}H}^T + \mathbf{error}_D\end{aligned} \tag{A17}$$

where the $\mathbf{error}_D$ is given as

$$\mathbf{error}_D = \{\mathbf{I} - \mathbf{H\widetilde{B}H}^T(\mathbf{H\widetilde{B}H}^T+\widetilde{\mathbf{R}})^{-1}\}\Delta(\mathbf{H\widetilde{B}H}^T) = \widetilde{\mathbf{R}}(\mathbf{H\widetilde{B}H}^T+\widetilde{\mathbf{R}})^{-1}\Delta(\mathbf{H\widetilde{B}H}^T) \tag{A18}$$

Again only one error term that is similar to the *MDJ* diagnostic, and its scalar version is given as,

$$error_D = \frac{\sigma_o^2\Delta}{1+\gamma} \tag{A19}$$

Error analysis for the diagnostics using passive observations can also be derived. For the passive *MDJ* diagnostics, Equation (21), we have similarly to (A14),

$$\begin{aligned}\mathrm{E}[(\hat{A}-B)_c(\hat{A}-B)_c] &= \mathbf{H}_c\widetilde{\mathbf{K}}\mathrm{E}[\mathbf{dd}^T]\widetilde{\mathbf{K}}\mathbf{H}_c^T \\ &= \mathbf{H}_c\widetilde{\mathbf{B}}\mathbf{H}^T(\mathbf{H\widetilde{B}H}^T+\widetilde{\mathbf{R}})^{-1}\mathbf{H\widetilde{B}H}_c^T + \mathbf{error}_{MDJ\_passive}\end{aligned} \tag{A20}$$

with a single error term given as,

$$\mathbf{error}_{MDJ\_passive} = \mathbf{H}_c\widetilde{\mathbf{B}}\mathbf{H}^T(\mathbf{H\widetilde{B}H}^T+\widetilde{\mathbf{R}})^{-1}\Delta(\mathbf{H\widetilde{B}H}_c^T) \tag{A21}$$

To express a scalar version of this equation we need to account for the background error correlation ρ between the active observation location and the passive observation location, and thus expressed as,

$$error_{MDJ\_passive} = \frac{\rho^2 \sigma_b^2 \Delta}{1+\gamma} \tag{A22}$$

Finally, the fundamental diagnostic of cross-validation Equation (7) does not depend explicitly on the innovation covariance consistency. However, attaining its true minimum by tuning only γ and L_c as would suggest Equation (22), does introduce some innovation in-consistency, which all other optimal diagnostics Equations (18)–(21) has to account for.

References

1. Ménard, R.; Robichaud, A. The chemistry-forecast system at the Meteorological Service of Canada. In Proceedings of the ECMWF Seminar Proceedings on Global Earth-System Monitoring, Reading, UK, 5–9 September 2005; pp. 297–308.
2. Robichaud, A.; Ménard, R. Multi-year objective analysis of warm season ground-level ozone and $PM_{2.5}$ over North-America using real-time observations and Canadian operational air quality models. *Atmos. Chem. Phys.* **2014**, *14*, 1769–1800. [CrossRef]
3. Robichaud, A.; Ménard, R.; Zaïtseva, Y.; Anselmo, D. Multi-pollutant surface objective analyses and mapping of air quality health index over North America. *Air Qual. Atmos. Health* **2016**, *9*, 743–759. [CrossRef] [PubMed]
4. Moran, M.D.; Ménard, S.; Pavlovic, R.; Anselmo, D.; Antonopoulus, S.; Robichaud, A.; Gravel, S.; Makar, P.A.; Gong, W.; Stroud, C.; et al. Recent advances in Canada's national operational air quality forecasting system. In Proceedings of the 32nd NATO-SPS ITM, Utrecht, The Netherlands, 7–11 May 2012.
5. Ménard, R.; Deshaies-Jacques, M. Evaluation of analysis by cross-validation. Part I: Using verification metrics. *Atmosphere* **2018**, in press.
6. Daley, R. The lagged-innovation covariance: A performance diagnostic for atmospheric data assimilation. *Mon. Weather Rev.* **1992**, *120*, 178–196. [CrossRef]
7. Daley, R. *Atmospheric Data Analysis*; Cambridge University Press: New York, NY, USA, 1991; p. 457.
8. Talagrand, O. A posteriori evaluation and verification of analysis and assimilation algorithms. In *Proceedings of Workshop on Diagnosis of Data Assimilation Systems, November 1998*; European Centre for Medium-Range Weather Forecasts: Reading, UK, 1999; pp. 17–28.
9. Todling, R. Notes and Correspondence: A complementary note to "A lag-1 smoother approach to system-error estimaton": The intrinsic limitations of residuals diagnostics. *Q. J. R. Meteorol. Soc.* **2015**, *141*, 2917–2922. [CrossRef]
10. Hollingsworth, A.; Lönnberg, P. The statistical structure of short-range forecast errors as determined from radiosonde data. Part I: The wind field. *Tellus* **1986**, *38A*, 111–136. [CrossRef]
11. Ménard, R.; Deshaies-Jacques, M.; Gasset, N. A comparison of correlation-length estimation methods for the objective analysis of surface pollutants at Environment and Climate Change Canada. *J. Air Waste Manag. Assoc.* **2016**, *66*, 874–895. [CrossRef] [PubMed]
12. Janjic, T.; Bormann, N.; Bocquet, M.; Carton, J.A.; Cohn, S.E.; Dance, S.L.; Losa, S.N.; Nichols, N.K.; Potthast, R.; Waller, J.A.; et al. On the representation error in data assimilation. *Q. J. R. Meteorol. Soc.* **2017**. [CrossRef]
13. Hollingsworth, A.; Lönnberg, P. The verification of objective analyses: Diagnostics of analysis system performance. *Meteorol. Atmos. Phys.* **1989**, *40*, 3–27. [CrossRef]
14. Desroziers, G.; Berre, L.; Chapnik, B.; Poli, P. Diagnosis of observation-, background-, and analysis-error statistics in observation space. *Q. J. R. Meteorol. Soc.* **2005**, *131*, 3385–3396. [CrossRef]
15. Ménard, R. Error covariance estimation methods based on analysis residuals: Theoretical foundation and convergence properties derived from simplified observation networks. *Q. J. R. Meteorol. Soc.* **2016**, *142*, 257–273. [CrossRef]
16. Kailath, T. An innovation approach to least-squares estimation. Part I: Linear filtering in additive white noise. *IEEE Trans. Autom. Control* **1968**, *13*, 646–655.

17. Marseille, G.-J.; Barkmeijer, J.; de Haan, S.; Verkley, W. Assessment and tuning of data assimilation systems using passive observations. *Q. J. R. Meteorol. Soc.* **2016**, *142*, 3001–3014. [CrossRef]
18. Waller, J.A.; Dance, S.L.; Nichols, N.K. Theoretical insight into diagnosing observation error correlations using observation-minus-background and observation-minus-analysis statistics. *Q. J. R. Meteorol. Soc.* **2016**, *142*, 418–431. [CrossRef]
19. Caines, P.E. *Linear Stochastic Systems*; John Wiley and Sons: New York, NY, USA, 1988; p. 874.
20. Cohn, S.E. The principle of energetic consistency in data assimilation. In *Data Assimilation*; Lahoz, W., Boris, K., Richard, M., Eds.; Springer: Berlin/Heidelberg, Germany, 2010.
21. Mitchell, H.L.; Daley, R. Discretization error and signal/error correlation in atmospheric data assimilation: (I). All scales resolved. *Tellus* **1997**, *49A*, 32–53. [CrossRef]
22. Mitchell, H.L.; Daley, R. Discretization error and signal/error correlation in atmospheric data assimilation: (II). The effect of unresolved scales. *Tellus* **1997**, *49A*, 54–73. [CrossRef]
23. Joiner, J.; da Silva, A. Efficient methods to assimilate remotely sensed data based on information content. *Q. J. R. Meteorol. Soc.* **1998**, *124*, 1669–1694. [CrossRef]
24. Migliorini, S. On the quivalence between radiance and retrieval assimilation. *Mon. Weather Rev.* **2012**, *140*, 258–265. [CrossRef]
25. Chapnik, B.; Desroziers, G.; Rabier, F.; Talagrand, O. Properties and first application of an error-statistics tunning method in variational assimilation. *Q. J. R. Meteorol. Soc.* **2005**, *130*, 2253–2275. [CrossRef]
26. Ménard, R.; Deshiaes-Jacques, M. Error covariance estimation methods based on analysis residuals and its application to air quality surface observation networks. In *Air Pollution and Its Application XXV*; Mensink, C., Kallos, G., Eds.; Springer International AG: Cham, Switzerland, 2017.
27. Skachko, S.; Errera, Q.; Ménard, R.; Christophe, Y.; Chabrillat, S. Comparison of the ensemlbe Kalman filter and 4D-Var assimilation methods using a stratospheric tracer transport model. *Geosci. Model Dev.* **2014**, *7*, 1451–1465. [CrossRef]
28. Efron, B. *An Introduction to Boostrap*; Chapman & Hall: New York, NY, USA, 1993; p. 436.

6

Chemical and Biological Characterization of Particulate Matter (PM 2.5) and Volatile Organic Compounds Collected at Different Sites in the Los Angeles Basin

Arthur K. Cho [1,2,*], Yasuhiro Shinkai [3], Debra A. Schmitz [1], Emma Di Stefano [1], Arantza Eiguren-Fernandez [4], Aline Lefol Nani Guarieiro [5], Erika M. Salinas [6], John R. Froines [1] and William P. Melega [2]

1 Department of Environmental Health Sciences, School of Public Health, University of California Los Angeles, Los Angeles, CA 90095-1772, USA; dschmitz@ucla.edu (D.A.S.); emmadeste@gmail.com (E.D.S.); jfroines@ucla.edu (J.R.F.)
2 Department of Molecular and Medical Pharmacology, University of California Los Angeles, Los Angeles, CA 90095-1772, USA; WMelega@mednet.ucla.edu
3 Environmental Biology Laboratory, Faculty of Medicine, University of Tsukuba, 1-1-1 Tennodai, Tsukuba, Ibaraki 305-8575, Japan; ya_shinkai@md.tsukuba.ac.jp
4 Aerosol Dynamics, Inc., Berkeley, CA 94710, USA; arantza@aerosol.us
5 Universidade Federal da Bahia, Instituto de Química, 40170290 Salvador-BA, Brasil; alinelefol@gmail.com
6 División de Ciencias Básicas e Ingeniería, Universidad Autónoma Metropolitana, Azcapotzalco, Av San Pablo Xalpa 150, Mexico; eri_salinas@hotmail.com
* Correspondence: acho@mednet.ucla.edu

Abstract: Background: Most studies on air pollution (AP) exposure have focused on adverse health effects of particulate matter (PM). Less well-studied are the actions of volatile organic compounds (VOCs) not retained in PM collections. These studies quantified chemical and biological properties of both PM2.5 and VOCs. Methods: Samples were collected near the Port of Los Angeles (Long Beach, LB), railroads (Commerce, CM), and a pollution-trapping topography-site (San Bernardino, SB). Quantitative assays were conducted: (1) chemical—prooxidant and electrophile content, (2) biological—tumor necrosis factor-α (TNF-α) and heme oxygenase-1 (HO-1) expression (3), VOC modulation of PM effects and (4), activation of the antioxidant response element (ARE) using murine RAW 264.7 macrophages. Results: SB site samples were the most potent in the chemical and biological assays, followed by a CM railroad site. Only PM2.5 exhibited significant proinflammatory responses. VOCs were more potent than PM2.5 in generating anti-inflammatory responses; further, VOC pretreatment reduced PM-associated TNF-α expression. VOCs significantly increased ARE activation compared to their corresponding PM2.5 which remained at background levels. Conclusion: Ambient VOCs are major contributors to adaptive responses that can modulate PM effects, in vitro, and, as such, need to be included in comprehensive assessments of AP.

Keywords: PM2.5; VOC; volatile organic compounds; prooxidants; electrophiles; tumor necrosis factor alpha; hemeoxygenase-1; ambient air; antioxidant response element; murine RAW 264.7 macrophages

1. Introduction

Epidemiological and clinical studies have firmly established associations between ambient air pollution (AP) levels and adverse health effects (see for examples, [1–4]). To advance our understanding of the relationships between AP exposure and disease processes, chemical and biological studies are needed to identify and characterize underlying molecular and cellular mechanisms. An essential element of those studies is the use of quantitative methodologies that allow for the comparison of results across other AP studies conducted in different local environments and geographical regions.

Our prior studies have been focused on characterizing reactive chemicals in AP that interact with cellular targets to elicit responses such as inflammation. In ambient air, these chemicals are differentially distributed between particle and vapor phases as obtained in AP sample collections. The particulate matter phase (typically PM2.5) contains mostly prooxidants such as redox active metals together with higher molecular weight quinones and humic-like substances (HULIS) [5,6] together with inorganic electrophiles such as arsenic and zinc. The vapor phase contains mostly volatile organic compounds (VOCs) that include quinones [7–9] and electrophilic carbonyls [10]. At the functional level, prooxidants catalyze the reduction of oxygen to reactive oxygen species and electrophiles react with nucleophilic functions, such as thiolate and amino groups, to form covalent bonds (see Graphical Abstract).

Surprisingly, in recent years, research on biological assessments of AP has largely focused only on the consequences of PM exposure without addressing the potential biological contributions of corresponding VOCs. To address this deficiency, the present studies were designed to generate quantitative data on both PM2.5 and VOCs of ambient AP samples for evaluating their distinct chemical reactivities and biological effect profiles.

In AP biological studies, the consequences of chemical exposure upon cellular components involves induction of oxidative stress and/or protein modifications that at low levels activate cytoprotective mechanisms but at higher levels initiate proinflammatory responses [11]. To understand the relationship between levels of reactive chemicals and this continuum of cellular responses, in this study we determined in PM2.5 and VOC fractions of ambient AP samples the prooxidant content based on their ability to transfer electrons from dithiothreitol to oxygen [12] and electrophile content, based on their ability to inactivate glyceraldehyde 3-phosphate dehydrogenase (GAPDH) [13]. The same fractions were then subjected to cellular assays designed to determine relative potencies of their biological responses.

The AP samples for this study were collected at three regions that neighbor railyards (Long Beach (LB), Commerce (CM), and San Bernardino (SB)) in the Los Angeles Basin with the objective of assessing regional differences in the chemical and biological actions of AP [14]. The chemical study results show distinct distributions of reactive chemicals: prooxidants in the PM2.5 and the electrophiles in the VOCs. The biological study results show that the samples with high reactive chemical content were the most potent in eliciting biological responses: PM2.5 promoting an inflammatory response and VOCs promoting an anti-inflammatory response.

Additionally, a second set of AP samples were collected at CM at sites 0.03–1 mile from its railyards to assess their chemical and biological actions as a function of proximity to the emission source. Those CM local samples exhibited profiles similar to the three regions' data, i.e., high chemical reactivity was associated with potent biological effects; furthermore, distance-dependent chemical and biological reactivities were also observed.

The pro- and anti-inflammatory biological responses to the PM2.5 and VOC samples raised the question of whether additive or antagonistic interactions occur following ambient AP exposures (i.e., PM2.5 + VOCs). We could not address this issue directly insofar as the concentrations of PM2.5 and VOC extracts precluded simultaneous exposure studies, however, we did conduct a two phase exposure in which RAW 264.7 macrophages were pre-exposed to VOCs, followed 24 h later by exposure to PM2.5. The results show that pre-exposure to VOCs reduces the magnitude of an inflammatory response to subsequent PM2.5. Based on prior studies showing that compounds contained in VOCs can activate the Nrf2-ARE pathway [15], the current samples were similarly tested. VOCs, but not PM2.5, activated this pathway, suggesting a cellular mechanism for their anti-inflammatory actions. Collectively, our in vitro studies show that VOCs contain reactive chemicals capable of inducing significant anti-inflammatory effects and, as such, need to be included with PM analyses to enable comprehensive studies of ambient AP.

2. Materials and Methods

2.1. *Materials*

Rabbit muscle glyceraldehyde-3-phosphate dehydrogenase (GAPDH), nicotinamide adenine dinucleotide (NAD^+), ethylenediaminetetraacetic acid (EDTA), glyceraldehyde-3-phosphate (GAP), dithiothreitol (DTT), 5,5′-dithiobisbis-(2-dinitrobenzoic acid (DTNB), and diethylenetriaminepentaacetic acid (DTPA) were purchased from Sigma-Aldrich (St. Louis, MO, USA). Other reagents were of the highest grade available and purchased from Fisher Scientific (Pittsburgh, PA, USA). Murine macrophage cell line RAW 264.7 cells were purchased from American Type Cell Culture (Manassas, VA, USA). Dulbecco's modified Eagle medium (DMEM) and penicillin/streptomycin were purchased from Life Technologies (Carlsbad, CA, USA). Fetal bovine serum was purchased from Gemini Bio-Products (Sacramento, CA, USA). The ELISA kits for TNF- α were purchased from BD Biosciences Pharmingen (San Diego, CA, USA) and the kits for HO-1 were purchased from Enzo Life Sciences (Farmington, NY, USA). Lipofectamine LTX transfection reagent was obtained from Life Technologies (Carlsbad, CA, USA).

2.2. *Sample Collection and Extraction*

2.2.1. PM2.5

Medium-volume samplers (Tish, Cleves, OH) were located in the selected locations; PM2.5 and vapor samples were collected continuously for 48 h as one time collections. The LB and SB samples were collected in early summer (June). The CM samples were collected continuously over a 5 d period in late spring (May). Teflon-coated glass fiber filters (Pall Corp, East Hills, NY) were used for PM2.5 collection and XAD-4 resin beds (Acros, Thermo Fisher Scientific) for the vapor-phase. Sampling details and matrix cleaning procedures have been previously published (16). Estimates of the volume equivalent for each vapor sample analyzed were based on the air volume collected divided by the fraction of the total extract used in each analysis. Aqueous suspensions of PM2.5 samples were prepared by sonicating filter punches in cell culture water for 20 min and the corresponding volume of air (m^3/cm^2) calculated to normalize the results. As the mass of the PM2.5 on the filter was not measured, the volume equivalent of air was used to describe the final concentration of the particles in the aqueous suspension, which was 2.2–6.0 m^3/mL.

2.2.2. VOCs

XAD-4 resin beds containing the trapped vapor phase organic components corresponding to each particle sample were extracted by sonication (30 min) with dichloromethane. The suspension was filtered through a 0.45 μm nylon filter (Millipore, Billerica, Massachusetts), the volume reduced, and solvent evaporated into a known volume of dimethyl sulfoxide (DMSO), so the concentration for analysis could be expressed as m^3 per mL of DMSO. The final concentration of the organic extract was approximately 300 m^3/mL of DMSO. Blank XAD-4 resin extracts were prepared as described previously and used as controls [16]. Highly polar compounds such as those with multiple hydroxyl moieties that are associated with the vapor phase would not be extracted with dichloromethane.

2.3. *Chemical Assays*

2.3.1. DTT Assay

This assay measured the prooxidant content of the sample from its ability to transfer electrons from DTT to oxygen [12–17]. In the procedure, aliquots of vapor-phase and PM2.5 water suspensions were incubated with DTT (Sigma Chemical Co., MO, USA) for times varying from 10 to 30 min. The reaction was quenched at specific times and after addition of DTNB to the complex with the remaining DTT, the absorbance at 412 nm was measured. Rates were calculated averaging duplicate runs, and were blank-corrected. The units were DTT nmol consumed per min per m^3, with 95% confidence intervals derived from regression analysis of rates.

2.3.2. GAPDH Assay

Electrophilic reactivity was measured from the sample's ability to inhibit or inactivate the thiolate enzyme GAPDH, through covalent bonding [13]. In brief, a mixture of 1 unit of rabbit GAPDH was incubated with aliquots of the organic extracts of vapors or water suspensions of particles under argon gas at 25 °C for 120 min. The reaction was then quenched by adding an equal volume of cold DTT solution; GAPDH activity, based on the rate of nicotinamide adenine dinucleotide (NADH) formation, was measured by its absorption at 340 nm. The ability to inactivate the enzyme was expressed in equivalents of N-ethylmaleimide (NEM), the standard electrophile, which was included in each assay as a control. Samples were run in triplicate and values reported as averages ± SEM. The assay provided a measure for electrophile content that was based on structures capable of interacting with the catalytic center of the enzyme and provided a quantitative estimate of the electrophilic content which could be used in comparison studies. The units used were the equivalents of NEM per m^3.

2.4. Cell Culture and Treatment

2.4.1. Cell Viability

Cell viability was determined by the 3-(4,5-dimethylthiazol-2yl)-2,5-triphenyl tetrazolium bromide (MTT) tetrazolium salt colorimetric assay [18]. A 5 mg/mL MTT solution was prepared in phosphate buffered saline and sterilized by filtration through a Steriflip. RAW264.7 cells were exposed to particle and vapor samples in 96-well plates for 16 h then treated with 10 μL of 5 mg/mL MTT for 2 h at 37 °C. The medium was removed and 100 μL DMSO was added to dissolve the formazan. Absorbance was measured at 540 nm on a Biotek Synergy 2 Multi-mode Microplate Reader. To determine suitable sample concentration ranges for these experiments, air sample toxicity was assessed by RAW 264.7 macrophages exposure to concentrations of 1 m^3/mL for 16 h. The loss of cells was normalized to that caused by a blank filter extract for the PM2.5 and a blank XAD resin extract for the vapors. The PM2.5 loss was 113.6% ± 3.9% (SEM for N = 8) and that for XAD was 94.7% ± 3.0% of their respective controls. Based on these results, all subsequent cell experiments used concentrations of 1 m^3/mL or less.

2.4.2. Cell Exposure

Murine RAW 264.7 cells were cultured in DMEM, supplemented by 1% penicillin-streptomycin and 10% FBS as described by Li et al. [11] with slight modifications. Cells were exposed to three concentrations (0.1 to 2.0 m^3 air equivalent/mL) of PM2.5 or VOC extracts in duplicate for 16 h, after which the medium and cells were separated and the cells subjected to lysis to obtain a cell extract. The cell extracts were used in the ELISA assays for HO-1 and the medium used to assay TNF-α. The results were analyzed by linear regression procedures (Graph Pad Prism, (San Diego CA, USA) to determine (a) concentration dependency of the response, and (b), if the slope was significant, its value, which is a measure of the potency of the sample.

2.5. Two Phase Study VOC Pretreatment/PM Exposure

In phase 1, cells were exposed to the VOC at a single concentration in triplicate (1 m^3/mL) together with the relevant controls for 24 h. In phase 2, the medium was removed and replaced with fresh medium containing the challenge agent, or PM2.5, also at 1 m^3/mL for 16 h. Cells and media were then processed for HO-1 and TNF-α analyses

2.5.1. ELISA Assays

The ELISA assays were performed following instructions provided by the manufacturers (HO-1; Enzo Life Sciences; TNF-α, BD Pharmingen). The results reported are the differences between the control and the experimental conditions. Values for HO-1 (ng/mg cell protein) and TNF-α (pg/mg cell protein) were normalized to cell protein.

2.5.2. ARE/EpRE Activation

DNA transfections were performed with Lipofectamine LTX transfection reagent (Life Technologies, Carlsbad, CA) following the manufacturer's instructions performing cell culture in 12-well plates. ARE-luciferase cDNA (1 μg/well) and pRL-TK cDNA (0.1 μg/well) or transfection reagent (2 μL/well) were mixed with serum-free media. Before addition to the cells, the DNA solution and transfection reagent solution were mixed together and incubated for 20 min at room temperature to allow the formation of complexes. The complexes were mixed with the culture media and incubated for 24 h to allow transfection. After transfection, the cells were exposed to the samples as described above and luciferase activity measured in cellular extracts according to the manufacturer's instructions (Dual-Luciferase reporter assay system; Promega, WI, USA) with a multi-mode microplate reader (Synergy 2, Bio-Tek, Winooski, VT, USA).

2.6. Collection Sites

The ambient air samples for this study were collected in three communities with railyards in the Los Angeles Basin: Commerce (CM), Long Beach (LB) and San Bernardino (SB) (Figure 1). CM, with two railyards, is located in the midtown area of Los Angeles. LB is southwest of CM and neighbors the Pacific Ocean and the Port of Los Angeles, which is the largest container port in the United States as measured by container volume and cargo value. Its emission sources would include both local railyards and the Port. The SB site is located in the eastern end of the Basin, approximately 80 miles from the coast and represents a receptor site that receives air parcels that have been subjected to modifications by photochemical and chemical reactions as they move from east to west across the Basin [19], together with emissions from a SB railyard. We previously observed that as the air mass moves east, a trend toward higher levels of 9,10-phenanthroquinone was found in the air samples, consistent with its formation from phenanthrene by atmospheric processes [7].

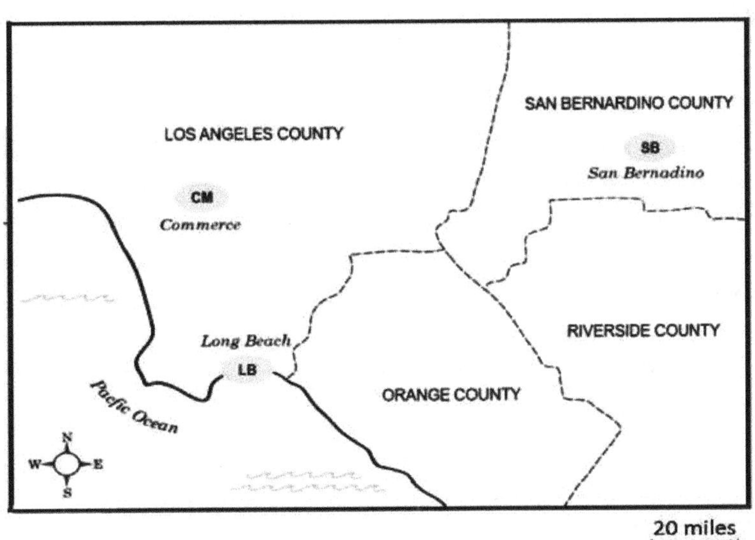

Figure 1. Map of the Los Angeles Basin with locations of the air collection sites: Long Beach (LB), Commerce (CM) and San Bernardino (SB). The Los Angeles Basin contains a coastal area bordered by the San Gabriel Mountains located in northern Los Angeles County and western San Bernardino County.

The second set of collections focused on CM at different distances nearby its railyards to assess ambient air properties as a function of distance from emission sources (Figure 2).

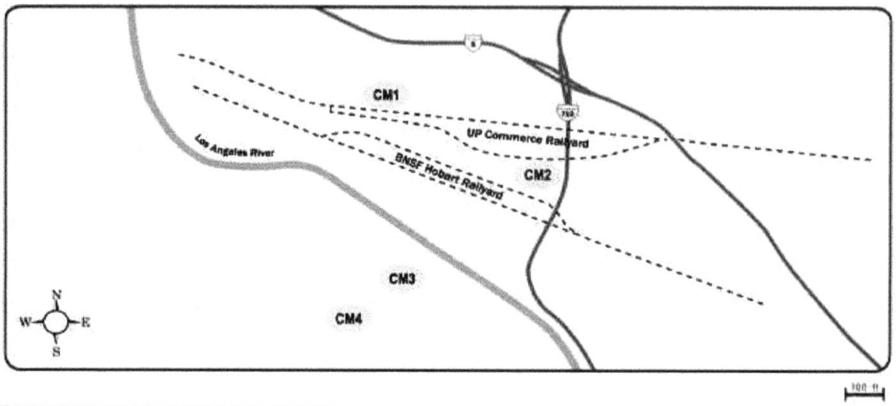

Figure 2. Map of the Commerce railyards with locations of air collection sites.

The sites, CM1 and CM2, were closest to the CM-Union Pacific railyard and downwind from the Burlington Northern and Santa Fe (BNSF) Hobart yard. Sites CM3 and CM4 were both upwind of the BNSF yard, but CM3 was closer to the yards and to a diesel truck processing center. CM4 was considered as a background site, minimally directly impacted by the railyard activities. It should be pointed out that as the samples were collected continuously 24-h/d for 5-d, changes in air movement during that period could have minimized differences due to changes from prevailing onshore to infrequent offshore winds. The distances from the railyards for the CM collections are summarized in Table 1.

Table 1. Location of collection sites at Commerce.

Site	Distance from Railyard (ft)	Emission Environment
CM1	155	Railyard
CM2	400	Railyard (between 2 tracks)
CM3	4440	Diesel truck loading site
CM4	7000	Background

2.7. Biological Reactivities

Two markers were used to assess cellular responses by the RAW 264.7 macrophage cell line: TNF-α as a proinflammatory response [20,21] and HO-1 as an anti-inflammatory response [22]. Cellular responses were determined at three concentrations (0.1, 0.5 and 1.0 m^3/mL) to ensure that the responses compared were linear over the concentrations used. The results were analyzed by linear regression to obtain slopes of the concentration vs. response curves. The results provide an assessment of the concentration-dependency of the response with the slope providing a quantitative assessment of potency in stimulating expression of the target markers, TNF-α and HO-1 (see Supplementary Materials).

2.8. Computational Procedures

All computations were performed with GraphPad Prism 8.12 (San Diego, CA, USA). Linear regression procedures were used to generate the best-fit values, the standard error of the slope and its 95% confidence intervals together with an assessment of the significance of the slope provided by its p value. These procedures generated Pearson correlation coefficients and their respective p values.

3. Results

3.1. Chemical and Biological Reactivities for Basin Sites

The PM and VOC samples from SB were the most reactive compared to those collected at CM and LB. For all three regions, the prooxidant content was mostly associated with the PM2.5 fraction (Figure 3).

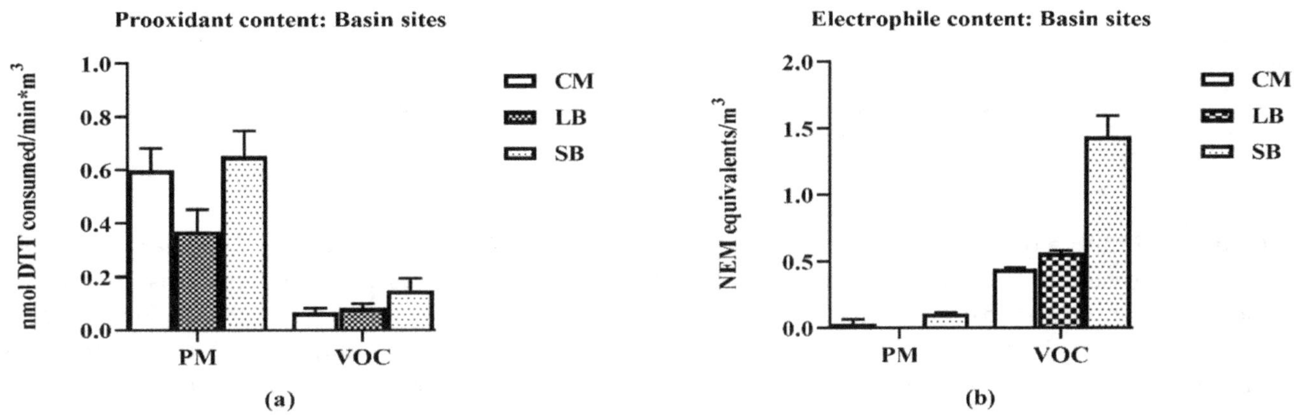

Figure 3. Chemical properties of particulate matter (PM) and volatile organic compound (VOC) air samples collected at CM, LB and SB. For each region's PM and VOC samples, nmol dithiothreitol (DTT) consumed (prooxidant content (**a**)) and N-ethylmaleimide (NEM) equivalents derived from a glyceraldehyde 3-phosphate dehydrogenase (GAPDH)-inhibition assay (electrophile content (**b**)) were measured. The VOCs represent the dichloromethane extracts of XAD resin traps of the vapor phases. Error bars indicate 95% CI (n = 3) for DTT results; SEM (n = 3) for the NEM results (values of the CIs are detailed in supplemental data—Table S1).

In contrast, electrophile content was mostly associated with the VOC fractions; electrophile content of the PM2.5 was close to background levels. SB VOC electrophile content was approximately three-fold higher than those for CM, LB. All PM2.5 samples increased TNF-α expression, with SB being 40-fold more potent than CM and LB (Figure 4). Increases in TNF-α expression were not detected in the VOC samples, i.e., they did not exhibit a concentration-dependent response (data not shown).

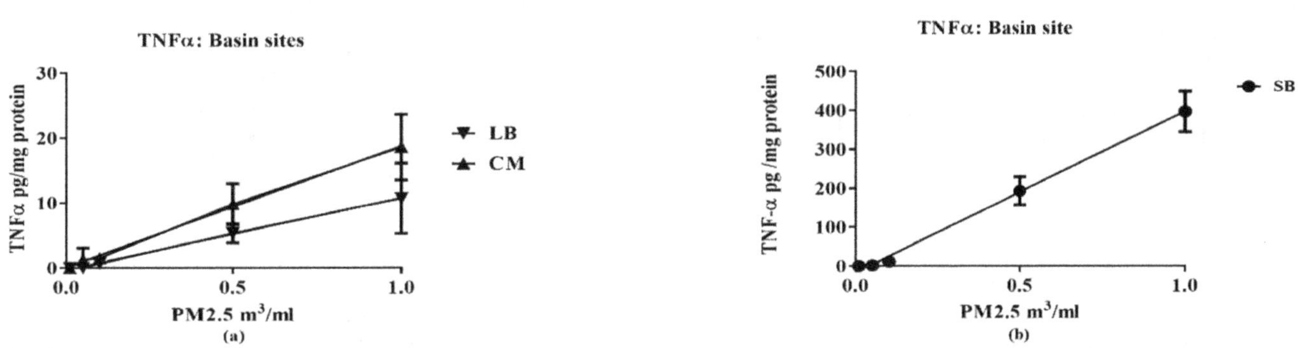

Figure 4. Relative potencies of air samples from CM, LB (**a**) and SB (**b**) to increase TNF-α in RAW 264.7 macrophages. TNF-α expression in the media was measured following exposure to the indicated PM 2.5 concentrations for 16 h. The solid lines represent the best line fit following linear regression analysis of the results (n = 3 for each data point) with potencies indicated by the slopes (values of the regression analysis are detailed in supplemental data—Table S4).

All VOC samples increased HO-1 expression, with SB being three-fold more potent than CM, LB, (Figure 5). The SB PM also increased HO-1 expression but the response was markedly lower than its VOC; concentration-dependent increases in HO-1 expression for CM PM2.5 and LB PM2.5 were not detected.

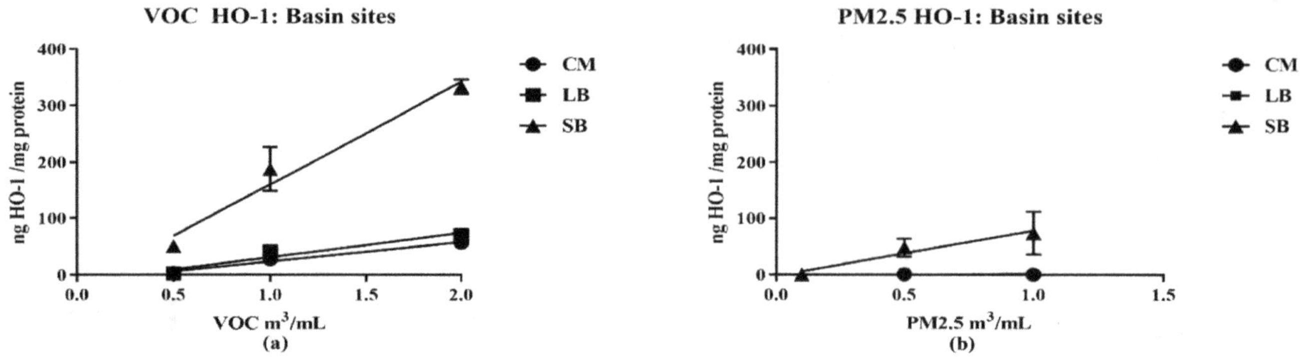

Figure 5. Relative potencies of VOC and PM 2.5 air samples from CM, LB (**a**) and SB (**b**) to increase HO-1 expression in RAW 264.7 macrophages. Cell extracts were measured following exposure to the indicated concentrations of either VOC or its corresponding PM for 16 h. The lines represent the best line fit following linear regression analysis of the results (n = 3 for each data point) with potencies indicated by the slopes (values of the regression analysis are detailed in supplemental data—Table S5).

3.2. Chemical and Biological Reactivities: CM Railroad Sites

The collection sites were located at different distances (0.03 to 1.3 mi from the railyards (Table 1) identified as CM1 to CM4 in increasing distance. The PM2.5 prooxidant and VOC electrophile content decreased as a function of distance from the railroad emission source (Figure 2) with the exception of CM2 which was adjacent to two railyards.

Prooxidant content was mostly associated with PM2.5 samples while electrophile content was mostly associated with the VOC samples. VOC prooxidant and PM2.5 electrophile contents remained close to background levels for all samples (Figure 6).

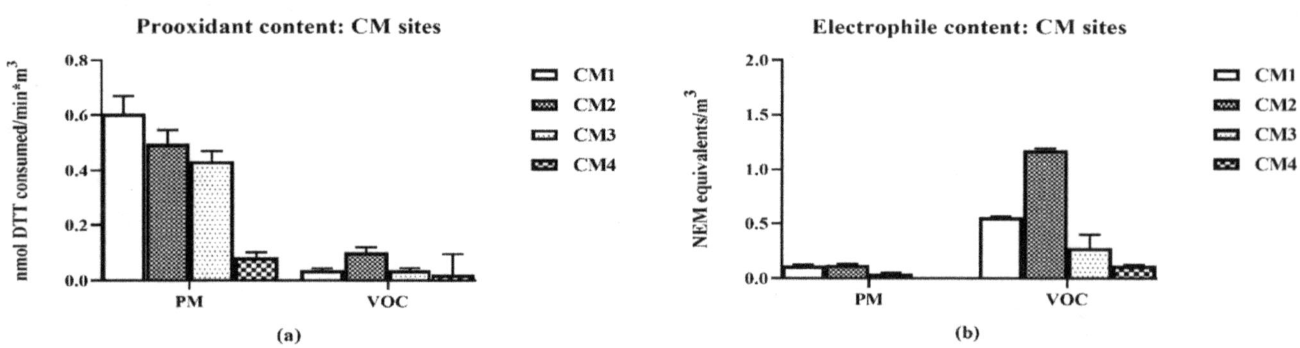

Figure 6. Chemical properties of PM and VOC air samples collected near to the CM railroad area (CM1—closest, CM4 most distant). For each PM and VOC local sample, the nmol DTT consumed (prooxidant content (**a**)) and NEM equivalents (electrophile content (**b**)) were measured. Error bars indicate 95% CI for DTT (n = 3) results and SEM (n = 3) for the GAPDH results; analysis methods as described in Figure 4, (values of the CIs are detailed in supplemental data—Tables S2 and S3).

The biological reactivities paralleled the profiles of the regional samples. Further, PM2.5 TNF-α and VOC HO-1 expression showed a 'distance' pattern, with CM2 being the most potent and CM4 (most distant) being least potent; PM2.5 HO-1 expression remained close to background levels for all samples (Figure 7).

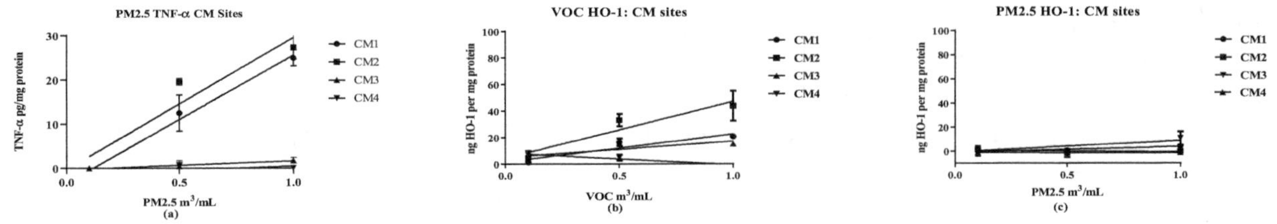

Figure 7. TNF-α expression (**a**) and HO-1 expression (**b**,**c**) in RAW 264.7 macrophages using air samples collected nearby CM railroads (CM1—closest, CM4 most distant). Macrophages were incubated with three concentrations of VOC or PM for 16 h. The lines represent the best line fit following linear regression analysis of the results (n = 3 for each data point) with potencies indicated by the slopes (values of the regression are detailed in supplemental data—Tables S4 and S5).

3.3. SB VOC Pretreatment Effects on SB PM2.5 2.5 Expression of TNF-α and HO-1 in RAW 264.7 Macrophages

The anti-inflammatory properties of the VOC suggested that VOC pretreatment may affect the attenuation of the pro-inflammatory effects of a subsequent PM2.5 challenge. We selected the most potent samples—SB VOC for the pretreatment and SB PM2.5 for the challenge. A sequential exposure experiment was performed in which cells were first exposed to the SB VOC for 24 h. The media was then replaced with fresh media containing control filter extracts or suspensions of the SB PM2.5 and the exposure continued for 16 h. Macrophages and media were analyzed for HO-1 and TNF-α content, respectively (Figure 8).

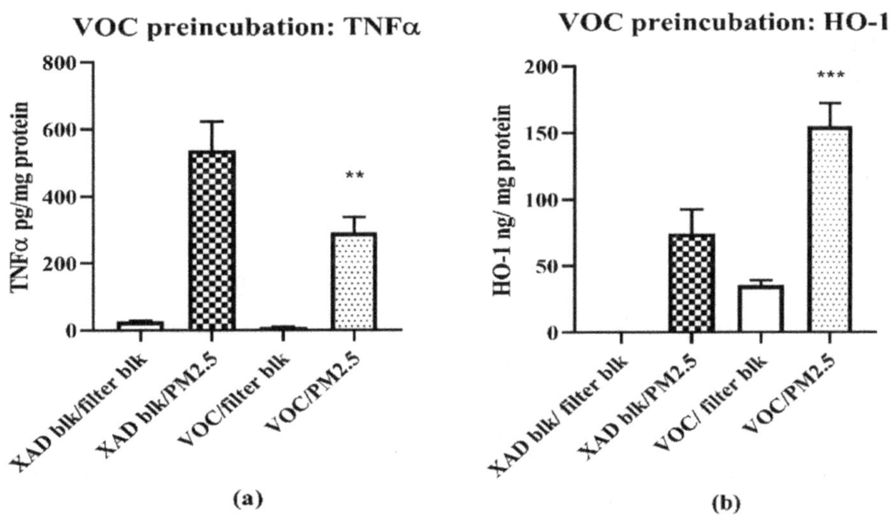

Figure 8. SB VOC pretreatment effects on SB PM 2.5 expression of TNF-α (**a**) and HO-1 (**b**) in RAW 264.7 macrophages. Cells were preincubated with VOC (1 m^3/mL) for 24 h followed by incubation with PM 2.5 (1 m^3/mL) for 16 h; TNF-α—XAD Blk/PM 2.5 vs. VOC/PM 2.5, n = 4; $p^{**} < 0.005$; HO-1—XAD Blk/PM 2.5 vs. VOC/PM 2.5, n = 4; *** $p < 0.001$.

The reason for the two-stage exposure was the limitation in the volume of sample solutions that could be added to the plate cells while maintaining the incubation conditions (see Methods for details). The results show a 54% decrease in the TNF-α response to the PM2.5 challenge following pretreatment to SB VOCs at 1 m^3/mL, n = 4; $p^{****} < 0.005$, together with a significant 200% increase in HO-1 expression (Blank XAD/SB vs. SB VOC/SB 2.5; (n = 4; $p < 0.001$).

3.4. Activation of the antioxidant Response Element (ARE) in RAW 264.7 Macrophages by Air Samples from CM, LB and SB

In prior studies, we measured an HO-1 response in VOC samples collected at Riverside CA (a sample site nearby SB). Those samples also showed an electrophilic effect on the ARE [23].

Accordingly, we evaluated in this study the three regional samples for their potential to activate the ARE. The results show that all VOC samples significantly increased ARE activation 1.5- to three-fold compared to their corresponding PM2.5 samples which remained at background levels. The SB VOC sample was approximately two-fold more potent than the LB and CM samples (Figure 9).

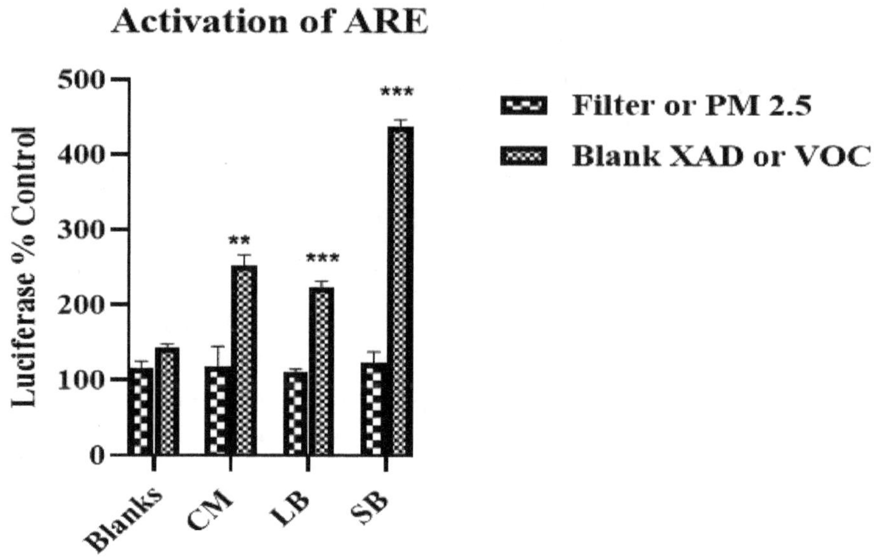

Figure 9. Activation of the antioxidant response element (ARE) in RAW 264.7 macrophages by air samples from CM, LB and SB. Cells transfected with ARE-luciferase cDNA were exposed to either PM2.5 and VOC from CM, LB, SB at 1 m^3/mL for 16 h and ARE activation assessed; (see Methods for ARE procedures). Each value is the mean ± SD, n = 4/sample site; site values normalized to corresponding blanks; CM VOC vs. CM PM2.5, ** $p < 0.005$; LB VOC vs. LB PM 2.5, SB VOC vs. SB PM2.5, *** $p < 0.00001$.

4. Discussion

In our prior AP studies on PM2.5 and VOCs, we characterized chemical species that can participate in two reactions associated with acute toxicity, (1) prooxidants that catalyze the formation of hydrogen peroxide and hydroxyl radical, and (2) electrophiles that form covalent bonds with nucleophilic groups such as amino and thiolate found in proteins. Furthermore, we showed that VOC electrophiles associated with AP can activate cellular signaling pathways relevant to adverse health effects [23]. However, VOC biological assessment has received minimal attention in conjunction with PM research despite its apparent contributions to AP toxicology, with the exception of adductomics studies by Rappaport and his associates (e.g., [24]).

For this study, analytical methods we previously developed and validated [14] were used to characterize chemical and biological properties of SB, CM and LB AP samples. The samples were obtained from (1) an aqueous suspension of the particulates, PM2.5, and (2) an organic extract of the VOCs trapped by polystyrene resins placed downstream of the PM2.5 filters which have been shown to contain the majority of the total electrophiles present in ambient air [14]. Based on our previous analyses of seasonal effects on AP properties [14,16], the samples for this study collected in late spring/early summer were likely have relatively higher prooxidant and VOC electrophile content compared to winter collections. Nonetheless, despite such quantitative differences, predominantly PM—TNF-α—and VOC electrophile responses were consistently observed for both seasonal collections. The quantitative nature of the regional data allowed for multiple comparisons between different parameters of AP chemical reactivities in terms of pro- and anti-inflammatory properties, and their distribution between PM 2.5 and VOC phases. Prooxidant reactivity and proinflammatory responses (TNF-α increases) were predominantly associated with PM samples while electrophile reactivity and anti-inflammatory responses (HO-1 increases and ARE activation) were predominantly associated with VOCs. Future studies, both in vitro and in vivo, are needed to determine whether the interactions of these responses are synergistic or antagonistic following concurrent exposure of PM and VOCs. Presently, evidence for a potential interaction is suggested from the results of the two-phase studies we conducted with RAW 264.7 macrophages in which the anti-inflammatory effects of the VOC suppressed the pro-inflammatory effects of the PM2.5.

The following analysis illustrates how the data we generated can be used to explain differences in the response profiles of the two most reactive and potent samples, SB and CM2.

Of particular note, the SB and CM2 samples have comparable chemical reactivities in the DTT assay (0 65, 0.50; Table 2), but the SB pro- and anti-inflammatory potencies as derived from the biological assays were significantly greater (slope values from linear regression analysis; Table 2B). Namely, the SB TNF-α and HO-1 responses were 14 and four times more potent, respectively, than those generated by CM2. We interpret these differences in terms of the chemical content in the samples being collected at different regions of the LA Basin. The CM2 sample consists mostly of railyard traffic emissions, whereas the SB sample has been subjected to atmospheric chemical reactions on the air mass as it moves approximately 80 miles across the Basin. These reactions convert, among other compounds, polynuclear aromatic hydrocarbons to their corresponding quinones as exemplified by 9,10-phenanthroquinone generation (PQ) [7] a potent biological prooxidant associated with PM [25]. Relatively higher levels of such organic prooxidants in the SB PM2.5 sample are suggested by the DTT assay data (Table 2B) that show a larger fraction of the SB PM activity not inhibited by metal chelation (i.e., DTPA-insensitive), in support of this phenomenon of atmospheric formation of organic prooxidants. Analogously, the generation of organic electrophiles such as 1,4-benzoquinone (BQ) likely contribute to the more potent SB anti-inflammatory HO-1 response in the VOC fraction (Table 3). Based on these considerations, we suggest that the higher pro- and anti-inflammatory actions of the SB sample compared to that of CM2 are due to its higher content of organic reactants generated by photochemical processes.

Table 2. Prooxidant content of SB and CM2 samples.

	DTT		DTT + DTPA	
	Mean	**95% CI**	**Mean**	**95% CI**
SB	0.65	0.55–0.75	0.16	0.11–0.20
CM2	0.50	0.47–0.55	0.08	0.03–0.13
SB/CM2 ratio	1.31		1.89	

Prooxidant content, defined by DTT activity (nmols DTT consumed/min*m^3), in the presence and absence of 20 μM, diethylenetriamine pentaacetic acid (DTPA), a metal chelator [16].

Table 3. Biological properties of SB and CM2 samples.

	PM2.5 TNF-α		VOC HO-1	
	Equation	**Slope: 95% CI**	**Equation**	**Slope: 95% CI**
SB	Y = 413.9*X − 16.7	319.9–507.9	Y = 185.2.4*X − 37.8	158.5–212.9
CM2	Y = 29.88*X − 0.34	22.58–37.18	Y = 42.74*X + 4.49	22.76–58.22
SB/CM2 ratio	13.8		4.3	

Slopes of cellular responses to three concentrations of PM2.5 from the indicated PM2.5 samples as a function of cellular protein concentration. Units: Y = TNF-α pg/mg protein; Y = HO-1 ng/mg protein/m^3; X = m^3/mL.

A second consideration in determining relative biological potencies of PM and VOCs is their cellular access upon exposure. PM/metal ions' entry into cells is regulated by endocytosis and transport processes [6,26], whereas quinones such as PQ and BQ are neutral and non-polar compounds that can readily penetrate membranes to promote oxidative stress-related reactions [27]. These biological differences underscore the complex chemical composition of PM, which limits the interpretive value of a general chemical assay such as the DTT assay and necessitate inclusion of parallel biological assays for more definitive evaluation of PM reactivities.

Direct extrapolation of these results to potential toxicological effects in humans, i.e., PM-induced oxidative stress and VOC-induced covalent adducts, remains to be established. In one of the very few direct comparisons of 'whole' diesel exhaust (DE) and VOC components, Campen et al. [28] compared the cardiovascular effects of these two samples in mice and observed major effects consistent with myocardial ischemia, attributable to VOC components. These VOC effects were also seen in vitro using isolated blood vessels. Although some of the chemical components were determined, they did not, however, assay for redox-active species or electrophiles. A second relevant study examined the role of electrophiles that included 1,2-naphthoquinone, a component of VOCs [25] in the actions of DE on mouse lungs [29]. In an intact lung model and in multiple cultured human lung cell models, they showed activation of transient receptor potential ankyrin-1, a pro-inflammatory mediator, on airway C-fibers by a variety of qualitatively and quantitatively different DE particles. These studies demonstrated the actions of VOCs underlying the toxicity of emissions and emphasized the need for further studies that model human exposure conditions.

5. Conclusions

We have shown that quantitative assessments of chemical and biological properties of ambient PM2.5 and VOCs can be used effectively to characterize, compare and contrast AP across different geographical regions, to account for effects of atmospheric modifications on air mass, and to evaluate exposure proximity to an emission source. Of major significance is the consistent characterization of VOCs in terms of their electrophilic chemical reactivity and biological modulation of HO-1 expression and ARE activation. Collectively, these results have human health implications. VOCs, being non-polar volatile species, will readily enter the lungs and penetrate membranes to access all body compartments and thus can potentially contribute to multiple organ effects. Furthermore, electrophilic properties of VOCs can affect the long-term inactivation of protein function resulting from their covalent binding with tissue nucleophiles. Consequently, the effects of low level VOC exposure in AP can accumulate over time and contribute to chronic adverse health effects.

Author Contributions: Conceptualization, A.K.C., J.R.F., W.P.M.; methodology, Y.S., D.A.S., E.D.S., A.E.-F., A.L.N.G., E.M.S.; software, A.K.C.; validation, A.K.C., W.P.M., D.A.S.; writing—original draft preparation, A.K.C., Y.S., W.P.M.; writing—review and editing A.K.C., W.P.M., Y.S.; supervision, A.K.C.; project administration, A.K.C., J.R.F.; funding acquisition, J.R.F., A.K.C. All authors have read and agreed to the published version of the manuscript.

References

1. Berhane, K.; Chang, C.C.; McConnell, R.; Gauderman, W.J.; Avol, E. Association of Changes in Air Quality With Bronchitic Symptoms in Children in California, 1993–2012. *JAMA* **2016**, *315*, 1491–1501. [CrossRef] [PubMed]
2. Cassee, F.R.; Heroux, M.E.; Gerlofs-Nijland, M.E.; Kelly, F.J. Particulate matter beyond mass: Recent health evidence on the role of fractions, chemical constituents and sources of emission. *Inhal. Toxicol.* **2013**, *25*, 802–812. [CrossRef] [PubMed]
3. Delfino, R.J.; Wu, J.; Tjoa, T.; Gullesserian, S.K.; Nickerson, B.; Gillen, D.L. Asthma morbidity and ambient air pollution: Effect modification by residential traffic-related air pollution. *Epidemiology* **2014**, *25*, 48–57. [CrossRef] [PubMed]
4. McConnell, R.; Wu, W.; Berhane, K.; Liu, F.; Verma, G. Inflammatory Cytokine Response to Ambient Particles Varies due to Field Collection. *Am. J. Respir. Cell Mol. Biol.* **2013**, *48*, 497–502. [CrossRef] [PubMed]
5. Ghio, A.J.; Soukup, J.M.; Madden, M.C. The toxicology of air pollution predicts its epidemiology. *Inhal. Toxicol.* **2018**, *30*, 327–334. [CrossRef]
6. Ghio, A.J.; Madden, M.C. Human lung injury following exposure to humic substances and humic-like substances. *Environ. Geochem. Health* **2018**, *40*, 571–581. [CrossRef]

7. Eiguren-Fernandez, A.; Miguel, A.; Lu, R.; Purvis, K.; Grant, B. Atmospheric formation of 9,10-phenanthroquinone in the Los Angeles air basin. *Atmos. Environ.* **2008**, *42*, 2312–2319. [CrossRef]
8. Cho, A.; Di Stefano, E.; Ying, Y.; Rodriguez, C.E.; Schmitz, D. Determination of Four Quinones in Diesel Exhaust Particles, SRM 1649a and Atmospheric PM2.5. *Aerosol Sci. Technol.* **2004**, *38*, 68–81. [CrossRef]
9. Jakober, C.A.; Riddle, S.G.; Robert, M.A.; Destaillats, H.; Charles, M.J. Quinone emissions from gasoline and diesel motor vehicles. *Environ. Sci. Technol.* **2007**, *41*, 4548–4554. [CrossRef]
10. Jakober, C.A.; Robert, M.A.; Riddle, S.G.; Destaillats, H.; Charles, M.J. Carbonyl emissions from gasoline and diesel motor vehicles. *Environ. Sci. Technol.* **2008**, *42*, 4697–4703. [CrossRef]
11. Li, N.; Kim, S.; Wang, M.; Froines, J.; Sioutas, C.; Nel, A. Use of a stratified oxidative stress model to study the biological effects of ambient concentrated and diesel exhaust particulate matter. *Inhal. Toxicol.* **2002**, *14*, 459–486. [CrossRef] [PubMed]
12. Cho, A.K.; Sioutas, C.; Miguel, A.H.; Kumagai, Y.; Schmitz, D.A. Redox activity of airborne particulate matter at different sites in the Los Angeles Basin. *Environ. Res.* **2005**, *99*, 40–47. [CrossRef] [PubMed]
13. Shinyashiki, M.; Rodriguez, C.E.; Di Stefano, E.W.; Sioutas, C.; Delfino, R.J. On the interaction between glyceraldehyde-3-phosphate dehydrogenase and airborne particles: Evidence for electrophilic species. *Atmos. Environ.* **2008**, *42*, 517–529. [CrossRef]
14. Eiguren-Fernandez, A.; Di Stefano, E.; Schmitz, D.A.; Guarieiro, A.L.N.; Salinas, E.M. Chemical reactivities of ambient air samples in three Southern California communities. *J. Air. Waste Manag. Assoc.* **2015**, *65*, 270–277. [CrossRef] [PubMed]
15. Shinkai, Y.; Nakajima, S.; Eiguren-Fernandez, A.; Stefano, E.D.; Schmitz, D.A. Ambient vapor samples activate the Nrf2-ARE pathway in human bronchial epithelial BEAS-2B cells. *Environ. Toxicol.* **2013**, *25*, 1222–1230. [CrossRef] [PubMed]
16. Eiguren-Fernandez, A.; Miguel, A.H.; Froines, J.R.; Thurairatnam, S.; Avol, E.L. Seasonal and Spatial Variation of Polycyclic Aromatic Hydrocarbons in Vapor-Phase and PM2.5 in Southern California Urban and Rural Communities. *Aerosol Sci. Technol.* **2004**, *38*, 447–455. [CrossRef]
17. Kumagai, Y.; Koide, S.; Taguchi, K.; Endo, A.; Nakai, Y. Oxidation of proximal protein sulfhydryls by phenanthraquinone, a component of diesel exhaust particles. *Chem. Res. Toxicol.* **2002**, *15*, 483–489. [CrossRef]
18. Denizot, F.; Lang, R. Rapid colorimetric assay for cell growth and survival. Modifications to the tetrazolium dye procedure giving improved sensitivity and reliability. *J. Immunol. Methods* **1986**, *89*, 271–277. [CrossRef]
19. Ito, T.; Bekki, K.; Fujitani, Y.; Hirano, S. The toxicological analysis of secondary organic aerosol in human lung epithelial cells and macrophages. *Environ. Sci. Pollut. Res. Int.* **2016**, *26*, 22747–22755. [CrossRef]
20. Sama, P.; Long, T.C.; Hester, S.; Tajuba, J.; Parker, J. The cellular and genomic response of an immortalized microglia cell line (BV2) to concentrated ambient particulate matter. *Inhal. Toxicol.* **2007**, *19*, 1079–1087. [CrossRef]
21. Park, H.S.; Kim, S.R.; Lee, Y.C. Impact of oxidative stress on lung diseases. *Respirology* **2009**, *14*, 27–38. [CrossRef] [PubMed]
22. Wu, L.; Wang, R. Carbon monoxide: Endogenous production, physiological functions, and pharmacological applications. *Pharmacol. Rev.* **2005**, *57*, 585–630. [CrossRef] [PubMed]
23. Iwamoto, N.; Nishiyama, A.; Eiguren-Fernandez, A.; Hinds, W.; Kumagai, Y. Biochemical and cellular effects of electrophiles present in ambient air samples. *Atmos. Environ.* **2010**, *44*, 1483–1489. [CrossRef]
24. Grigoryan, H.; Edmands, W.; Lu, S.S.; Yano, Y.; Regazzoni, L. Adductomics Pipeline for Untargeted Analysis of Modifications to Cys34 of Human Serum Albumin. *Anal. Chem.* **2016**, *88*, 10504–10512. [CrossRef] [PubMed]
25. Eiguren-Fernandez, A.; Miguel, A.; Di Stefano, E.; Schmitz, D.; Cho, A. Atmospheric Distribution of Gas- and Particle-Phase Quinones in Southern California. *Aerosol Sci. Technol.* **2008**, *42*, 854–861. [CrossRef]
26. Su, R.; Jin, X.; Li, H.; Huang, L.; Li, Z. The mechanisms of $PM_{2.5}$ and its main components penetrate into HUVEC cells and effects on cell organelles. *Chemosphere* **2020**, *241*, 125127. [CrossRef]
27. Taguchi, K.; Fujii, S.; Yamano, S.; Cho, A.K.; Kamisuki, S. An approach to evaluate two-electron reduction of 9,10-phenanthraquinone and redox activity of the hydroquinone associated with oxidative stress. *Free Radic. Biol. Med.* **2007**, *43*, 789–799. [CrossRef]

28. Campen, M.J.; Babu, N.S.; Helms, G.A.; Pett, S.; Wernly, J. Nonparticulate components of diesel exhaust promote constriction in coronary arteries from ApoE-/- mice. *Toxicol. Sci.* **2005**, *88*, 95–102. [CrossRef]
29. Deering-Rice, C.E.; Memon, T.; Lu, Z.; Romero, E.G.; Cox, J. Differential Activation of TRPA1 by Diesel Exhaust Particles: Relationships between Chemical Composition, Potency, and Lung Toxicity. *Chem. Res. Toxicol.* **2019**, *32*, 1040–1050. [CrossRef]

7

Interpreting Mobile and Handheld Air Sensor Readings in Relation to Air Quality Standards and Health Effect Reference Values: Tackling the Challenges

George M. Woodall [1,*], Mark D. Hoover [2], Ronald Williams [1], Kristen Benedict [1], Martin Harper [2,†], Jhy-Charm Soo [2], Annie M. Jarabek [1], Michael J. Stewart [1], James S. Brown [1], Janis E. Hulla [3], Motria Caudill [4], Andrea L. Clements [1], Amanda Kaufman [1], Alison J. Parker [5], Martha Keating [1], David Balshaw [6], Kevin Garrahan [7], Laureen Burton [7], Sheila Batka [8], Vijay S. Limaye [8], Pertti J. Hakkinen [9] and Bob Thompson [1]

1 Environmental Protection Agency, Research Triangle Park, NC 27711, USA; williams.ronald@epa.gov (R.W.); benedict.kristen@epa.gov (K.B.); jarabek.annie@epa.gov (A.M.J.); stewart.michael@epa.gov (M.J.S.); brown.james@epa.gov (J.S.B.); clements.andrea@epa.gov (A.L.C.); kaufman.amanda@epa.gov (A.K.); keating.martha@epa.gov (M.K.); thompson.bob@epa.gov (B.T.)

2 National Institute for Occupational Safety and Health, Morgantown, WV 26505, USA; mhoover1@cdc.gov (M.D.H.); mharper@zefon.com (M.H.); jsoo@cdc.gov (J.-C.S.)

3 Army Corps of Engineers, Sacramento, CA 95814, USA; janis.e.hulla@usace.army.mil

4 Agency for Toxic Substances and Disease Registry, Atlanta, GA 30329, USA; caudill.motria@epa.gov

5 ORISE Fellow hosted by U.S. Environmental Protection Agency, Washington, DC 20004, USA; parker.alison@epa.gov

6 National Institute for Environmental Health Sciences, Research Triangle Park, NC 27709, USA; balshaw@niehs.nih.gov

7 Environmental Protection Agency, Washington, DC 20004, USA; garrahan.kevin@epa.gov (K.G.); burton.laureen@epa.gov (L.B.)

8 Environmental Protection Agency, Chicago, IL 60605, USA; batka.sheila@epa.gov (S.B.); limaye.vijay@epa.gov (V.S.L.)

9 National Library of Medicine, Bethesda, MD 20892, USA; hakkinenp@mail.nlm.nih.gov

* Correspondence: woodall.george@epa.gov.

† Retired. Current affiliation with Zefon International, Inc., Ocala, FL 34474, USA.

Abstract: The US Environmental Protection Agency (EPA) and other federal agencies face a number of challenges in interpreting and reconciling short-duration (seconds to minutes) readings from mobile and handheld air sensors with the longer duration averages (hours to days) associated with the National Ambient Air Quality Standards (NAAQS) for the criteria pollutants-particulate matter (PM), ozone, carbon monoxide, lead, nitrogen oxides, and sulfur oxides. Similar issues are equally relevant to the hazardous air pollutants (HAPs) where chemical-specific health effect reference values are the best indicators of exposure limits; values which are often based on a lifetime of continuous exposure. A multi-agency, staff-level Air Sensors Health Group (ASHG) was convened in 2013. ASHG represents a multi-institutional collaboration of Federal agencies devoted to discovery and discussion of sensor technologies, interpretation of sensor data, defining the state of sensor-related science across each institution, and provides consultation on how sensors might effectively be used to meet a wide range of research and decision support needs. ASHG focuses on several fronts: improving the understanding of what hand-held sensor technologies may be able to deliver; communicating what hand-held sensor readings can provide to a number of audiences; the challenges of how to integrate data generated by multiple entities using new and unproven technologies; and defining best practices in communicating health-related messages to various audiences. This review summarizes the challenges, successes, and promising tools of those initial ASHG efforts and Federal agency progress on crafting similar products for use with other NAAQS pollutants and the HAPs. NOTE: The opinions expressed are those of the authors and do not necessary represent the opinions of their Federal Agencies or the US Government. Mention of product names does not constitute endorsement.

Keywords: air pollutants; ambient air; indoor air; citizen science; toxic chemicals

---

1. Introduction

The *Air Sensors 2013: Data Quality and Applications* workshop, held in Research Triangle Park, N.C., highlighted the substantial advances in the development of portable air sensors capable of providing real-time measurements of ambient air pollution [1]. One anticipated benefit for the use of air sensors is the potential to expand upon the already well-established network of air quality monitors for key air pollutants. Sensor manufacturers introduced a number of new, relatively low-cost, portable air sensors capable of continuously measuring ambient levels of ozone (O_3) and nitrogen dioxide (NO_2). In addition, considerable progress was described regarding the development of air sensors capable of measuring ambient concentrations of particulate matter (PM), and total volatile organic compounds (VOCs). Since that 2013 workshop, the field of portable air quality sensing technology has continued to evolve at a rapid pace, with commercially available products now available for O_3, NO_2, PM, VOCs, as well as for other pollutants, both alone and in complex mixtures. While the potential for these technologies continues to be great, significant challenges in their application remain. Most notably, there is a wide range of data quality differences between different sensor manufactures and models [2], and there is a significant question about how air quality data collected on timescales as short as 1-min should be interpreted in comparison to the available health reference values for air pollutants which are typically based on exposure durations of several hours to many years [3]. For example, commercially available portable air sensors for O_3 can provide the user with minute-by-minute O_3 concentrations; however, interpreting these very short-term measurements with respect to potential adverse health effects is difficult. The health-based National Ambient Air Quality Standard (NAAQS) for O_3 is based on an 8-h average concentration which is backed by thousands of studies from numerous independent researchers and is established through a rigorous process to be scientifically defensible and legally enforceable; no such standards have been developed for these shorter duration exposures. Similar challenges exist when applying Occupational Safety and Health Administration (OSHA) standards. The most comparable OSHA exposure standards include the Permissible Exposure Limit (PEL), which is an 8-h time-weighted average, and the Short Term Exposure Limit (STEL), which is a 15-min time-weighted average.

Detection and monitoring of most non-NAAQS environmental chemicals (including the hazardous air pollutants), toxins and pathogens still largely involves identifying each individual agent, which often requires sending samples to a remote analytical laboratory for analyses. Delivery of laboratory results may take days, weeks or even months. Although the deployment of portable direct-reading instruments, such as photon ionization detectors (PIDs) for total VOCs, can provide some real-time information, these screening instruments lack the sensitivity or selectivity delivered by analytical laboratories and thus, are unable to fully inform a user with a critical need for high-specificity (e.g., rapid response decision-makers). New technologies such as miniaturized light emitting diodes, ultra-violet light detectors and functionalized graphene resistors have enabled the development of chemical detectors capable of delivering laboratory-quality analyses in near real-time. Coupling these types of sensors with global positioning and cell phone technologies may enable the detection, quantification, and visual monitoring of environmental contamination in real-time from remote locations. These promising technologies are expensive, however, and will likely remain out of reach for all but the most dedicated citizen scientist. As described in the following sections, understanding and advancing these types of applications has been a focus of the ASHG: beginning with defining the relevance of air sensors; facing the challenges of communicating across diverse key audiences; dealing with data validity issues; and moving toward a future with more reliable and useful sensor readings.

Formation of the Air Sensors Health Group (ASHG)

Recognizing the potential widespread use of portable air sensors, the likely data interpretation challenges these sensors would present to state and local governments, and the opportunity for collaboration across Federal agencies, the ASHG was formed in 2013. The ASHG is a multi-institutional collaboration of Federal agencies devoted to keeping abreast of new sensor technologies and to assist in the proper interpretation of sensor data as potential indicators of air quality. ASHG consists of experts in a number of areas, including toxicology, public health, engineering, monitoring and sampling, ambient air, indoor air, and occupational health, to name a few.

The ASHG monitors the state of the science at each institution and aims to find common ground on how sensors might effectively be used to meet a wide range of research needs in the occupational, indoor air, and ambient air settings. The ASHG also aims to be a resource for state, regional, and tribal organizations, as well as for citizen scientists and community members considering the use of portable air sensors for air pollution research and decision-making. To date, the ASHG has focused on multiple fronts: assisting EPA Program Offices in developing tools and message statements regarding the potential for adverse health effects from the short-duration air sensor readings for PM and O_3; providing analysis comparing short-term readings to longer-duration averages from existing official monitoring stations used in determining compliance to the NAAQS; and developing prototype visual tools to assist in communicating appropriate interpretive messages. These ASHG contributions have been incorporated into projects and programmatic products which are discussed in later sections of this paper. This review summarizes the challenges, and successes of those initial ASHG efforts, and Federal agency progress on crafting similar products for use with other pollutants.

2. Challenges

2.1. Relevance of Sensor Measurements

A critical concern for interpreting readings from air quality sensors is to have a realistic understanding of what the sensors are actually measuring; how that relates to the expectations of users (including citizen scientists, researchers, regulatory agencies, etc.) for sensitivity, specificity, and robustness of the intended application; and the extent to which those measurements of exposure can appropriately be used to communicate potential hazard and manage risks to public health. The magnitude of this concern varies greatly across the types of pollutants purportedly being measured and is related to a number of factors, such as the chemical and physical nature of the pollutant, the reactivity of the pollutant, concentration and compositional changes over time and location, and the influences with various atmospheric conditions. EPA guidance on achieving high-quality data through systematic planning using the data quality objectives process can be found online [4]. Standards for data quality can differ slightly depending on the context; additional standards are discussed as those contexts are examined in the ensuing sections of this review. The ASHG recognized early in their discussions that this ideal for collecting high-quality data may not be readily obtainable for all potential users of air pollutant sensors.

Low-cost air sensors on the market to date are predominantly of three types; optical, electrochemical, or metal oxide. Generally, these sensor types are not as specific and do not incorporate the front end conditioning or selection that is present in Federal Reference Method (FRM) or Federal Equivalent Method (FEM) measurement devices. Therefore, low-cost sensor measurements may suffer from the influence of co-responsive pollutants, environmental conditions, and even sensor component production variations. The need to understand what the sensor is measuring is of prime importance. In order to better understand the measurement and potential confounding influences, an informed user must understand the physical and chemical nature of a pollutant of interest and how it responds to the environment as well as the sensor measurement technology itself. Field testing may help a user understand the influence of confounding factors and could identify co-responsive pollutants. The interpretation and use of low-cost air sensor data will be enhanced by simultaneous collection of co-responsive pollutant concentrations and environmental data including temperature, relative humidity, wind, and weather (i.e., precipitation, fog), as well as observations about local pollution sources. In this context, chemical sensors must be validated for chemical specificity and sensitivity under the environmental conditions expected in the field prior to any consideration for demonstrating regulatory compliance.

2.1.1. Interpreting Sensor Readings

Managing expectations about what sensor measurement data can and cannot be used for is paramount in communicating with both citizen scientists and the general public at large, who for a variety of reasons (e.g., personal health, general air quality interest) may, in the future, routinely consult a sensor. In anticipation of this interest on the part of the public, the U.S. EPA began a pilot effort in June 2016, to begin addressing the interpretation of short-term sensor readings in the context of air quality [5]. There are several challenges to interpreting these data. Among these challenges are sensor performance and the short-term, sometimes instantaneous output from a sensor. Short-term sensor data come from instruments of unknown performance quality and, importantly, these short-term concentrations cannot be compared to the NAAQS to draw conclusions about what these nearly instantaneous exposures may mean in terms of health impacts. The NAAQS are based on longer exposure durations (e.g., 8-h or 24-h averages) consistent with the health evidence from the reviews of these standards. This health evidence does not support linking 1-min (or shorter) ozone or $PM_{2.5}$ concentrations to adverse health effects, thus a 1-min sensor reading is not directly comparable to the NAAQS, or to the related Air Quality Index (AQI) categories.

To help the public understand the implications of these readings, scientists at EPA have piloted a color-coded Sensor Scale that might be used in conjunction with the AQI to help the public better understand what their short-term sensor data mean in the context of local and regional air quality and consequently make behavioral decisions about outdoor activities. Statistical approaches were used to understand the relationship between short-term ozone and fine particulate matter measurements with longer term averages. The Sensor Scales and explanatory background materials are housed in the Air Sensor Toolbox on the EPA website [6] and have also been described elsewhere. To test the effectiveness of these messages and their visual presentation, EPA conducted focus groups in 2015 and 2017, before and after the deployment of the pilot respectively. Analysis of the outcomes of these focus groups is underway. After considering the input from the focus groups, EPA will refine the messages as appropriate and consider outreach to air quality sensor developers on these focus group findings. Additional work is also on-going to develop similar message schemes for selected HAPs.

In addition to the NAAQS, EPA's Integrated Risk Information System (IRIS) provides inhalation reference concentration (RfC) values for HAPs and other key pollutants important to many EPA Programs with a focus on chronic exposure durations (from years to a lifetime). The Agency for Toxic Substances and Disease Registry (ATSDR), a federal public health agency of the U.S. Department of Health and Human Services, develops similar substance-specific minimal risk level (MRL) values for acute (1–14 days), intermediate (15–364 days), and chronic (365 days and longer) exposure durations [7]. ATSDR's work is focused on Superfund sites, but environmental health specialists apply the MRLs in a wide range of investigations. Many state agencies also develop inhalation health effect reference values in support of their programmatic needs. More generically, the term reference value is used in this text to include all of the various values referred to as standards, guideline values, toxicity values, health benchmarks, etc., and includes values developed for use in emergency response, occupational exposure monitoring, and those protective of the general public. Additional reports are available for a more complete comparison of the available systems of health effect reference values [3,8], and between values for specific pollutants [3,9–13].

2.1.2. Occupational Versus Environmental Exposures

One issue likely to be resolved on a case-by-case basis is the overlap between the environment and the workplace. An employee may be exposed to a chemical in the workplace at a higher concentration than in an outdoor environment. In addition, because the worker is assumed to be fit and healthy and to have periods of recovery from exposure, the levels of concentration deemed tolerable are also higher. A citizen wearing a sensor outdoors, may well carry it into their workplace, and could then experience problems reconciling the levels of pollutants considered acceptable in these different milieus. For example, a sensor designed to help citizens avoid pollution from motor vehicles might measure carbon monoxide. EPA sets a primary NAAQS of 9 ppm averaged over an 8-h period and 35 ppm averaged over a 1-h period, not to be exceeded more than once per year. In the workplace, the situation is not quite as straightforward, but all the limit values which might be applied are higher than those under the NAAQS. The OSHA Permissible Exposure Limit is 50 ppm, while the National Institute for Occupational Safety and Health (NIOSH) Recommended Exposure Limit is 35 ppm, and the American Conference of Governmental Hygienists (ACGIH) Threshold Limit Value is 25 ppm, all averaged over an 8-h period, as with the 9 ppm NAAQS standard. If the monitor alarm is set for the 8-h average NAAQS value, or possibly even at the one-hour average value, the alarm might easily be triggered in the workplace, which might be, for example, a bus garage or foundry. The ensuing discussion with the workplace safety manager or the employer regarding the acceptability of the exposure situation might be difficult for both parties in the absence of well-thought out responses. Nevertheless, in a holistic vision of the exposome, which should consider all exposures over all life-stages, it may not be appropriate to regard ambient, indoor and workplace exposures as somehow "different", to be always measured and assessed separately [14], and so it is to be hoped that it will not be necessary to turn off ambient continuous air monitors "at the factory gates".

2.1.3. Global/International Perspectives

Low cost air quality sensors are of interest worldwide. Multiple studies have distributed sensors in an effort to apply them to local environmental research [15–18]. One of the most notable efforts involving multiple countries and metropolitan areas was the Citi-Sense project [19,20]. This diverse research project involved technology developers, citizen scientists, academics, and professional research organizations using low cost sensor technologies with a purposeful intent [21,22]. The pan-European initiative, EuNetAir, has goals of developing harmony in sensor selection and deployment strategies as well as coordinating sensor evaluation protocols to ensure the timely integration of air quality sensors into monitoring networks [23–25]. The Clean Air Asia consortium, a partnership of multiple Asian-based cities, sensor enthusiasts, academics, and air quality professionals are attempting to improve air quality and improve the overall living conditions in some of the most polluted cities in the world [26].

Even with the apparent world-wide enthusiastic use of low cost sensors to inform public awareness of environmental conditions, there is also a call to ensure the data being collected from these devices across the globe are accurate enough to be used in a purposeful manner [27]. This call for an adequate understanding of sensor performance is not only a reasonable approach but one that must be pursued with the same enthusiasm as those wishing to disseminate low cost sensors to a global population.

A multitude of pseudo air quality index messaging applications are available on the internet from sources around the world [28]. Some of these applications are using low cost sensors to collect air quality data, potentially not accounting for the accuracy of the measurement or the environmental setting in which the measurements are taking place (e.g., indoor, outdoor, near source categories). Furthermore, use of subjective data messaging on health impacts of such sensor measurements without a scientific basis has the potential of confusing the public-at-large and their understanding of sound air quality awareness indices.

The World Health Organization (WHO) describes air pollution as a "major environmental risk to health". The WHO Air Quality Guidelines (AQG) provide an assessment of air pollution health effects and recommend reference values for health-harmful pollution levels of ozone, PM_{10}, $PM_{2.5}$, SO_2, and NO_2 [29]. ATSDR has considered AQGs in the screening process of health evaluations for multiple site investigations in recent years [30–33]. The European Commission (EC) has developed air quality standards for the six US criteria pollutants, as well as benzene, polycyclic aromatic hydrocarbons (PAHs), and three metals: arsenic, cadmium, and nickel.

2.2. Communicating Across Audiences

Portable sensors present a great deal of promise for identifying personal exposure to toxicants present in the environment. However, communication between government or public health agencies, sensor manufacturers, researchers, employers and employees, and citizen scientists is particularly important given the varying degrees of accuracy across sensors, the complexities of environmental exposure, and the difficulties in interpreting potential health risk based on readings from a portable sensor. Thus, the subsections below summarize the various efforts that agencies represented on the ASHG membership have put toward communicating with these groups.

2.2.1. Citizen Scientists and Communities

Over the last decade, members of the public have become increasingly engaged in taking measurements of their environments and otherwise contributing to scientific research. Enabled by the rapid pace of growth of sensor technology, many citizen scientists collect data on air quality in their local environments, both individually and as a part of organized projects. The growth of these technologies and a surge in enthusiasm for these approaches has pushed the boundaries of traditional institution-driven research. Often called "citizen science", these efforts are also referred to as civic or community science, community-based monitoring, crowdmapping, participatory science, open science, or crowdsourcing. In *citizen science*, the public participates voluntarily in the scientific process, addressing problems in ways that may include formulating research questions, conducting scientific experiments, collecting and analyzing data, interpreting results, making new discoveries, developing technologies and applications, and solving complex problems [34]. Of particular relevance to air quality research and air sensors, *community science* or *community citizen science* is "collaboratively led scientific investigation and exploration to address community-defined questions, allowing for engagement in the entirety of the scientific process. Unique in comparison to citizen science, community science may or may not include partnerships with professional scientists, emphasizes the community's ownership of research and access to resulting data, and orients toward community goals and working together in scalable networks to encourage collaborative learning and civic engagement" [35].

The growth of citizen science offers significant opportunities and challenges for federal agencies. Citizen science increases public understanding and community and civic engagement with science and environmental issues, especially locally. Citizen science can connect agencies to the public, provide opportunities for working together towards common goals, support innovation, and make science more accessible and available. It provides data that would otherwise be inaccessible, helps generate a more comprehensive understanding of variation over space and time, and can increase our understanding of social science and human behavior. At the same time, low cost sensor technologies and citizen science introduce challenges to federal agencies, such as communicating risk to project participants, increased public pressure for actions like regulations and enforcement, and communication of data quality. In December 2016, the National Advisory Council for Environmental Policy and Technology provided EPA with advice and recommendations for how to maximize the benefits of citizen science and respond to the corresponding challenges. Recommendations include building technical capacity, providing guidance and communicating data quality needs for different data uses, and integrating citizen science into the full range of EPA's work [34].

EPA is collaborating with citizen scientists to share tools and technology, conduct air monitoring studies, and interpret sensor data. The Air Sensor Toolbox website [6] was created in 2014 to provide resources and tools to citizens interested in learning more about conducting successful air monitoring projects. Topics include how to use sensors, interpreting sensor data, information about EPA air monitoring projects, funding resources, and local air monitoring examples across the United States. The Air Sensor Guidebook [36] provides a comprehensive overview of air pollutants, sensors, study design, data collection, and data interpretation. Other resources include sensor evaluation reports highlighting performance of various sensors on the market, standard operating procedures for sensors, fact sheets, blogs, and training videos from the EPA-sponsored Community Air Monitoring Training in 2015.

Another way EPA engages communities is through collaborations. One such collaboration was conducted in the Ironbound community in Newark, New Jersey through a Regional Applied Research Effort (RARE) grant [37]. EPA scientists trained citizen scientists from the Ironbound Community Corporation (ICC) (The Ironbound section of Newark is a multicultural, multiracial mosaic whose population of 50,000 reflects the diversity and the challenges in urban America. ICC impacts the lives of nearly 1000 people daily and thousands annually. The majority of ICC's 3000 annual clients are from very low to low income households with low literacy and English proficiency and multiple family stressors.) to operate sensor pods developed by the Office of Research and Development (ORD). The collaboration allowed for joint decisions regarding study design, sensor siting, and collection, validation and interpretation of sensor readings. Following the Ironbound project, a similar RARE grant collaborative project was performed in Ponce, Puerto Rico. EPA scientists used lessons learned from the first project to attempt to improve study design topics such as roles and responsibilities and data validation and interpretation. Tasks such as managing, validating, and interpreting large datasets can be an obstacle for community groups involved in air monitoring projects.

ATSDR and NIOSH have partnered with University of Cincinnati, Georgia Tech, and University of Texas Arlington to pilot the use of real-time sensors as sentinels that trigger sample collection via more conventional devices. For example, a low-budget hydrogen sulfide detector can be programmed such that, once a certain threshold is exceeded, a VOC canister will be filled and then sent for laboratory analysis. ATSDR is developing these projects either to characterize peak exposures or as the first stage in assessing the need for a full-blown traditional air monitoring study.

Sensor collocation is an important step to perform before embarking on any monitoring project, regardless of who is conducting the monitoring. This process involves siting sensors in line with regulatory monitors (the gold standard) for a period of time in order to compare the two datasets and establish a regression equation to normalize sensor data and make it more accurate. EPA scientists recently created tools designed to assist citizens in this process, including an easy-to-use Excel™ macro that allows one to compare two datasets, such as sensor and reference data, and with one click provides a regression equation, comparative data graphs, and descriptive statistics. A training document that explains collocation and why it is important, and how to perform a proper collocation, accompanies the macro. EPA is partnering with the Clean Air Carolina community group and the Eastern Band of Cherokee Indians to conduct their own collocation projects, pilot test these resources and provide input on how to improve them for use by the general public.

2.2.2. State, Local and Tribal Agencies

State, tribal and local health and regulatory agencies were identified as a primary target audience early in discussions within the ASHG. The more proximately located agencies and offices are more likely to be called by the public and press, and may be more resource-challenged to interpret data generated using sensors. The EPA Regional Offices were identified to be the most likely first contact point for EPA. Similarly, ATSDR has Regional Offices that are another resource for health agencies.

Health agencies at all levels of government are called upon by the public and press to interpret air monitoring data. City and county health departments, depending on their size and community needs, may have environmental health specialists on staff who are familiar with air monitoring methods. Frequently, local agencies will defer to their respective state health department, many of which have specialists that are supported by the ATSDR Cooperative Agreement Program. State or local agencies may in turn also contact ATSDR or its sister agency, the National Center for Environmental Health (NCEH), which is part of the Centers for Disease Control and Prevention (CDC). As noted earlier in Section 2.1.1, there are several reference value systems which have been applied in interpreting exposures to non-criteria air pollutants, including the RfC and MRL values. Health agencies also frequently make site-specific exposure dose calculations and may evaluate health risks for shorter averaging periods by adapting the dose-response information in toxicology studies on which the reference values are based.

In order to address these needs, EPA Regional Offices have initiated outreach to state, tribal, and local environmental agencies to facilitate information sharing amongst stakeholders and to develop an inventory of sensor-based community air quality investigations. EPA Regional Office staff have highlighted EPA's Air Sensor Toolbox for Citizen Scientists [38] in this outreach, facilitating exchanges on sensor operation and maintenance, funding opportunities, and tools for data interpretation. For example, the Minnesota Pollution Control Agency provided input to the EPA Regional Office in Chicago (Region 5) on its own experience managing an air sensor loan program, and shared documentation it had developed to inform community members about proper sensor operation.

Monitoring for the NAAQS pollutants is delegated to state agencies with oversight from EPA Regional Offices and EPA's Office of Air Quality Planning and Standards (OAQPS); there is a structured regulatory program to determine compliance status with each NAAQS. Emissions of HAP are regulated at the source of the emissions, and monitoring is mostly initiated at the local level (i.e., state, tribal and local agencies), with federal guidance and some federal funding. There are very specific regulatory requirements in monitoring for the NAAQS pollutants but the States and Tribes have more leeway in monitoring for other purposes (e.g., for HAPs). EPA's Superfund Program uses various technologies for emergency response, and new sensors may be useful to inform decisions in these scenarios. Federally recognized Tribes interface with EPA and ATSDR on a government-to-government basis; many receive EPA grants or ATSDR technical assistance to do monitoring and other environmental health projects.

2.2.3. Sensor Manufacturers

Low cost air quality sensors have exploded onto the commercial market in the last decade. Development and manufacturing of these devices is not isolated to a single sector or industry. Some of the earliest developers were design teams associated with academic industrial/design arts programs, telecommunication research groups, and non-profit citizen scientists [39]. In particular, these developers took advantage of inexpensive sensor components and their individual areas of expertise to craft air quality sensors to meet a wide variety of needs [40]. More traditional instrumentation manufacturers have recently started to invest research capital into the low cost sensor area [41]. Their presence is the result of the growing citizen science movement and an obvious market niche.

The existence of a multitude of sensor manufacturers has resulted in both obvious benefits and deficits. On the plus side, there are numerous low cost sensors in a price range typically well under $1000 available for the interested user for both particle as well as gas phase pollutant monitoring. Many offer user friendly features such as immediate data visualization through smart phone applications with wireless data transmission. Some concerns associated with low cost sensor production is that often manufacturers may lack the technical know-how or capabilities to adequately test or calibrate their devices prior to releasing them to the market [27]. Other concerns include reliability issues between replicate copies of the same sensors attributed to either a poor overall manufacturing process associated with the fully assembled sensor or issues with the individual components themselves [42,43].

Recent sensor evaluation efforts by recognized institutions, including the U.S. EPA and others [6,44] would appear to be having a positive impact upon sensor manufacturing. In particular, work conducted in 2012 resulted in some of the first reported performance tests of sensors from a wide variety of manufacturing sources [2]. The evaluations resulted in an almost immediate update of the technology by many of the manufacturers to overcome issues first revealed in the tests (e.g., battery failures, low detection sensitivities, poor telecommunications protocols). Some of those early manufacturers would appear to have left the market while other new ones have joined. These observations suggest that it is likely that more mature instrument manufacturers will continue to invest in this area and capture more of the market share as start-up groups continue to develop unique devices to meet a particular niche.

2.3. Calibration and Validation

As mentioned previously, the calibration procedures conducted by most low-cost sensor manufacturers are often severely inadequate or non-existent and fail to establish the sensor's performance characteristics, especially when used in the ambient environment. End users and research organizations have often taken the lead on establishing the performance of these emerging technologies and sharing their results. Noted groups establishing sensor performance include the European Union's Joint Research Center [45], the U.S. EPA [46], and the South Coast Air Quality Management District's AQ-SPEC laboratory [47]. Results show that while nephelometric (light scattering) devices might perform well in direct chamber-based evaluations, they often reveal significant departures from a true reference grade instrument response under real-world (ambient) conditions [1]. In like manner, while excellent chamber-based response relationships have been observed for select gas phase pollutants [2], ambient test results have been less promising or inconsistent [42]. There are of course, exceptions to these observations and certain devices and sensors appear to show more promise relative to their performance features [48].

Gas detection in the workplace has a long history of being driven by flammability concerns, beginning with the miners' safety lamp, but direct-reading real-time instruments have also been used to determine toxic gases for almost 100 years; colorimetric detector tubes patented for detecting carbon monoxide in the 1930's are examples. More recent developments of infra-red analyzers and gas chromatography detectors (with or without an associated gas chromatograph) have been portable, but not personal, although current research into miniaturization may alter that situation. Electrochemical cells have been developed for specific acute hazards, such as carbon monoxide, hydrogen sulfide and chlorine. These are used in process safety applications, but have also been adapted for use in personal dosimeters. The quality of data used for safety monitoring and personal exposure assessment in the workplace is of paramount importance, particularly in relation to acutely hazardous substances. The European Union countries have developed guidance for the performance, testing, selection, installation, use and maintenance of electrical apparatus used for the direct detection and direct concentration of toxic gases and vapors. Other Standards-setting organizations have published similar products, and a joint Working group of the International Organization for Standardization (ISO) and the International Electrotechnical Commission (IEC) are working on umbrella standards to replace these (these Standards are detailed in Appendix A). In the USA, NIOSH has developed guidance for ensuring data quality for gas and vapor detection, including during emergency response [49,50]. The American Industrial Hygiene Association, through their Gas and Vapor Detection Systems Committee (now Real-Time Detection Systems Committee) has developed guidance for manufactures to report specifications for electronic real-time gas and vapor detection equipment [51].

Recognizing the need to calibrate instrument response under real-world conditions, variable performance based on changing conditions (e.g., aerosol size-distribution), and deteriorating sensor response as due to age, new research efforts have been focused on the idea of auto-calibration methods [52]. Such methods may be able to maintain the accuracy of a node-based distributed sensor system for longer term measurements.

Extending beyond the need for validation in controlled laboratory settings to "real world" conditions and environmental epidemiological and citizen science applications, the National Institute for Environmental Health Sciences (NIEHS) released a phased field validation program in 2013 [53]. Through this effort, teams of engineers, exposure scientists and environmental epidemiologists are working together to test sensor performance at the pilot scale, including iterative improvements in sensor system design and performance and then scaling up the application to a full-sized epidemiological study to demonstrate the added value of temporally and spatially resolved air pollution exposure assessment relative to existing 'citizen' measures.

3. Related Projects and Programs

Although the ASHG has not been a prime mover in development of key products or programs, it has served as an advisory group and has contributed to a number of projects and programs, as noted in the sections below.

3.1. Village Green

The U.S. EPA's Village Green project has provided a test-bed for long-term evaluation and application of emerging air quality sensors in a variety of community settings [54]. While the sensor technologies for particulate matter and ozone are not considered truly low cost (~$6000/each), they do represent mid-tier technologies [55] which are showing good to excellent capabilities for certain attributes. In particular, they are capable of providing extended periods (months) of ambient air quality monitoring using sustainable energy (solar power) with little or no technical support and often with a high degree of agreement with local reference monitoring [56]. Another unique feature of the Village Green is that it was designed to stream data continuously to the public via its web-based data portal. It was in fact, the first real-time public reporting of air quality data for environmental awareness purposes provided to the general public by the US EPA. Since its conception in 2014, Village Green stations have now been deployed in a total of eight (8) U.S. metropolitan areas and involve a variety of air quality sensor and emerging technologies (variable sustainable power supplies, data microprocessing features, etc.).

At the time of its development as a technology test bed and community air quality awareness tool, its capabilities to meet purposeful air quality data analyses were not established. EPA's investigations into short time interval sensor data messaging provided an opportunity to use the Village Green for such an analysis due to its extensive database containing 1-min measurements of both ozone and fine particulate matter. Analysis of that database resulted in a pilot Sensor Scale associated with potential short time interval air quality measurements [5]. EPA is launching a pilot project to test a new tool for making instantaneous outdoor air quality data useful for the public. The new "Sensor Scale" is designed to be used with air quality sensors that provide data in short time increments–often as little as one minute [5]. EPA developed the scale to help people understand the 1-min data the stations provide and how to use those data as an additional tool for planning outdoor activities.

3.2. The E-Enterprise Advanced Monitoring Team (EEAMT)

Under the direction of the E-Enterprise Leadership Council, a joint EPA/State Advanced Monitoring team was formed in April 2015 to address the challenges and opportunities presented by rapidly changing air and water technologies. Data interpretation is one of five priority projects identified in the path forward for EPA, States, and Tribes [57]. In order to provide context and interpretation of advanced monitoring data in formats relevant and understandable to users, the team was charged with advancing (1) statistical analyses to understand the relationship between continuous data and data collected over longer-term averaging times or via discrete (e.g., bi-weekly) sampling;

(2) development of visualization tools (e.g., interactive maps) and websites with appropriate messaging; and (3) development of outreach and communication materials.

Conducting statistical analyses to understand the relationship between short term measurements and longer term standards or discrete measurements is a blanket need across media and pollutants. While messaging already exists to alert the public about air or water quality conditions experienced over specific time periods (e.g., 1-h, 8-h, or 24-h), the same messaging should not be used to translate short term (e.g., 1-min) measurements. Two articles discussing an approach to relate short term ozone and particulate matter ($PM_{2.5}$) measurements with longer term averages have been published [58,59]. Other needs identified by the team include data analysis and messaging for SO_2, NO_2, CO, PM_{10}, benzene, total volatile organic compounds (VOCs), and other specific VOC compounds.

In order to create visualization tools that make continuous and discrete state/federal data more accessible and understandable, EPA and States have determined it is important to present real-time (e.g., hourly concentration values) along with appropriate caveats. Examples of caveats include marking data as "raw", "provisional", or "final" and distinguishing between different forms of data (e.g., regulatory vs. peer reviewed). A need exists to display continuous data along with discrete measurements while giving the data context including geographical and meteorological information. Mockups were drafted as part of a thought process to address these issues and allow for EPA and States to provide guidance or expertise to big data developers desiring to stream, collect, or use regulatory and sensor data. Once short term messaging has been developed, it is important to develop appropriate outreach and communication materials that EPA and States can use for consistent messaging including frequently asked questions (FAQs) and a standardized, centralized repository of metrics that break down toxicity/health information by pollutant for both short term and long term effects (e.g., gaseous, metals, PM, pH, toxins, dissolved oxygen, etc.). Outreach material should describe the limitations of sensors and sensor networks including information on what we do not know and cannot measure with confidence.

3.3. *Homeland Security Applications*

Air sensors have many important applications relating to homeland security, but two roles are especially important: (1) serving as sentinels to detect the release of dangerous substances into the air, and (2) providing measurements of the amount of dangerous substances present in the air so that health risks can then be assessed and appropriate decisions made to protect public health.

Serving as sentinels to provide early warning of the release of dangerous substances is a very critical function for real-time air sensors. In this capacity, air sensors are designed to monitor background levels of various substances (dangerous chemicals or/or their indicators) and detect any significant increases in their presence. Air sensor networks have already been deployed extensively as sentinels in several major cities, chemical plants, military installations, and during major public events. These networks have been deployed on roof tops, in underground transportation systems, and in various types of mobile ground vehicles and aircraft.

In addition to their role as sentinels to detect the release of dangerous chemicals, air sensors may be used to measure concentrations of harmful chemicals which can potentially provide an estimate of health risks to persons breathing the air. Estimating health risks from acute or short-term exposure durations is a challenging task, as has already been noted in this paper. Most of the tools used to assess health risks are more centered on long-term, chronic, or lifetime exposure to chemicals. For example, as EPA's premier program for assessing health risks from chemicals, the Integrated Risk Information System (IRIS) has focused on dose-response data for chronic exposures. Health effect reference values derived to assess long-term health risks have very little, if any, value for assessing health risks from acute, short-term, or intermittent exposure scenarios, and few have been developed to assess those health risks to the general public.

A number of chemical release accidents, including the December 1984 accidental release of methyl isocyanate from a Union Carbide chemical plant in Bhopal, India, led to the formation of the Acute

Exposure Guideline Level (AEGL) program in 1995 [60]. Standing Operating Procedures for deriving AEGL values were developed by the National Research Council (NRC) [61] for use in both planning for and during a catastrophic chemical release event, using values covering inhalation exposures from 10-min to 8-h.

Shortly after the events of 11 September 2001, the need for additional sets of values for use for remediation in the aftermath of such events led to the development of another set of health metrics, called Provisional Advisory Levels (PALs). PALs were derived by building upon the AEGL methodology developed by the NRC [61], to create values for use at 1-day, 30-days, 90-days, and 2-years durations [62]. Under the PALs program, EPA has drafted over 3000 numeric values for over one hundred priority chemical agents, two routes of exposure (ingestion and inhalation), three levels of harmful effects, and four relatively short exposure durations. These values, in addition to those already published by many other organizations [3], can provide a context on which to compare sensor-derived exposure levels.

3.4. Indoor Air–Non-Industrial

Indoor air quality (IAQ), which is the air quality within buildings (i.e., homes, schools, offices and other non-industrial buildings) and other enclosed spaces, can affect the health, comfort and ability to perform for occupants, from infants to senior citizens. IAQ involves many factors including: outdoor air contributions; chemical, micro-biological, and particulate contaminants; and, characteristics of the indoor climate such as temperature, humidity and airflow. Americans spend about 90 percent of their time indoors [63–66], where pollutant levels, like some VOCs, may be two to five times higher—and occasionally 100 times higher than outdoors [67–69]. EPA traditionally has focused its IAQ efforts to identify sources and developing guidance to reduce human exposure to unhealthy indoor air and to provide low-cost mitigation strategies consistent with public health practices. In addition, many international professional, trade, and standards organizations have dedicated committees addressing a broad range of indoor air quality issues. With the increased availability of low-cost sensor technology, however, the resources we use to assess indoor air have been expanding.

One challenge for sensor work within indoor environments is the fact that the several pollutants important in the indoor environment cannot be detected accurately by current low-cost sensor technology. Indoor air sensors have the same challenges that ambient air sensors encounter when attempting to evaluate short-term sensor readings. Most of the indoor air pollutants also share the challenge associated with HAPs, the lack of enforceable or agreed upon health based standards to with which to compare sensor readings. As with HAPs, the available health effect reference values may be inappropriate for direct use with indoor air sensor readings. There are several other challenges unique to using sensors indoors—for example, unlike the ability of ambient air sensors to be collocated with regulatory monitors to help assess accuracy, there is no similar availability for indoor air.

While guidance development around IAQ best practices will continue within EPA, the potential for increased use of sensor technology for indoor air quality continues to evolve. Sensors that work with occupants' health-assessment technologies and those that can actuate building systems or individual appliances automatically may help enhance IAQ efforts in the built environment. There is a great potential that, as low-cost sensor technology improves, it may be a viable complement to comprehensive building system approaches currently used by the IAQ community and would help inform and improve IAQ stakeholders' ability (both professional and consumer level) to use IAQ management tools, interpret air quality data for their space, and assess the benefits of IAQ-related actions.

It is important that we not only continue the engineering "fixes" that are constantly improving source control, as well as ventilation and filtration/air cleaning systems, but also work to improve sensor technology as a tool to help optimize building performance and increase the accessibility of these tools as a complement to integrated strategies for improving IAQ for all parts of society. Within the context of IAQ, this may involve creating future systems that can be automated, improving the integration of these low-cost tools with building and health-assessment systems, establishing guidance that creates IAQ best practices that integrates these tools into existing public health practices and provides appropriate messaging for the public to better understand their health in their indoor environments.

4. Maximizing the Usefulness of Sensor Readings

As discussions within the ASHG have progressed over the years, approaches have been considered on how sensors may improve our understanding of real-world dosimetry, and how we might better organize the process of developing and using sensors through life-cycle analysis. Both of these approaches are more immediately relevant to research-grade sensor technologies; however, as more improvements in sensitivity and specificity are made to the low-cost sensors available to citizen scientists, these approaches will apply to a more universal set of sensor users.

4.1. Transitioning Sensors into Dosimeters and Future of Exposure Science

Radiation biology provides a paradigm for transitioning sensor-derived data into estimates of internal dose. Biological effect data from early animal studies modeled radiation exposures to define internal dose and algorithms were developed so that the readout from portable detector could be translated into the dose delivered to internal target organs.

It is the internal dose (the concentration of the agent at the cell/organ/tissue being affected) and not the environmental exposure level that is more closely related to toxicological effects. Health-based standard setting and development of health effect reference values are moving to apply emerging advances in exposure and toxicological sciences aimed at a more seamless integration of exposure and internal dose metrics, together with the incorporation, coherent alignment, and coordination of novel data streams [70–72]. Elucidation of the linkages between exposures and adverse effects in humans and the ecosystem will result in a better understanding on which to develop effective management strategies.

Constructing a robust context for this integration calls for coordinated research with human-health and ecologic-health scientists to identify, collect, and evaluate data that capture internal and external markers of exposure in a format that improves the analysis and modeling of exposure–response relationships and links to emerging methods for hazard identification such as high-throughput test systems (HTS). Figure 1 depicts selected scientific and technologic advances for measuring and monitoring considered in relation to a conceptual integration of exposure and dose as espoused by the National Research Council [71].

Mattingly et al. [73] developed an exposure ontology (ExO) designed to address the lack of exposure information required to elucidate environmental contributions to diseases, translate molecular insights from new technologies such as HTS, and aid assessment of human and ecological risks. The ExO formalized definition such as "exposure receptor" and centralized the role of exposure science with the intent to extend its ability to integrate and analyze exposure information within the broader context of environmental health. The exposure receptor can be an organ, tissue or cell, and the exposure stressor can be a biological, physical, or psychosocial agent, so that exposure assessment may include estimating the magnitude, frequency, and duration of an exposure, along with characteristics of the specific receptor. This concept of the exposome is consistent with the life-cycle approach articulated above regarding sensors. Such integration will expand the impact of exposure data and inform existing environmental health data by providing associated real-world exposure context.

On the response side of the exposure-dose-response linkage, adverse outcome pathways (AOP) are emerging as an important construct for the integration of toxicological effects across various levels of biological organization (e.g., genomic, cellular, target tissue, individual, population) based on HTS and identification of molecular initiating events [74,75]. This construct is also completely consistent with the ExO, and more recently Teeguarden et al. [76] have proposed formal linkage of exposure science with the AOP by means of the aggregate exposure pathway (AEP).

Therefore, emerging technologies now provide a method whereby measures of environmental exposure can be translated to an internal dose which can then be used in applications such as biomonitoring of populations or in cell systems. These approaches essentially provide a scalable platform with which to depict both exposure and internal dose. Alignment of exposures across experimental toxicity-testing systems can be achieved by understanding, measuring, and applying this information on the processes that control the time course of concentrations and delivery of chemicals and particles to target cells in various test systems at different scales (e.g., scale of ecological epidemiological studies, target tissue dose for in vivo animal studies, delivered concentration in HTS) that may be used as the basis for developing health-based values to which sensor data may be compared. Such exposure and internal dosimetry considerations can provide a context for the interpretation of emerging sensor data and may also inform future considerations of the form of various health-based reference values.

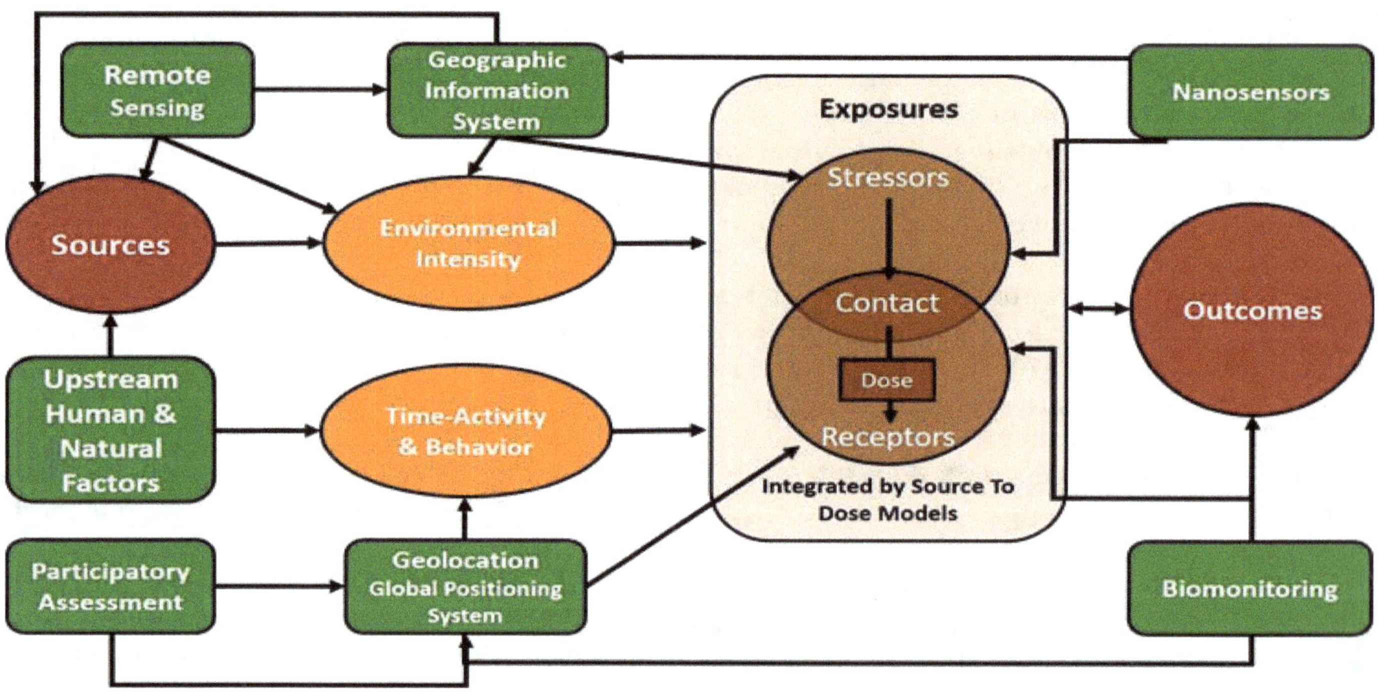

Figure 1. Conceptual integration of exposure and dose (from National Research Council).

4.1.1. Dosimeters in the Workplace

The Army projects aimed at developing a naphthalene dosimeter are an example of how biomarkers of exposure are being integrated with chemical sensor-derived data [77]. The first in the series of naphthalene dosimeter projects was funded through the Army's SBIR program. It developed an instrument capable of measuring naphthalene from air every three minutes [78]. A second project, underway at NIOSH, is an independent evaluation of the performance of the instrument. A third project lead by the U.S. Army Research Institute of Environmental Medicine is deploying the prototype instrument on military fuel-handlers to measure the concentration of naphthalene in their breathing zone. Concurrent to collecting the personal real-time exposure data, biomarkers of naphthalene exposure are being collected from the exhaled breath, urine and skin of the individual wearing the air sensor. Following the radiation dosimetry paradigm, the concurrently collected exposure and biomarker data will be used in a fourth project that will inform development of a model to estimate internal dose [79].

4.1.2. Dosimetry for NAAQS Air Pollutants

Portable sensors available for some NAAQS pollutants (namely, PM, O_3, and NO_2) have been evaluated by EPA. There are not specific metabolites that can be used to quantify the general relationship between exposure and dose. However, unlike naphthalene which enters the body by both dermal absorption and inhalation, at least these three NAAQS pollutants enter the body predominately by inhalation. In general, greater than 80% of inhaled ozone is absorbed in the respiratory tract (see Table 5-1 of U.S. EPA, 2013 [80]). Thus, dose can be approximated as the product of O_3 concentration, minute ventilation, and duration of exposure. Conceivably, an estimate of minute ventilation could be derived using an accelerometer in combination with data on the mass and gender of an individual. A more exact linkage to predicted decrements in lung function due to O_3 exposure, at least in healthy individuals, could be calculated by linking dosimeter data with an existing model [81]. This type of approach could be applied to both occupational and ambient ozone exposures or adapted for other compounds where health endpoints are closely related to inhaled dose.

Development of a dosimeter for PM is complicated due to the dependence of particle deposition on inhaled particle size, route of breathing, tidal volume, breathing rate, and lung size [82]. A couple of PM samplers that mimic the deposition in adults during nasal breathing and light exercise based on the International Commission on Radiological Protection (ICRP) model [83] have been developed. First, Koehler et al. [84] employed the use of a foam plug in which particle deposition efficiencies are similar to the ICRP [83] predicted total respiratory tract deposition (average of adult males and females) for particles between 0.05 and 2 μm. Total deposited dose in the respiratory tract can be determined either gravimetrically or by digesting the foam and extracting metals or organs for quantification by other means. Second, TSI has developed a Nanoparticle Surface Area Monitor (Model 3550) which can be used to determine particle surface area depositing in the tracheobronchial and alveolar regions of the lung for particles between 0.01 and 1 μm. The Model 3550 is designed to match predicted deposition in an adult male. Its operation is based on the diffusion charging of particles followed by detection of the aerosol using an electrometer. Particle doses can be assessed by the second, computed as a time-weighted average, or cumulative total deposed particle surface area. Available portable PM sensors typically use light scattering to estimate PM mass or number concentration in air for micron-sized or larger particle fractions such as $PM_{2.5}$. Unless an underlying PM size distribution is known or assumed, estimates of dose cannot be derived based on data from these portable PM sensors.

The utility of a dosimeter for ambient NO_2 is questionable. For NO_2, one of the critical endpoints is asthma exacerbation via an increase in airways responsiveness, as discussed in Section 5.2.2 of the Integrated Science Assessment (ISA) for NO_2 [85]. Since the increase in airway responsiveness does not appear to be associated with NO_2 dose for NO_2 concentrations between 100 and 600 ppb [86], it may be sufficient to monitor NO_2 concentration. Additionally, ambient NO_2 concentrations are generally only elevated near roadways, as discussed in Section 2.5.3 of the 2016 ISA [85]. An elevated NO_2 sensor reading could serve as a warning for a person with asthma to take actions to reduce exposure, e.g., by using recirculation of air in an automobile or avoiding outdoor activities in close proximity to major roadways or during periods of increased traffic.

4.1.3. Dosimeters for Hazardous Air Pollutants (HAPs)

There are 187 air pollutants designated as HAPs, including several VOC, that are emitted by point sources and under the purview of Section 112 of the Clean Air Act [87]. These pollutants are known or suspected to cause cancer or other serious health effects, such as reproductive effects or birth defects, or adverse environmental effects and are often also encountered in the workplace. As sensor technologies emerge to address these pollutants, our recommendations for life-cycle characterization and dosimetry as discussed above will be essential to create proper context for interpretation of sample measurements with comparisons to appropriate health effect reference values, when such values are available [3].

4.2. Adopting a Life-Cycle Approach

An overarching life-cycle framework and decision-making process that the ASHG has encouraged as an ideal for air quality sensor applications is illustrated in Figure 2. The life-cycle concept was originally developed for use in emergency response situations [88], adapted and applied for radioactive air sampling and instrumentation [89], adopted as a systematic way to organize the framework of the White House's signature initiative on Nanotechnology for Sensors and Sensors for Nanotechnology [90], and most recently expanded to meet all manner of the emerging sensor needs for safety, health, well-being and productivity [91]. The lifecycle begins with a clear and complete identification of the purpose of the measurement, including what needs to be measured, under which conditions it needs to be measured, and how well it needs to be measured. The lifecycle guides the research and development, prototype testing, qualification type testing, production control testing, and training needs for the sensor system. The lifecycle further defines procedures for acceptance testing, initial calibration, functional checks, conduct and evaluation of operational experience, maintenance and recalibration, and periodic performance testing to confirm continued successful use of the sensor system. Effective following of the life-cycle process ensures that the sensor methods and instrumentation will work as intended under realistic conditions. Documentation and continuous improvement are essential at each step.

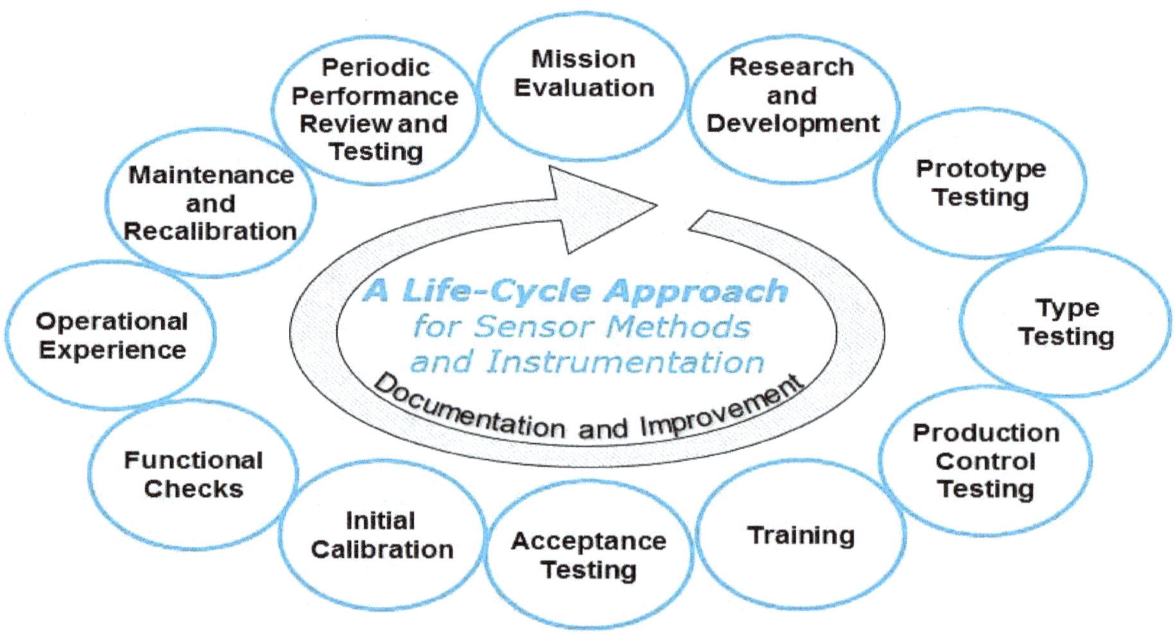

Figure 2. The life-cycle approach for sensor methods and instrumentation [88–91].

Use of the lifecycle supports an approach to sensor methods and instrumentation that is consistent with the roles served by resources such as *My Air, My Health: An HHS/EPA Challenge* [92] to develop and validate new methods, and the online *AirNow* [93] maps and forecast data that are collected and shared using federal reference or equivalent monitoring techniques or techniques approved by the state, local or tribal monitoring agencies.

4.2.1. A Working Definition of Air Quality Sensors and Health Informatics

A proposed working definition of *Air Quality Sensors and Health Informatics* might be "the science and practice of determining which information is relevant to meeting air sensor and health objectives; developing and implementing effective mechanisms for collecting, validating, storing, sharing, analyzing, modeling, and applying the information; and then confirming that appropriate decisions were made and that desired mission outcomes were achieved" [94]. The additional steps in the informatics lifecycle include "conveying experience to the broader community, contributing to generalized knowledge, and updating standards and training" [95]. Successful informatics endeavors will apply all of the steps in the process. ASHG members have also assisted with the development of an attempt to define "data readiness levels", which could help with relevance and reliability issues for the collection, sharing, and application of air sensor data [96].

4.2.2. Roles and Responsibilities of Sensor and Data Customers, Creators, Curators, and Analysts

In the context of our working definition of informatics for air sensors and health, the roles and responsibilities of the myriad individuals who are engaged in the development and application of air quality sensors can be viewed as fitting into four categories: sensor and data customers (who specify the sensors and data needs for their intended purposes), sensor and data creators (who will develop relevant and reliable sensors and data to meet the customer needs), sensor and data curators (who will maintain and ensure the quality of the sensors and sensor data), and sensor data analysts (who will develop and apply models for data analysis and interpretation that are consistent with the quality and quantity of the data and that those data meet the customers' needs). In some instances, the same individuals may perform all roles, and in the larger global reality the individuals and their roles may extend over significant distances, organizations, and time periods. As shown in Figure 3, effective communication across the many customer, creator, curator, and analyst interfaces is essential, and that communication across each of the six interfaces must work effectively in both directions [95]. This vision follows the views of Hendren et al. [97] on a collaborative approach to assessing, evaluating, and advancing the state of the field for data curation in the emerging field of nanomaterials and nanotechnology.

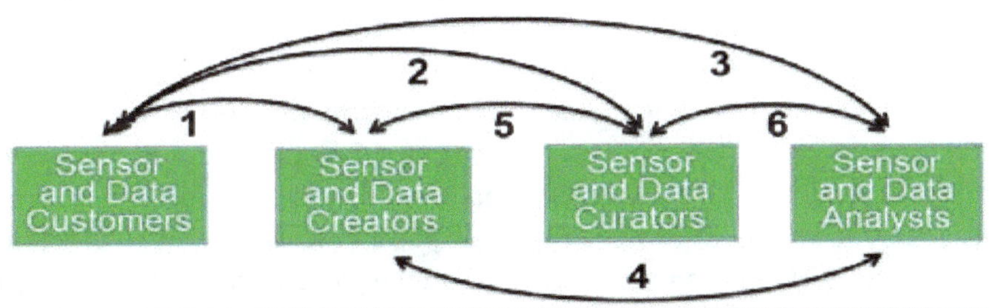

	Set Mission Objectives	Determine Relevance	Collect	Validate	Store	Share	Analyze and Model	Apply	Confirm Effectiveness	Convey Experience	Generalize	Update Guidance
Customers	X	X						X	X	X	X	X
Creators		X	X	X					X			X
Curators		X		X	X	X			X			X
Analysts		X		X			X		X		X	X

Figure 3. Communication interfaces, roles, and responsibilities for air quality sensor and health data customers, creators, curators, and analysts [95].

5. Forecast of Advancing Technologies

Major areas of opportunity to advance the state of the art and application of sensor technologies include: strengthening and sustaining infrastructures for investments and collaborations, improving the sharing of information, integrating data and a coherent interpretation across various venues to construct a cumulative accounting of exposures, and advancing breakthroughs in miniaturization of sensor systems. At this writing, there are efforts across multiple organizations (public and private; large and small; formal and informal) with additional involvement by independent inventors. Entrepreneurial organizations such as Aclima have become involved in collecting data from mobile sensors attached to Google street-view cars [98] and are leading the way to making those data accessible.

Equally important are opportunities to increase the linkage of measurements of exposure via sensor readings to additional parameters useful to estimate dosimetry: biometric data (e.g., pulse rate, breath rate, and the like, which are now available from devices such as FitBit); geospatial measurements of location and daily movement patterns; and of the physical environment (temperature, humidity, etc.). These capabilities are most actively being developed within the occupational exposure arena but migration of these technologies to the low-cost sensor market is anticipated as value-added features. Properly informing and providing useful guidance to ensure these technologies are appropriately engaged will be another challenge for groups like the ASHG.

5.1. *Infrastructure Needs*

5.1.1. Forecast and Statement of Needed Investment

When real-time sensing is combined with concurrent GPS coordinates, the derived data are multidimensional. Today, sensor-derived data include streaming of latitude, longitude, altitude in addition to the chemical data. However, the hardware and software to fully capture the potential usefulness is lacking. For example, visualization of multidimensional data might be imaged as a video of a multi-colored topographical map that flexes as an individual moves through different levels of the positional and chemical data. Depth of color might represent various concentrations of chemicals detected by the sensors. If the full potential is to be captured, investment is essential. Infrastructure is needed to not only capture multidimensional data streams but also analyze it as it is being produced. This is especially relevant to homeland security and for rapid response following a catastrophic event. The application of sensor technology to the field of epidemiology can also be cited as an investment opportunity. New evidence of cause-and-effect will become evident by developing the ability to retrieve archived multidimensional sensor-derived data and overlaying it with geographical disease prevalence information. Challenges of scalability across dimensions of time and geography must be addressed to properly interface with available health-based standards and reference values.

As described in Section 3.1, there is great interest, and investment, in transforming toxicity testing with the application of high throughput analyses using cell cultures [70]. Generating evidence to demonstrate the validity of these testing methods is an ongoing challenge. In this context, we have an opportunity to transform systematic hazard identification and risk assessment processes by combining real-time exposure monitoring with real-time image analysis. Again, investment in the data capture, analytics and archiving capabilities is essential if this opportunity is to be realized.

5.1.2. Scientific Literature Collection and Coordination

Another essential investment is needed to network the fields of Health Science, Electrical Engineering and Computer Science. A first step might be to encourage sensor-related journals to petition the National Library of Medicine for incorporation into the database. The National Library of Medicine's PubMed database [99] provides the ability to search for peer-reviewed publications. This includes research about the use of sensors to assess indoor and outdoor air. As an example, PubMed includes air sensor-related publications from a journal called "Environmental Monitoring and Assessment" such as this 2017 one about "Public engagement on urban air pollution: an exploratory study of two interventions" [100]. However, not all scientific journals are indexed in PubMed and this is especially true for journals specializing in emerging areas and technologies Editors and publishers can submit their journals for inclusion in PubMed; however, not all journals will be accepted based on NLM's selection criteria [101]. The relatively new PubMed Commons allows for the sharing of opinions and information about PubMed citations [102]. This could include comments about the techniques used in a publication or alerting readers to consider looking at publications with more current approaches and findings.

5.2. Miniaturization

Miniaturization has been essential to recent successes in moving laboratory-scale technologies to the field. For example, to meet the demands for personal exposure monitoring of the size characteristics and spatial distribution of ultrafine particles, Fierz et al. [103] have developed a compact, real-time instrument called the diffusion size classifier (DiSC) which provides particle size information that is in good agreement with the much larger and more expensive laboratory-based aerosol spectrometers. Another personal monitoring device recently released into the market is the portable aerosol mobility spectrometer (PAMS) which simultaneously measures the number-weighted size distribution of submicrometer aerosol, including the nanoparticle fraction [104]. In developing the PAMS, NIOSH investigators used miniaturization to overcome the prohibitive size, weight, and cost limitations of the previous technology, and they also eliminated burdensome regulatory and administrative limitations for record-keeping and transportation of the instrument by replacing the traditionally used radioactive source for particle charge conditioning with a nonradioactive bipolar diffusion charger. Compared to traditionally used instruments, these innovations provided reductions in size (by a factor of 20), weight (by a factor of 10–15), and cost (by a factor of 4), along with improvements in analytical performance. However, further improvements in instrument size, weight, cost, and performance are still needed to make these technologies more widely affordable, deployable, and able to operate in the wide range of particle number concentrations that can be encountered in both workplaces and ambient environments. Needed improvements include smaller, more reliable power sources, including the possibility of body-heat or motion-driven power sources, and more compact data collection, processing, and memory capabilities. As described in the perspective article by Fadel et al. [105], advances in nanotechnologies could enable and accelerate the development of inexpensive, portable devices for the broad detection, identification, and quantification of biological and chemical substances, including sensors for air quality and health applications.

6. Summary and Conclusions

The goals of the ASHG are to provide an open dialogue from the multiple disciplines and agencies represented on a number of related issues: (1) improving the understanding of what sensor technologies may be able to deliver for these agencies to meet their missions; (2) communication of what small-scale sensor readings can provide to a number of audiences; and most centrally, (3) best practices in communicating health-related messages to numerous audiences.

In this paper, we have attempted to cross-reference other related projects and provide additional resources to an interested reader to pursue additional information from credible sources. Additional aims were to provide an update on the advances made to date under the auspices of these various programs, to forecast potential applications of rapidly emerging sensor technologies and to foster a collaborative response to challenges involved in their application to research and support of decisions related to air quality management.

Acknowledgments: Funding for publication costs were provided by the U.S. Environmental Protection Agency, Washington, DC USA, This project was supported in part by an appointment to the Internship/Research Participation Program at the U.S. Environmental Protection Agency, administered by the Oak Ridge Institute for Science and Education through an interagency agreement between the U.S. Department of Energy and EPA, with additional support provided by the National Institutes of Health, National Library of Medicine.

Author Contributions: George Woodall, Mark Hoover, Ronald Williams and Janis Hulla were the primary architects of the outline for this review paper; all authors contributed in writing the paper.

Appendix A

Standards for equipment used to detect and determine toxic gases and vapors in a workplace or similar situation.

A.1. Published

Europe:

1. EN 45544:2000 Workplace atmospheres. Electrical apparatus used for the direct detection and direct concentration measurement of toxic gases and vapours. Part 1: General requirements and test methods; Part 2: Performance requirements for apparatus used for measuring concentrations in the region of limit values; Part 3: Performance requirements for apparatus used for measuring concentrations well above limit values; Part 4: Guide for selection, installation, use and maintenance.

USA:

1. Underwriters Laboratory: UL 2075 Gas and Vapor Detectors and Sensors
2. American National Standards Institute/International Safety Association: ANSI/ISA-92.00.01-2010 Performance Requirements for Toxic Gas Detectors; ANSI/ISA 92.00.02-2013 Installation, Operation, and Maintenance of Toxic Gas-Detection Instruments
3. American Society for Testing and Materials: ASTM E2885-13 Standard Specification for Handheld Point Chemical Vapor Detectors (HPCVD) for Homeland Security Application

International Electrotechnical Commission:

1. IEC 60079-29-1:2007 Explosive atmospheres—Part 29-1: Gas detectors—Performance requirements of detectors for flammable gases
2. IEC 60079-29-2:2007 Explosive atmospheres—Part 29-2: Gas detectors—Selection, installation, use and maintenance of detectors for flammable gases and oxygen

Other:

1. Australian/New Zealand Standard: AS/NZS 4641:2007 Electrical apparatus for the detection of oxygen and other gases and vapours at toxic levels—General requirements and test methods.
2. ISO database: https://www.iso.org/committee/52702/x/catalogue/p/0/u/1/w/0/d/0
3. CEN Database: https://standards.cen.eu/dyn/www/f?p=204:7:0::::FSP_ORG_ID:6245&cs=178094E67E1897102F190938A48C7A285

A.2. Standards Proceeding through Process

ISO/IEC (IEC 62990-1) Workplace Atmospheres—Part 1: Gas detectors—Performance requirements of detectors for toxic gases

ISO/IEC (IEC 62990-2) Work-place Atmospheres—Part 2: Gas detectors—Selection, installation, use and maintenance of detectors for toxic gases and vapours and oxygen sensors.

References

1. Williams, R. Findings from the 2013 EPA Sensors Workshop. Available online: https://www.epa.gov/air-research/findings-2013-epa-air-sensors-workshop (accessed on 5 May 2017).
2. Williams, R.; Long, R.; Beaver, M.; Kaufman, A.; Zeiger, F.; Heimbinder, M.; Hang, I.; Yap, R.; Acharya, B.; Ginwald, B.; et al. *Sensor Evaluation Report*; U.S. Environmental Protection Agency: Washington, DC, USA, 2014.

3. U.S. Environmental Protection Agency. *Graphical Arrays of Chemical-Specific Health Effect Reference Values for Inhalation Exposures;* U.S. Environmental Protection Agency: Research Triangle Park, NC, USA, 2009; EPA/600/R-09/061. Available online: http://cfpub.epa.gov/ncea/cfm/recordisplay.cfm?deid=211003 (accessed on 5 May 2017).
4. U.S. Environmental Protection Agency. Guidance on Systematic Planning Using the Data Quality Objectives Process. Available online: https://www.epa.gov/fedfac/guidance-systematic-planning-using-data-quality-objectives-process (accessed on 5 May 2017).
5. U.S. Environmental Protection Agency. Air Sensor Toolbox: What Do My Sensor Readings Mean? Sensor Scale Pilot Project. Available online: https://www.epa.gov/air-sensor-toolbox/what-do-my-sensor-readings-mean-sensor-scale-pilot-project (accessed on 2 May 2017).
6. U.S. Environmental Protection Agency. Air Sensor Toolbox for Citizen Scientists, Researchers and Developers. Available online: https://www.epa.gov/air-sensor-toolbox (accessed on 2 May 2017).
7. ATSDR. Agency for Toxic Substances and Disease Registry Minimal Risk Levels (MRLs) 2016. Available online: https://www.atsdr.cdc.gov/mrls/index.asp (accessed on 12 June 2017).
8. Woodall, G.; Lipscomb, J.; Taylor, M. Review of health-based reference values for inhalation exposures. 2017; in preparation.
9. U.S. Environmental Protection Agency. *Inhalation Health Effect Reference Values for Toluene (CASRN 108-88-3);* U.S. Environmental Protection Agency: Research Triangle Park, NC, USA, 2012. Available online: http://oaspub.epa.gov/eims/eims-comm.getfile?p_download_id=512650 (accessed on 2 May 2017).
10. U.S. Environmental Protection Agency. *Chemical-Specific Reference Values for Benzene (CASRN 71-43-2);* U.S. Environmental Protection Agency: Research Triangle Park, NC, USA, 2012; EPA/600/R-12/047F1. Available online: http://oaspub.epa.gov/eims/eimscomm.getfile?p_download_id=512648 (accessed on 2 May 2017).
11. U.S. Environmental Protection Agency. *Inhalation Health Effect Reference Values for Ethylbenzene (CASRN 100-41-4);* U.S. Environmental Protection Agency: Research Triangle Park, NC, USA, 2012; EPA/600/R-12/047F2. Available online: http://oaspub.epa.gov/eims/eimscomm.getfile?p_download_id=512649 (accessed on 2 May 2017).
12. U.S. Environmental Protection Agency. *Inhalation Health Effect Reference Values for Xylene—All Isomers (CASRNs Mixed Isomers—1330-20-7; m-xylene—95-47-6; o-xylene—108-38-3; p-xylene—106-42-3);* U.S. Environmental Protection Agency: Research Triangle Park, NC, USA, 2012; EPA/600/R-12/047F4. Available online: http://oaspub.epa.gov/eims/eimscomm.getfile?p_download_id=512651 (accessed on 2 May 2017).
13. U.S. Environmental Protection Agency. *Inhalation Health Effect Reference Values for Manganese (CASRN 7439-96-5—Manganese) and Compounds (CASRN 1344-43-0; 1317-35-7; and 1129-60-5);* U.S. Environmental Protection Agency: Research Triangle Park, NC, USA, 2012; EPA/600/R-12/047F5. Available online: http://oaspub.epa.gov/eims/eimscomm.getfile?p_download_id=512652 (accessed on 2 May 2017).
14. Harper, M.; Weis, C.; Pleil, J.D.; Blount, B.C.; Miller, A.; Hoover, M.D.; Jahn, S. Commentary on the contributions and future role of occupational exposure science in a vision and strategy for the discipline of exposure science. *J. Expo. Sci. Environ. Epidemiol.* **2015**, *25*, 381–387. [CrossRef] [PubMed]
15. Fishbain, B.; Lerner, U.; Castell, N.; Cole-Hunter, T.; Popoola, O.; Broday, D.M.; Iñiguez, T.M.; Nieuwenhuijsen, M.; Jovasevic-Stojanovic, M.; Topalovic, D.; et al. An evaluation tool kit of air quality micro-sensing units. *Sci. Total Environ.* **2017**, *575*, 639–648. [CrossRef] [PubMed]
16. Bart, M.; Williams, D.E.; Ainslie, B.; McKendry, I.; Salmond, J.; Grange, S.K.; Alavi-Shoshtari, M.; Steyn, D.; Henshaw, G.S. High density ozone monitoring using gas sensitive semi-conductor sensors in the lower Fraser Valley, British Columbia. *Environ. Sci. Technol.* **2014**, *48*, 3970–3977. [CrossRef] [PubMed]
17. Gao, M.L.; Cao, J.J.; Seto, E. A distributed network of low-cost continuous reading sensors to measure spatiotemporal variations of PM2.5 in Xi'an, China. *Environ. Pollut.* **2015**, *199*, 56–65. [CrossRef] [PubMed]
18. Mead, M.I.; Popoola, O.A.M.; Stewart, G.B.; Landshoff, P.; Calleja, M.; Hayes, M.; Baldovi, J.J.; McLeod, M.W.; Hodgson, T.F.; Dicks, J.; et al. The use of electrochemical sensors for monitoring urban air quality in low-cost, high-density networks. *Atmos. Environ.* **2013**, *70*, 186–203. [CrossRef]
19. Jovasevic-Stojanovic, M. CITI-SENSE Development of Sensor-Based Citizens' Observatory Community for Improving Quality of Life in Cities. Available online: http://www.citi-sense.eu/Portals/106/Documents/Dissemination%20material/CITI-SENSE_Newsletter%20No4.pdf (accessed on 19 September 2017).

20. CITI-SENSE. Deliverable D9.16 Project Overview for the Lay People. 2016. Available online: http://co.citi-sense.eu/TheProject/Publications/Deliverables.aspx (accessed on 2 May 2017).
21. CITI-SENSE. Development of Sensor-Based Citizens' Observatory Community for Improving Quality of Life in Cities. Available online: http://www.citi-sense.eu/ (accessed on 2 May 2017).
22. Aspuru, I.; García, I.; Herranz, K.; Santander, A. Citi-Sense: Methods and tools for empowering citizens to observe acoustic comfort in outdoor public spaces. *Noise Mapp.* **2016**, *3*, 37–48. [CrossRef]
23. European Commission. Air Quality Standards. 8/6/16. Available online: http://ec.europa.eu/environment/air/quality/standards.htm (accessed on 2 May 2017).
24. Borrego, C.; Costa, A.M.; Ginja, J.; Amorim, M.; Coutinho, M.; Karatzas, K.; Sioumis, T.; Katsifarakis, N.; Konstantinidis, K.; De Vito, S.; et al. Assessment of air quality microsensors versus reference methods: The EuNetAir joint exercise. *Atmos. Environ.* **2016**, *147*, 246–263. [CrossRef]
25. Penza, M.; Consortium, E. COST Action TD1105: Overview of sensor-systems for air-quality monitoring. *Procedia Eng.* **2014**, *87*, 1370–1377. [CrossRef]
26. Clean Air Asia. Clean Air Asia India. Available online: http://cleanairasia.org/india/ (accessed on 2 May 2017).
27. Lewis, A.; Edwards, P. Validate personal air-pollution sensors. *Nature* **2016**, *535*, 29–31. [CrossRef] [PubMed]
28. BreezoMeter. BreezoMeter: Hyperlocal Air Quality Data, Available Globally. Available online: https://breezometer.com/ (accessed on 2 May 2017).
29. WHO. *Ambient (Outdoor) Air Quality and Health. Fact Sheet*; WHO: Geneva, Switzerland, 2016. Available online: http://www.who.int/mediacentre/factsheets/fs313/en/ (accessed on 2 May 2017).
30. ATSDR. *Health Consultation: Evaluation of Particulate Matter, Bennett Landfill Fire, Chester, South Carolina*; ATSDR: Atlanta, GA, USA, 2015.
31. ATSDR. *Health Consultation: Assessing the Public Health Implications of the Criteria (NAAQS) Air Pollutants and Hydrogen Sulfide as Part of the Midlothian Area Air Quality Petition Response, Midlothian, Ellis County, Texas*; ATSDR: Atlanta, GA, USA, 2016. Available online: https://www.atsdr.cdc.gov/sites/midlothian/health_consultations.html (accessed on 2 May 2017).
32. ATSDR. *Health consultation: Brooklyn Township PM2.5, Brooklyn Township, Susquehanna County, Pennsylvania*; ATSDR: Atlanta, GA, USA, 2016; Cost Recovery Number: 3A4K00. Available online: https://www.atsdr.cdc.gov/HAC/pha/BrooklynTownship/BrooklynTwnsp_pm2-5_HC_Final_04-22-2016_508.pdf (accessed on 2 May 2017).
33. ATSDR. *Health Consultation: Review of Analysis of Particulate Matter and Metal Exposures in Air, KCBX (AKA, Chicago Petroleum Coke sites), Cook County, IL, USA*; ATSDR: Atlanta, GA, USA, 2016.
34. NACEPT. *NACEPT 2016 Report: Environmental Protection belongs to the Public, a Vision for Citizen Science at EPA*; U.S. Environmental Protection Agency: Washington, DC, USA, 2016; EPA 219-R-16-001. Available online: https://www.epa.gov/faca/nacept-2016-report-environmental-protection-belongs-public-vision-citizen-science-epa (accessed on 2 May 2017).
35. Dosemagen, S.; Gehrke, G. Civic Technology and Community Science: A New Model for Public Participation in Environmental Decisions in Confronting the Challenges of Public Participation: Issues in Environmental, Planning, and Health Decision-Making. In Proceedings of the Iowa State University Summer Symposia on Science Communication, Ames, IA, USA, 30 May–1 June 2013.
36. Williams, R.; Kilaru, V.; Snyder, E.; Kaufman, A.; Dye, T.; Rutter, A.; Russell, A.; Hafner, H. *Air Sensor Guidebook*; U.S. Enviromental Protection Agency: Research Triangle Park, NC, USA, 2016; EPA/600/R-14/159. Available online: https://cfpub.epa.gov/si/si_public_record_report.cfm?direntryid=277996 (accessed on 2 May 2017).
37. U.S. Environmental Protection Agency. Citizen Science in Newark, New Jersey. 2015. Available online: https://www.epa.gov/sciencematters/citizen-science-newark-new-jersey (accessed on 12 July 2017).
38. U.S. Environmental Protection Agency. Air sensor toolbox: Resources and funding. 2016. Available online: https://www.epa.gov/air-sensor-toolbox/air-sensor-toolbox-resources-and-funding#TER (accessed on 5 June 2017).
39. Heimbinder, M. Apps & Sensors for Air Pollution. Taking Space: The HabitatMap & AirCasting Blog 2012 4/9/2012. Available online: http://www.takingspace.org/epa-apps-sensors-for-air-pollution-workshop/ (accessed on 2 May 2017).

40. Snyder, E.G.; Watkins, T.H.; Solomon, P.A.; Thoma, E.D.; Williams, R.W.; Hagler, G.S.; Shelow, D.; Hindin, D.A.; Kilaru, V.J.; Preuss, P.W. The changing paradigm of air pollution monitoring. *Environ. Sci. Technol.* **2013**, *47*, 11369–11377. [CrossRef] [PubMed]
41. MacDonnell, M.; Raymond, M.; Wyker, D.; Finster, M.; Chang, Y.; Raymond, T.; Temple, B.; Scofield, M.; Vallano, D. *Mobile Sensors and Applications for Air Pollutants*; U.S. Environmental Protection Agency: Washington, DC, USA, 2013.
42. Jiao, W.; Hagler, G.; Williams, R.; Sharpe, R.; Brown, R.; Garver, D.; Judge, R.; Caudill, M.; Rickard, J.; Davis, M.; et al. Community Air Sensor Network (CAIRSENSE) project: Evaluation of low-cost sensor performance in a suburban environment in the southeastern United States. *Atmos. Meas. Tech.* **2016**, *9*, 5281–5292. [CrossRef]
43. Smith, K.; Edwards, P.M.; Evans, M.J.J.; Lee, J.D.; Shaw, M.D.; Squires, F.; Wilde, S.; Lewis, A.C. Clustering approaches that improve the reproducibility of low-cost air pollution sensors. *Faraday Discuss.* **2017**, *200*, 621–637. [CrossRef] [PubMed]
44. Kaufman, A.; Brown, A.; Barzyk, T.; Williams, R. The Citizen Science Toolbox: A One-Stop Resources for Air Sensor Technology. In *EM: Air and Waste Management Association's Magazine for Environmental Managers*; Air & Waste Management Association: Pittsburgh, PA, USA, 2014.
45. Spinelle, L.; Gerboles, M.; Aleixandre, M.; Bonavitacola, F. Evaluation of metal oxides sensors for the monitoring of O3 in ambient air at Ppb level. *Chem. Eng. Trans.* **2016**, *54*, 319–324.
46. Williams, R. *Evaluation of Elm and Speck Sensors*; U.S. Environmental Protection Agency: Washington, DC, USA, 2015.
47. Air Quality Sensor Performance Evaluation Center (AQ-SPEC). South Coast Air Quality Management District. Available online: http://www.aqmd.gov/aq-spec/ (accessed on 2 May 2017).
48. Lin, C.; Gillespie, J.; Schuder, M.D.; Duberstein, W.; Beverland, I.J.; Heal, M.R. Evaluation and calibration of Aeroqual series 500 portable gas sensors for accurate measurement of ambient ozone and nitrogen dioxide. *Atmos. Environ.* **2015**, *100*, 111–116. [CrossRef]
49. NIOSH. *Components for Evaluation of Direct-Reading Monitors for Gases and Vapors*; NIOSH: Cincinnati, OH, USA, 2012; Publication No. 2012-12-162. Available online: https://www.cdc.gov/niosh/docket/archive/pdfs/niosh-220/2012-162.pdf (accessed on 2 May 2017).
50. NIOSH. *Addendum to Components for Evaluation of Direct-Reading Monitors for Gases and Vapors: Hazard Detection in First Responder Environments*; NIOSH: Cincinnati, OH, USA, 2012; Publication No. 2012-163. Available online: https://www.cdc.gov/niosh/docket/archive/pdfs/niosh-220/2012-163.pdf (accessed on 2 May 2017).
51. AIHA. Reporting Specifications for Electronic Real Time Gas and Vapor Detection Equipment Fact Sheet. Available online: https://www.aiha.org/governmentaffairs/PositionStatements/Reporting%20Specifications%20for%20Real%20Time%20Detection_Final.pdf (accessed on 2 May 2017).
52. Moltchanov, S.; Levy, I.; Etzion, Y.; Lerner, U.; Broday, D.M.; Fishbain, B. On the feasibility of measuring urban air pollution by wireless distributed sensor networks. *Sci. Total Environ.* **2015**, *502*, 537–547. [CrossRef] [PubMed]
53. National Institute of Environmental Health Science. Validation and Demonstration of Devices for Environmental Exposure Assessment (R21/R33). 2013. Available online: https://grants.nih.gov/grants/guide/rfa-files/RFA-ES-13-013.html (accessed on 2 May 2017).
54. U.S. Environmental Protection Agency. Village Green Project. 4/12/17. Available online: https://www.epa.gov/air-research/village-green-project (accessed on 2 May 2017).
55. U.S. Environmental Protection Agency. (2013) DRAFT Roadmap for Next Generation Air Monitoring. Available online: https://www.epa.gov/sites/production/files/2014-09/documents/roadmap-20130308.pdf (accessed on 12 July 2017).
56. Wan, J.; Hagler, G.; Williams, R.; Sharpe, B.; Weinstock, L.; Rice, J. Field assessment of the Village Green Project: An autonomous community air quality monitoring system. *Environ. Sci. Technol.* **2015**. [CrossRef]
57. Hindin, D.; Grumbles, B.; Wyeth, G.; Beneditc, K.; Watkins, T.; Aburn, G., Jr.; Ulrich, M.; Lang, S.; Poole, K.; Dapolito Dunn, A. Advanced monitoring technology: Opportunities and challenges a path forward for EPA, states, and tribes. *EM Environ. Manag.* **2016**, *11*, 16–21.
58. Keating, M.; Benedict, K.; Evans, R.; Jenkins, S.; Mannshardt, E.; Stone, S.L. Interpreting and communicating short-term air sensor data. *EM Environ. Manag.* **2016**, *11*, 22–25.

59. Mannshardt, E.; Benedict, K.; Jenkins, S.; Keating, M.; Mintz, D.; Stone, S. Analysis of short-term ozone and PM2.5 measurements: Characteristics and relationships for air sensor messaging. *J. Air Waste Manag. Assoc.* **2016**, *67*, 462–474. [CrossRef] [PubMed]
60. U.S. Environmental Protection Agency. History of Acute Exposure Guideline Levels (AEGLs). Available online: https://www.epa.gov/aegl/history-acute-exposure-guideline-levels-aegls (accessed on 12 July 2017).
61. NRC. *Standing Operating Procedures for Developing Acute Exposure Guideline Levels (Aegls) for Hazardous Chemicals*; National Academy Press: Washington, DC, USA, 2001.
62. Young, R.A.; Bast, C.B.; Wood, C.S.; Adeshina, F. Overview of the Standing Operating Procedure (SOP) for the development of Provisional Advisory Levels (PALs). *Inhal. Toxicol.* **2009**, *21*, 1–11. [CrossRef] [PubMed]
63. Klepeis, N.E.; Nelson, W.C.; Ott, W.R.; Robinson, J.P.; Tsang, A.M.; Switzer, P.; Behar, J.V.; Hern, S.C.; Engelmann, W.H. The National Human Activity Pattern Survey (NHAPS): A resource for assessing exposure to environmental pollutants. *J. Expo. Anal. Environ. Epidemiol.* **2001**, *11*, 231–252. [CrossRef] [PubMed]
64. Ott, W. *Human Activity Patterns: A Review of the Literature for Estimating Time Spent Indoors, Outdoors, and In-Transit*; U.S. Environmental Protection Agency, Office of Research and Development: Las Vegas, NV, USA, 1989.
65. Robinson, J.P.; Thomas, J. *Time Spent in Activities, Locations, and Microenvironments: A California National Comparison*; U.S. Environmental Protection Agency: Las Vegas, NV, USA, 1992.
66. U.S. Environmental Protection Agency. *Child-Specific Exposure Factors Handbook (Final Report) 2008*; U.S. Environmental Protection Agency: Washington, DC, USA, 2008; EPA/600/R-06/096F. Available online: http://cfpub.epa.gov/ncea/cfm/recordisplay.cfm?deid=199243 (accessed on 12 July 2017).
67. Wallace, L.A. *Project Summary: The Total Exposure Assessment Methodology (TEAM) Study*; U.S. Environmental Protection Agency: Washington, DC, USA, 1987.
68. SAB. *Integrated Human Exposure Committee Commentary on Indoor Air Strategy*; U.S. Environmental Protection Agency: Washington, DC, USA, 1998.
69. Nehr, S.; Hosen, E.; Tanabe, S.I. Emerging developments in the standardized chemical characterization of indoor air quality. *Environ. Int.* **2017**, *98*, 233–237. [CrossRef] [PubMed]
70. NRC. *Toxicity Testing in the 21st Century: A Vision and a Strategy*; National Academies Press: Washington, DC, USA, 2007.
71. NRC. *Exposure Science in the 21st Century: A Vision and a Strategy*; National Academies Press: Washington, DC, USA, 2012.
72. National Academies of Sciences, Engineering, and Medicine. *Using 21st Century Science to Improve Risk-Related Evaluations*; National Academies Press: Washington, DC, USA, 2017.
73. Mattingly, C.J.; Mckone, T.E.; Callahan, M.A.; Blake, J.A.; Hubal, E.A. Providing the missing link: The exposure science ontology ExO. *Environ. Sci. Technol.* **2012**, *46*, 3046–3053. [CrossRef] [PubMed]
74. Villeneuve, D.L.; Crump, D.; Garcia-Reyero, N.; Hecker, M.; Hutchinson, T.H.; Lalone, C.A.; Landesmann, B.; Lettieri, T.; Munn, S.; Nepelska, M.; et al. Adverse outcome pathway (AOP) development I: Strategies and principles. *Toxicol. Sci.* **2014**, *142*, 312–320. [CrossRef] [PubMed]
75. Villeneuve, D.L.; Crump, D.; Garcia-Reyero, N.; Hecker, M.; Hutchinson, T.H.; Lalone, C.A.; Landesmann, B.; Lettieri, T.; Munn, S.; Nepelska, M.; et al. Adverse outcome pathway development II: Best practices. *Toxicol. Sci.* **2014**, *142*, 321–330. [CrossRef] [PubMed]
76. Teeguarden, J.G.; Tan, Y.M.; Edwards, S.W.; Leonard, J.A.; Anderson, K.A.; Corley, R.A.; Kile, M.L.; Simonich, S.M.; Stone, D.; Tanguay, R.L.; et al. Completing the link between exposure science and toxicology for improved environmental health decision making: The aggregate exposure pathway framework. *Environ. Sci. Technol.* **2016**, *50*, 4579–4586. [CrossRef] [PubMed]
77. Hulla, J.; Snawder, J.E.; Proctor, S.P.; Chapman, G.D. DOD impact assessment and management of naphthalene-related risks. *Toxicologist* **2010**, *114*, 400.
78. Hug, W.F.; Bhartia, R.; Reid, R.D.; Reid, M.R.; Oswal, P.; Lane, A.L.; Sijapati, K.; Sullivan, K.; Hulla, J.E.; Snawder, J.; et al. Advanced environmental, chemical, and biological sensing technologies IX. In *Wearable Real-Time Direct-Reading Naphthalene and VOC Personal Exposure Monitor*; Vo-Dinh, T., Lieberman, R.A., Gauglitz, G., Eds.; SPIE: Bellingham, WA, USA, 2012; Volume 8366.
79. Hulla, J.; Proctor, S.; Snawder, J.E. The naphthalene dosimeter—Vanguard technology for improved health protection. *Toxicologist* **2015**, *144*, 110.

80. U.S. Environmental Protection Agency. *Integrated Science Assessment for Ozone and Related Photochemical Oxidants*; U.S. Environmental Protection Agency, Office of Research and Development, National Center for Environmental Assessment-RTP Division: Research Triangle Park, NC, USA, 2013.
81. Mcdonnell, W.F.; Stewart, P.W.; Smith, M.V.; Kim, C.S.; Schelegle, E.S. Prediction of lung function response for populations exposed to a wide range of ozone conditions. *Inhal. Toxicol.* **2012**, *24*, 619–633. [CrossRef] [PubMed]
82. Brown, J.S.; Gordon, T.; Price, O.; Asgharian, B. Thoracic and respirable particle definitions for human health risk assessment. *Part. Fibre Toxicol.* **2013**, *10*, 12. [CrossRef] [PubMed]
83. ICRP. *Human Respiratory Tract Model for Radiological Protection: A Report of a Task Group of the International Commission on Radiological Protection*; Pergamon Press: New York, NY, USA, 1994; Volume 24, pp. 1–482.
84. Koehler, K.A.; Clark, P.; Volckens, J. Development of a sampler for total aerosol deposition in the human respiratory tract. *Ann. Occup. Hyg.* **2009**, *53*, 731–738. [CrossRef] [PubMed]
85. U.S. Environmental Protection Agency. *Integrated Science Assessment for Oxides of Nitrogen-Health Criteria (Final Report)*; U.S. Environmental Protection Agency, Office of Research and Development, National Center for Environmental Assessment: Research Triangle Park, NC, USA, 2016.
86. Brown, J.S. Nitrogen dioxide exposure and airway responsiveness in individuals with asthma. *Inhal. Toxicol.* **2015**, *27*, 1–14. [CrossRef] [PubMed]
87. U.S. Congress. Clean Air Act as Amended in 1990, Section 112 (b) (1) Hazardous Air Pollutants. 1990. Available online: https://www.epa.gov/history/epa-history-clean-air-act-amendments-1990 (accessed on 12 July 2017).
88. Hoover, M.D.; Cox, M. Public protection from nuclear, chemical, and biological terrorism: Health Physics Society 2004 summer school. In *A Life-Cycle Approach for Development and Use of Emergency Response and Health Protection Instrumentation*; Brodsky, A., Johnson, R.H., Groans, R.E., Eds.; Medical Physics: Madison, WI, USA, 2004; pp. 317–324.
89. Hoover, M.D.; Cox, M. Radioactive air sampling methods. In *A Life-Cycle Approach to Development and Application of Air Sampling Methods and Instrumentation*; Maiello, M.L., Hoover, M.D., Eds.; CRC Press/Taylor & Francis: Boca Raton, FL, USA, 2011; pp. 43–52.
90. NNI. *Nanotechnology for Sensors and Sensors for Nanotechnology: Improving and Protecting Health, Safety, and Environment*; NNI: Washington, DC, USA, 2012.
91. Hoover, M.D.; Debord, D.G. Turning numbers into knowledge: Sensors for safety, health, well-being, and productivity. *Synergist* **2015**, *26*, 22–26.
92. InnoCentive. My Air, My Health: An HHS/EPA Challenge. Available online: https://www.innocentive.com/ar/challenge/9932947 (accessed on 12 July 2017).
93. U.S. Environmental Protection Agency. AirNow. Available online: https://www.airnow.gov/ (accessed on 12 July 2017).
94. De la Iglesia, D.; Harper, S.; Hoover, M.D.; Klaessig, F.; Lippelli, P.; Maddux, B.; Morse, J.; Nel, A.; Rajan, K.; Reznik-Zellen, R.; et al. Nanoinformatics 2020 Roadmap. National Nanomanufacturing Network 2011. Available online: http://eprints.internano.org/607/ (accessed on 20 June 2017).
95. Hoover, M.D.; Myers, D.S.; Cash, L.J.; Guilmette, R.A.; Kreyling, W.G.; Oberdörster, G.; Smith, R.; Cassata, J.R.; Boecker, B.B.; Grissom, M.P. Application of an informatics-based decision-making framework and process to the assessment of radiation safety in nanotechnology. *Health Phys.* **2015**, *108*, 179–194. [CrossRef] [PubMed]
96. NNI. *Nanotechnology Signature Initiative: Nanotechnology Knowledge Infrastructure (NKI) Data Readiness Levels Discussion Draft*; NNI: Washington, DC, USA, 2013. Available online: https://www.nano.gov/node/1015 (accessed on 20 June 2017).
97. Hendren, C.O.; Powers, C.M.; Hoover, M.D.; Harper, S.L. The Nanomaterial Data Curation Initiative: A collaborative approach to assessing, evaluating, and advancing the state of the field. *Beilstein J. Nanotechnol.* **2015**, *6*, 1752–1762. [CrossRef] [PubMed]
98. Google. Mapping the Invisible: Street View Cars Add Air Pollution Sensors. Available online: https://environment.google/projects/airview/ (accessed on 12 July 2017).
99. U.S. National Library of Medicine. PubMed. Available online: https://www.ncbi.nlm.nih.gov/pubmed (accessed on 17 August 2017).
100. Oltra, C.; Sala, R.; Boso, A.; Asensio, S.L. Public engagement on urban air pollution: An exploratory study of two interventions. *Environ. Monit. Assess.* **2017**, *189*, 296. [CrossRef] [PubMed]

101. U.S. National Library of Medicine. Fact Sheet: MEDLINE® Journal Selection. 30 June 2017. Available online: https://www.nlm.nih.gov/pubs/factsheets/jsel.html (accessed on 17 August 2017).
102. U.S. National Library of Medicine. PubMed Commons. Available online: https://www.ncbi.nlm.nih.gov/pubmedcommons/ (accessed on 17 August 2017).
103. Fierz, M.; Houle, C.; Steigmeier, P.; Burtscher, H. Design, Calibration, and field performance of a miniature diffusion size classifier. *Aerosol Sci. Technol.* **2011**, *45*, 1–10. [CrossRef]
104. Kulkarni, P.; Qi, C.; Fukushima, N. Miniature differential mobility analyzer for compact field-portable spectrometers. *Aerosol Sci. Technol.* **2016**, *50*, 1167–1179. [CrossRef] [PubMed]
105. Fadel, T.R.; Farrell, D.F.; Friedersdorf, L.E.; Griep, M.H.; Hoover, M.D.; Meador, M.A.; Meyyappan, M. Toward the responsible development and commercialization of sensor nanotechnologies. *ACS Sens.* **2016**, *1*, 207–216. [CrossRef] [PubMed]

8

Overview of the Model and Observation Evaluation Toolkit (MONET) Version 1.0 for Evaluating Atmospheric Transport Models

Barry Baker [1,2,*] and Li Pan [1,2]

1 NOAA Air Resources Laboratory, College Park, MD 20740, USA; li.pan@noaa.gov
2 Cooperative Institute for Climate and Satellites, University of Maryland at College Park, College Park, MD 20740, USA
* Correspondence: Barry.Baker@noaa.gov.

Abstract: This paper describes the development and initial applications of the Model and Observation Evaluation Tool (MONET) v1.0. MONET was developed to evaluate the Community Multiscale Air Quality Model (CMAQ) for the NOAA National Air Quality Forecast Capability (NAQFC) modeling system. MONET is designed to be a modularized Python package for (1) pairing model output to observational data in space and time; (2) leveraging the pandas Python package for easy searching and grouping; and (3) analyzing and visualizing data. This process introduces a convenient method for evaluating model output. MONET processes data that is easily searchable and that can be grouped using meta-data found within the observational datasets. Common statistical metrics (e.g., bias, correlation, and skill scores), plotting routines such as scatter plots, timeseries, spatial plots, and more are included in the package. MONET is well modularized and can add further observational datasets and different models.

Keywords: CMAQ; evaluation; air quality; software; visualization; statistics

---

1. Introduction

Ozone (O_3) and particulate matter smaller than 2.5 μm in diameter ($PM_{2.5}$) are among a handful of criteria air pollutants—pollutants the Clean Air Act requires to be monitored and regulated—that are primarily responsible for adverse impacts on human health [1]. Breathing these pollutants is recognized as major causes of acute and chronic respiratory and cardiovascular diseases, and premature mortalities associated with air pollution [2].

In relation to health hazards caused by air pollutants, air quality also has a direct impact on the economy. Trasande et al. [3] found the cost of medical care for preterm births attributable to $PM_{2.5}$ exposure to be between $2.43 and $9.66 billion. Ghude et al. [4] demonstrates the economic cost of poor air quality in India to crop yields is estimated to be around $1.26 billion annually. Tong et al. [5] reveals a similar influence to crop yields in the United States (U.S.).

Evaluating model simulations is critical for the development and implementation for forecasting [6]. Evaluation of meteorological and air quality parameters are essential in validating and improving model simulations.

Recently, air quality simulations increased from running for days or weeks to months or years, modeling domains increased resolution, and the use of ensembles greatly increased the amount of model output analyzed. The Community Multiscale Air Quality (CMAQ) modeling simulations, used for air quality and air composition modeling, currently run spanning modeling domains of regional to hemispheric scale and modeling times of days to years resulting in terabytes of model output available for analysis. Conventional methods of data analysis, such as spreadsheets, are not suited for a task this large. There are several model evaluation tools available for evaluating meteorological model simulations but few for evaluating air quality models [7]. Although there are many software packages, such as AMET, or visualization tools, such as Verdi, pycmbs, or Panoply, MONET is built using a single open-source language that retrieves, downloads, and analyses both model and observations on a regional scale. MONET is available on all major computing platforms, i.e., Mac, Linux, and PC, and provides the power of a dedicated programming language such as IDL, MATLAB, or ferret if needed.

The National Oceanic and Atmospheric Administration (NOAA) Air Resources Laboratory (ARL) developed the Model and Observation Evaluation Tool (MONET) to aid in the assessment of the National Air Quality Forecasting Capability (NAQFC) [8,9]. MONET reads, interpolates, and organizes model results to observation sites in both space and time resulting in a fast and flexible method to evaluate air quality modeling simulations. Although MONET was originally created to only evaluate CMAQ simulations, it can easily be expanded to include different model outputs (e.g., the Weather Research and Forecasting model, Next Generation Global Prediction System, and Comprehensive Air Quality Model with Extensions), along with adding additional observational sources (e.g., different ground based networks, satellite observations, and other in-situ observations).

This paper describes the structure and functionality of the MONET Python package (version 2.7). A broad description of MONET will be provided, followed by detailed description of how MONET works. Examples of the analysis products available from MONET will be presented and described. Finally, a discussion and future directions of MONET will be provided.

2. Tool Description

MONET pairs observations and gridded model prediction in space and time. This evaluates the model's performance for a set of predicted or diagnosed fields such as aggregating nitrous oxide (NO) and nitrogen dioxide (NO_2) into nitrogen oxides (NO_x). MONET is built in Python v2.7+ and is intended to follow object oriented concepts. Each model and observation has a specific object intended to be used for its specific cases. Stemming from such object structures, a verification object that inherits observation and model objects can be created allowing for targeted verification between sets of models and observations. MONET runs interactively or with a simple adaptable Python script.

Most of the dependencies can be obtained using commonly available Python packages, such as the Anaconda Python Package or the Enthought Canopy distribution. MONET uses the pandas Python package [10,11] to enable fast and efficient data manipulation using meta-data available in the observational datasets. Pairing of model and observations is done using the pyresample package, (available online: https://pyresample.readthedocs.io/en/latest/), which uses pykdTree to interpolate. Several interpolation methods are available; nearest neighbor, Gaussian, elliptical weighted averaging, or a user defined method, such as inverse distance weighting. Analysis is done using scipy, numpy, and scikit-learn Python functions, while plotting is achieved with Basemap (for spatial plotting), matplotlib, and seaborn [12,13]. Figure 1 presents a flowchart of the MONET software.

MONET runs in four phases; (1) creation of a model object that determines the model grid, variable names, and run duration; (2) creation of an observational object that parses observational data; (3) combines and interpolates model results to observations; and (4) plotting and statistical comparisons.

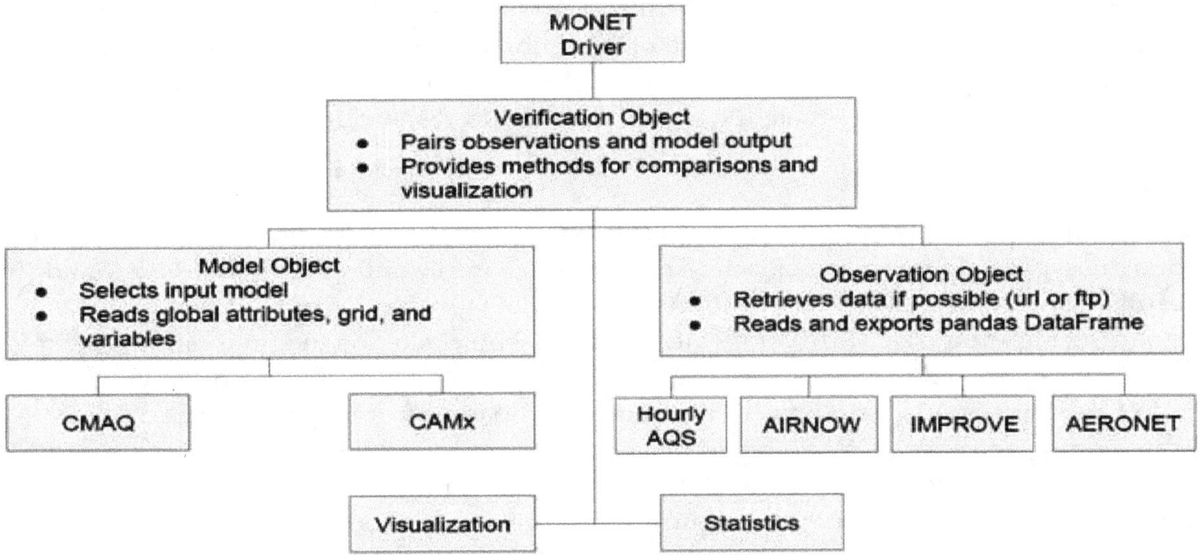

Figure 1. Flow chart of Model and ObservatioN Evaluation Tool (MONET).

2.1. Creation of Model and Observation Objects

MONET is designed to be modular in that each observational network or model used has its own set of specialized functions to handle the differences in each dataset. Currently, MONET can handle the hourly Environmental Protection Agency AirData (EPA AQS; available online: https://aqsdr1.epa.gov/aqsweb/aqstmp/airdata/download_files.html#Raw), including the Chemical Speciation Network (CSN) and AirNow (available online: https://www.airnow.gov) datasets, along with the Interagency Monitoring of Protected Visual Environments (IMPROVE; available online: http://vista.cira.colstate.edu/improve) and Aerosol Robotic Network (AERONET; available online: http://aeronet.gsfc.nasa.gov) datasets. Datasets can be readily downloaded through an FTP or http by using an array of datetime objects given to the observation object. Other networks, such as IMPROVE, require that the data be downloaded manually.

Each observation network is designed to monitor specific sets of air quality parameters, from individual pollutants to derived physical parameters (e.g., AOD). MONET makes it easy to compare individual pollutants or to aggregate model results to pair with the monitor dataset. Examples of this are NO_X ($NO + NO_2$) in AQS or AIRNOW or particle sulfate, an aggregate of the Aitken, accumulation, and coarse mode contribution to the sulfate particle concentration within CMAQ, from AQS and IMPROVE. MONET includes a set of standard aggregations for different pollutants but is versatile enough to allow user defined aggregations.

The observation objects require that a time range is given as an array of datetime objects to the object instance attribute. This is used to tailor data extraction from a pre-downloaded file covering the prescribed time range or otherwise expand the data retrieval. Then, observation objects read the raw datafile and process it into a pandas DataFrame, thus assigning it to an object instance with common columns of observational data (Obs), date (datetime), local date (datetime_local), latitude (Latitude), and longitude (Longitude). Depending on the network, more meta-data may be available, meaning that an observational object in MONET can be used as a standalone method to analyze observational data without the need for pairing. If adding a new network, only the latitude, longitude, observations, and date or timestamp is needed for the interpolation.

For model objects, the only information needed is the output file(s). In the case of overlapping simulations, the model results with the latest creation time stamp will be used. For instance, if two 48 h simulations with 24 h of overlap are given to MONET, then the first 24 h from the first simulation and the 48 h of the second day are used. MONET assigns the model object file to a class instance attribute and creates corresponding helper functions, such as that defining map projections for spatial plotting and retrieving and aggregating data. In the future, the model object will be done with the xarray Python package to keep a consistent view with observations. The xarray package is an N-Dimensional implementation of the pandas library that is highly suitable for scientific data that allows for larger then memory computations and efficient data extraction and resampling.

2.2. Pairing Observations and Model Results

MONET has a built-in driver for each model and observation pair, meaning that the model and observation are separately imported objects. This creates a simple and efficient method for adding new model results—once a model and observational dataset has its own Python object it can easily follow the same pairing algorithms. The driver creates the model object, reads the time range, and creates the model instance. The model time range is then provided to the observational object, where the data is downloaded and processed into a pandas DataFrame or loaded from a preprocessed file.

At this point, each species available in observations and model prediction are paired. MONET has the ability to calculate or aggregate data, such as the calculation of $PM_{2.5}$ (particulate matter with diameter less than 2.5 μm) mass concentration, from several modeled chemical species. Herein, a helper function retrieves the data in the model object to derive the needed data array. MONET then interpolates the model results for each timestamp that is available in the observational network to the observational site and merges them into a common pandas DataFrame containing pairings of model results and observations for all specified observation locations.

The Pyresample library (available online: https://pyresample.readthedocs.io/en/latest/) interpolates the model to observations. Pyresample uses (k-dimensional) KDTrees to transform data defined in one geometrical grid to another. Resampling uses the nearest neighbor, Gaussian, or a customized method, such as inverse distance weighting. MONET allows the user to define the radius of influence and number of neighbors to be used for data re-gridding.

After re-gridding, depending on the dataset, MONET resamples in time and then assigns these processed data to a separate pandas DataFrame for deriving certain species, such as the 8 h max ozone or the daily $PM_{2.5}$ concentration. Timing for opening model results, processing of observation data, and pairing takes less than 3 min with AIRNOW for a 48 h simulation. At this point, MONET proceeds to analyze the dataset with the observation and model result pairs.

3. Example of Tool Applications

MONET is a new software package developed for verifying the NAQFC and more generally the CMAQ model. It has been newly developed as described above and is available on GitHub. Detailed examples and installation instructions are available at https://github.com/noaa-oar-arl/MONET. Once MONET makes the verification object, all of the non-spatial plots, such as timeseries, scatter plots, and histograms, are executed through as a single line-command with flags specifying the user's selection of geographical extent and variables to be plotted. For instance, a user can specify that the geographical extent of their plot be within the U.S. EPA defined Conterminous United States (CONUS) region. MONET performs verification within the entire domain, a region, a state, a county, a metropolitan area, or a specific site. MONET also compares multiple simulations on a single figure simply by creating two verification objects and passing the figure handle. Figures 2–6 showcases some of the plotting routines found within MONET.

As an example of how easily MONET does a quick pairing and analysis, the following example shows how to use MONET to pair AirNow data and a CMAQ simulation. Begin by entering an interactive python session.

ipython –pylab (1)

Import the MONET object. This is the python object interface to the model and observation objects.

$$import\ monet \tag{2}$$

Set the concentration file or files, in this case concfiles, and the gridcro2d file, gridcro.

$$concfiles = \text{'/path/to/concentration/files/ *.ncf'} \tag{3}$$

$$gridcro = \text{'/path/to/gridcro2dfile.ncf'} \tag{4}$$

Use the MONET object to pair the AirNow data and CMAQ simulation by creating an instance of the verified airnow object, which imports both the CMAQ and AIRNOW objects.

$$m = monet.vairnow\ (concobj = concfiles, gridobj = gridcro) \tag{5}$$

MONET retrieves the observations if not in the current directory and the model is interpolated in space to the observational data points. The paired data can be accessed through the pandas DataFrame contained within the monet object, m.

$$m.df.head() \tag{6}$$

Then simple commands can be used to create different plots. For example, MONET creates a timeseries plot averaged over all of the observations found within the modeling domain with a simple one line execution as follows:

$$m.compare\_param\ (param = \text{'OZONE'}, timeseries = True) \tag{7}$$

Figure 2a–d shows a time series of the average concentration over the northeastern United States during the summer of 2016. The average value is shown in solid lines, observations in black, and model results in blue and purple. Shaded areas display one standard deviation from the mean. The footer illustates forecasting performance statistical measures specified by the user. The footer also shows the start date, end date, number of sites, and the number of measurements. Figure 2b presents an example of time series root mean square error plot comparing how two simulations performed for the EPA AQS sites in the U.S. EPA CONUS domain. Likewise, Figure 2c displays a mean bias time series of the two simulations. Figure 2d provides an example of speciated $PM_{2.5}$ sulfate using the IMPROVE network over the Pacific region. With the IMPROVE network daily average measurement every three days, MONET interpolates the hourly CMAQ results to each individual site and then averaged to create a daily concentration as if were an IMPROVE measurement. Then, MONET merges the daily averaged CMAQ results into the IMPROVE DataFrame using a mysql like merge found within Pandas.

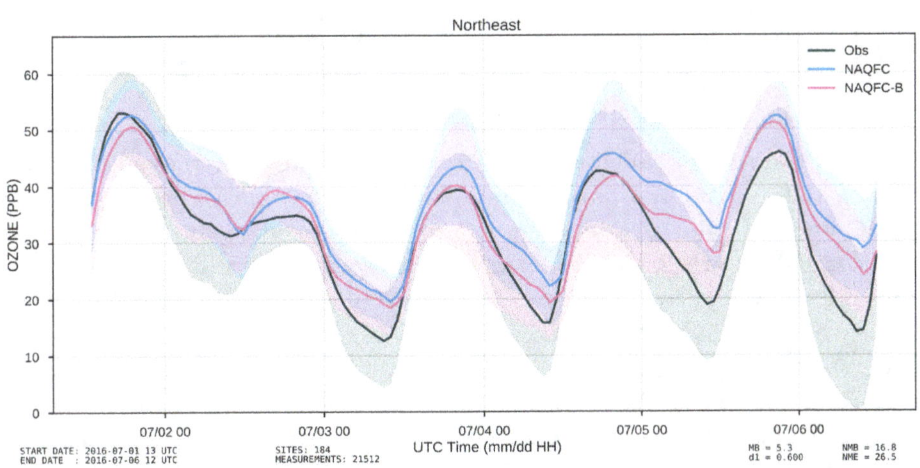

(a)

Figure 2. *Cont.*

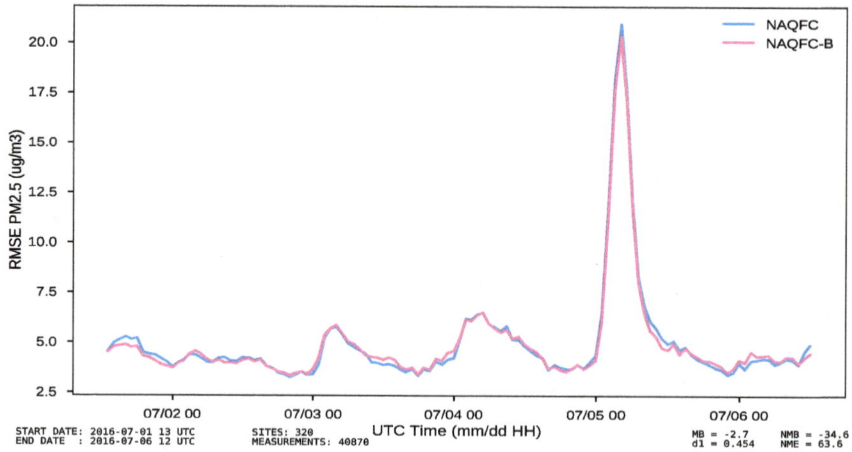

(**b**)

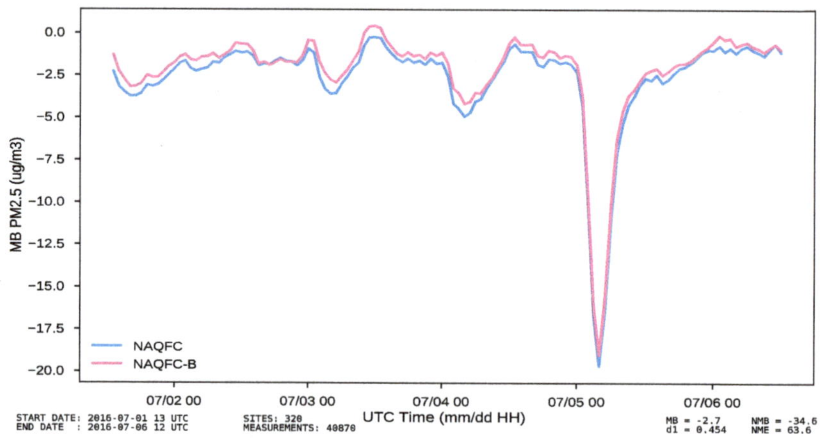

(**c**)

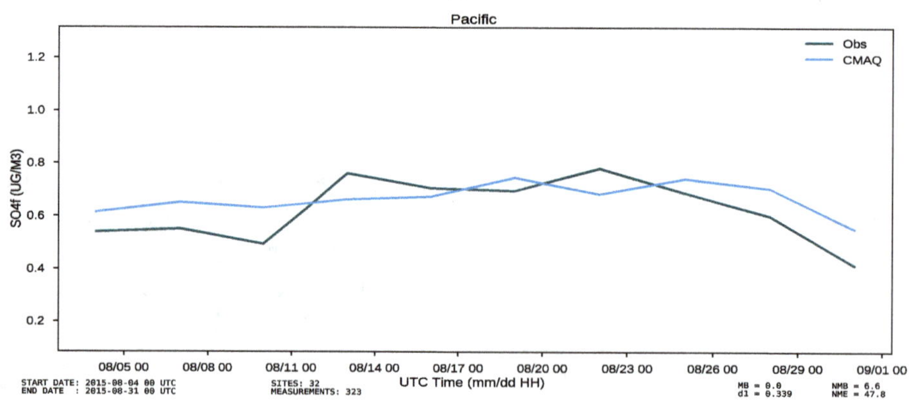

(**d**)

Figure 2. Examples of time-series (**a**) Hourly averaged values are shown in solid lines for EPA AQS observations (black), and two different simulations, National Air Quality Forecast Capability (NAQFC) (blue) and NAQFC-Beta (purple). Shaded regions represent one standard deviation for each hourly averaged time series; (**b**) Hourly averaged values of the root mean square error (RMSE) from EPA AQS observations are shown for two different simulations of $PM_{2.5}$, NAQFC (blue) and NAQFC-Beta (purple); (**c**) Hourly averaged values of the RMSE from EPA AQS observations are shown for two different simulations of $PM_{2.5}$, NAQFC (blue) and NAQFC-Beta (purple); (**d**) Daily averaged values are shown in solid lines for IMPROVE observations (black), and two different simulations, NAQFC (blue) and NAQFC-Beta (purple) in the Pacific region. Shaded regions represent one standard deviation for each daily averaged time series.

Figure 3a presents an example of a kernel density estimation (KDE) comparison. The black line shows an observational KDE and the blue represents the CMAQ model results. This type of plot (Figure 3a) is useful for inspection for distributional discrepancy between the model and observations. Using conditional distributions, such as conditioning on time of day or other meteorological factors like boundary layer height or wind speed, one can provide a more in depth analysis using MONET. Figure 3b displays the difference in the KDE between model and observation results. Difference KDE are important to show model biases in the most probable value.

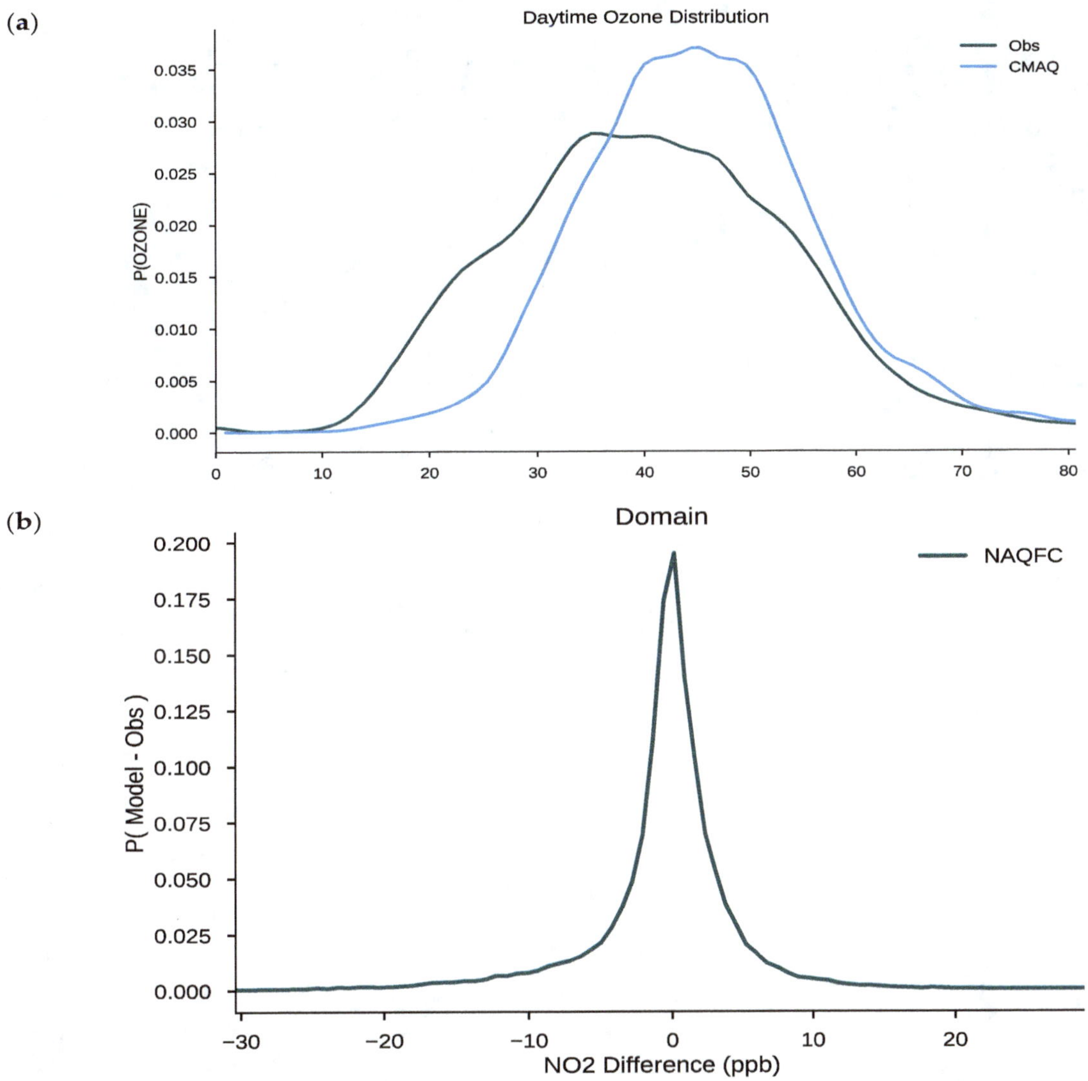

Figure 3. Examples of kernel density estimations (KDEs) (**a**) Daytime kernel density estimations for O_3 concentrations over the continental United States (U.S.) for both observations (black) and a Community Multiscale Air Quality Model (CMAQ) simulation (blue); (**b**) A kernel density estimation of the difference between the simulated NO_2 concentrations and observations.

Figure 4a shows an example of a spatial plot of model results against overlaid observations. The colorbar is customizable by passing colormaps that can be user defined or predefined in matplotlib. The colorbar levels can be tailored individually to any single observation. By default, the colorbar is scaled automatically based on the model data and a discrete colorbar with 15 value bins using the 'viridis' colormap. The 'viridis' colormap is perceptually uniform and sensitive for colorblind individuals. Keyword arguments for the matplotlib.pyplots.imshow can be passed to the MONET function to give additional control over the spatial plot. Spatial plots are useful to give quick glimpses of spatial patterns found within the model. Figure 4b shows an example of a spatial difference plot which is useful for providing spatial assessments for absolute discrepancy. The same type of plot can be created for different metrics such as for RMSE or correlation. In this particular example, it is easy to pertain that over large cities such as Los Angeles or Las Vegas there is a significant under prediction of ozone.

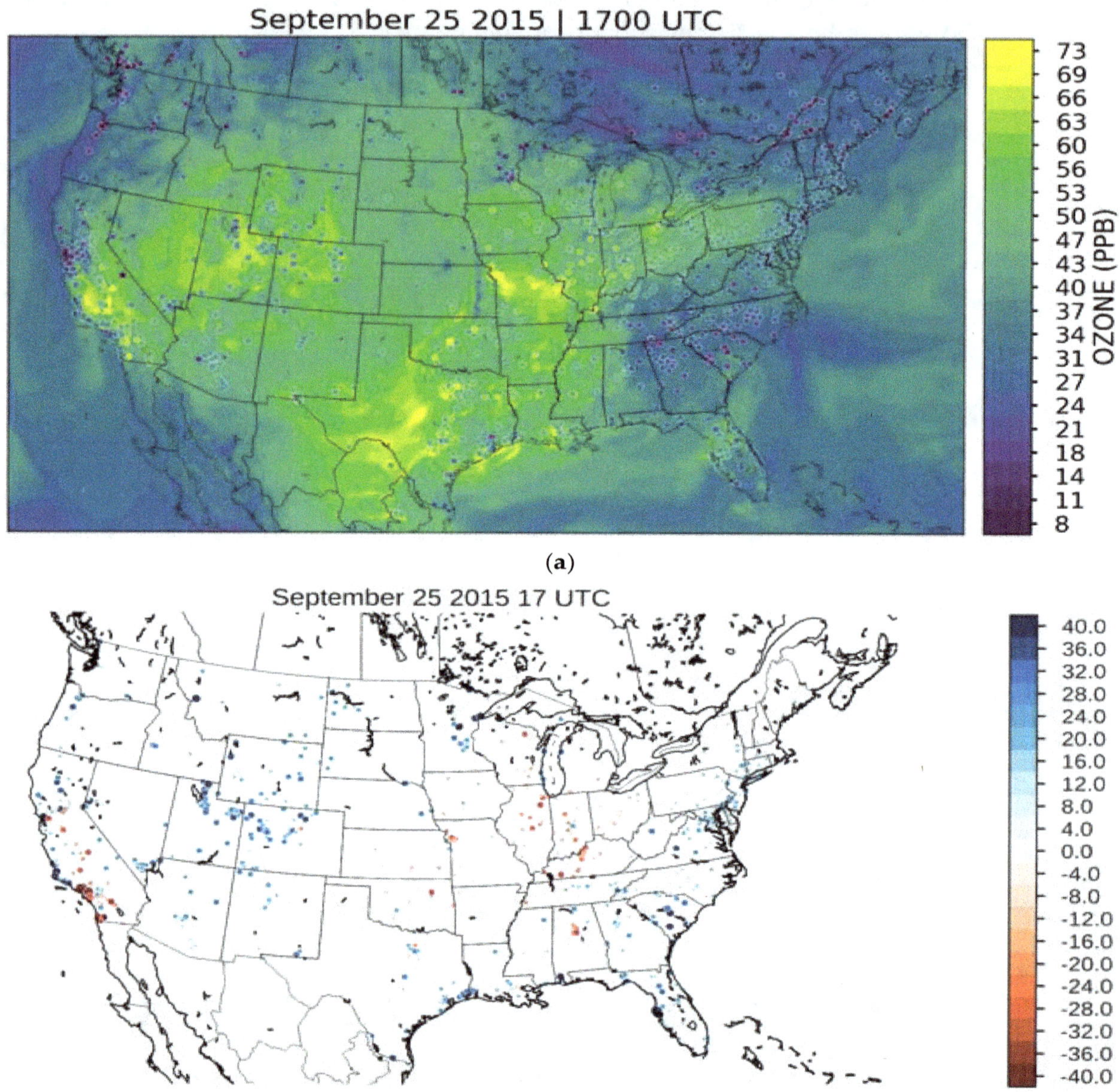

Figure 4. Examples of spatial plots. (**a**) Example of a spatial contour map with a discrete colorbar and observations overlaid; (**b**) Displayed is the normalized spatial O_3 bias (model–observation) at all of the monitors within the domain. The marker size is dependent on the absolute value of the normalized bias.

Figure 5a presents a scatter plot, while Figure 5b shows a difference scatter plot. Observations are displayed on the x-axis and model results on the y-axis. A one to one line is automatically created along with a line of best fit. The scipy linregress function creates the line of best fit. Currently, there is not a more complicated fitting algorithm available in MONET. The full power of scipy and scikit-learn is available however if a more in-depth analysis is needed and hopefully will be included in a future version of MONET. This type of plot is useful for determining the correlation of simulation to observational results.

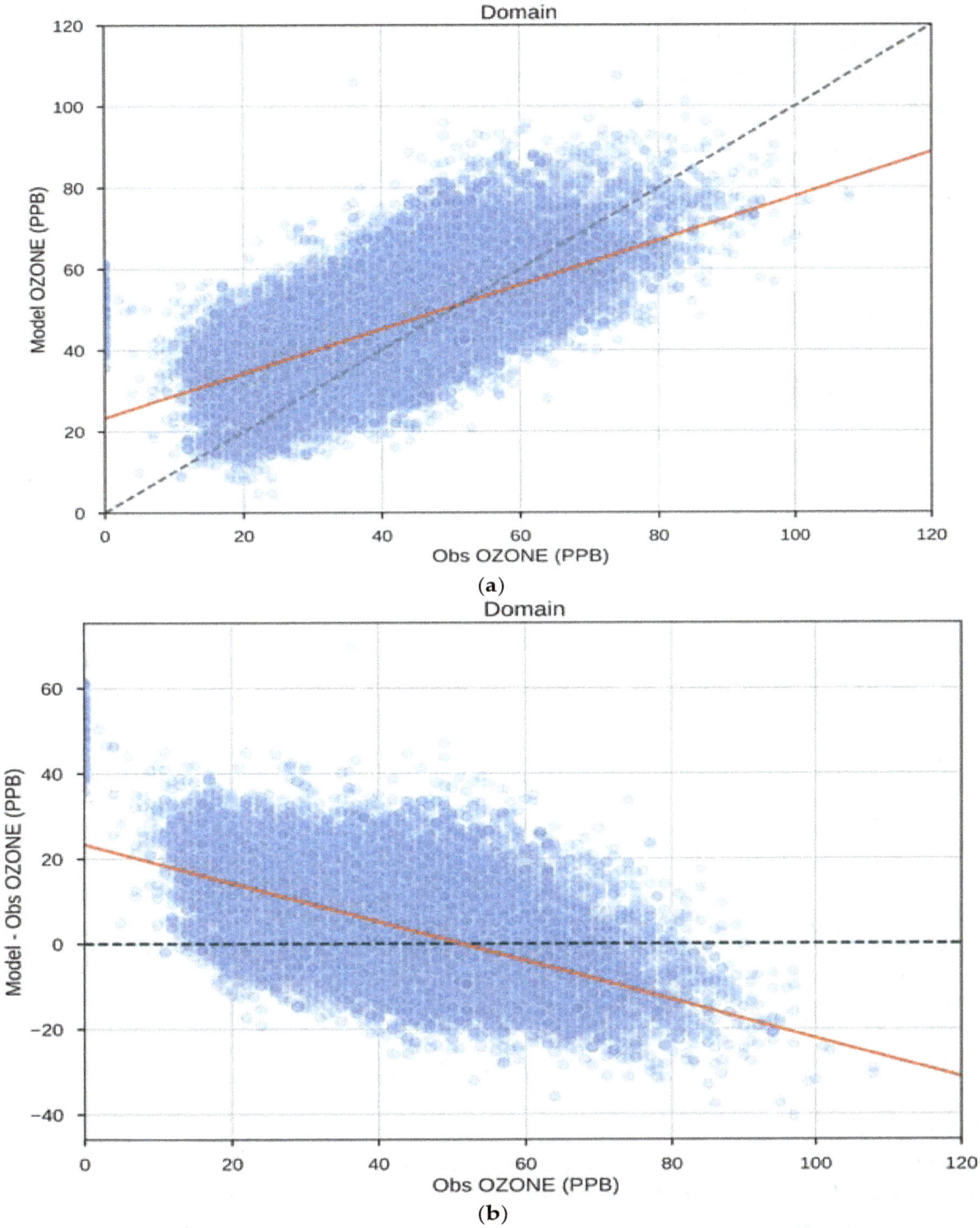

Figure 5. Examples of scatter plots in MONET (**a**) Scatter plot example of model versus ozone; (**b**) Scatter plot example of the difference in model and observation versus observations.

MONET uses the correlation, centered root mean square error, and the standard deviation measures to graphically display on its Taylor diagrams [14]. Taylor diagrams are an efficient way to graphically summarize model performance with observations. Currently, it is up to the user to ensure that data is close to a Gaussian distribution.

The relative merits of various model simulations can be inferred from Figure 6. Simulations that best compare to observations lie closest to the x-axis and have a low RMSE, while simulations that lie above the x-axis have a greater RMSE and lower correlation. The concentric arcs around the observation point show RMSE (in this graph at approximately 5.9 on the x-axis). Dispersion or variation within the simulations are graphically displayed by the distance from the concentric arcs. Simulations closest to the dotted arc have a standard deviation similar to observations.

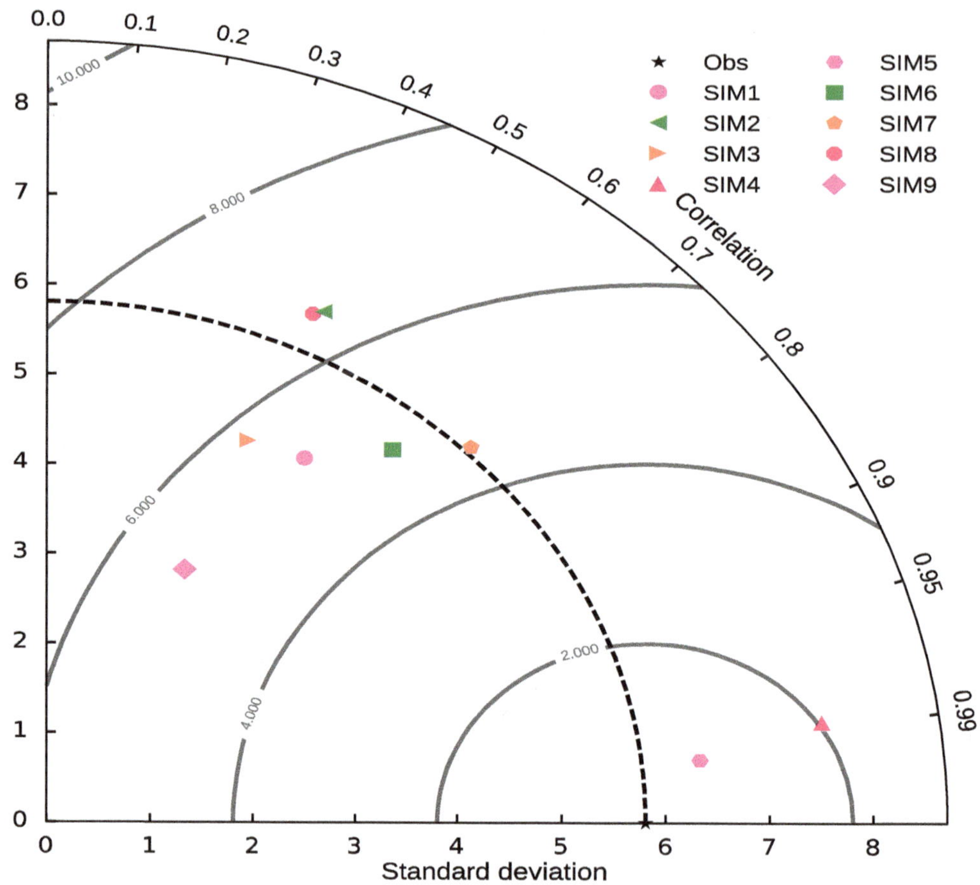

Figure 6. Taylor diagram example available in MONET. Taylor diagrams provide a concise visualization of multiple statistics on a single graph. In this example, the closer the simulation to the x-axis the better the correlation. The closer to the dotted line the better the model captured the variability of the model. The RMSE values follow the contours emanating from the x-axis.

In this example, SIM5 performed the best as it has the highest RMSE, correlation, and a respectable standard deviation as compared to observations. SIM9 has the lowest standard deviation and SIM2 and SIM8 have the lowest RMSE.

Different visualizations are being developed for future use. Specifically, cumulative distributions, spectral density analysis, and principle component analysis will be added. The addition of different observational data is also a high priority for use with MONET, including data from radiosondes, aircraft, ceilometers, and satellite data. MONET is a simple, easy to use object oriented approach for analyzing model results initially from NAQFC and more generally the CMAQ model.

4. Software and Data Availability

MONET is available on GitHub (https://github.com/noaa-oar-arl/MONET) as a free download. The verification package is written in Python v2.7+. The authors recommend the Anaconda Python package, available for free at https://www.continuum.io/downloads, to install supporting packages described in the readme file included at GitHub. As mentioned previously, MONET uses the Pandas python package to manipulate and store intermediate files. For one day of paired data, the file size is approximately 100 mb for all of the variables and meta-variables in AirNow within the NAQFC domain. MONET uses freely available observational data from the U.S. EPA, U.S. NOAA, U.S. NASA, and U.S. NPS.

5. Conclusions

MONET is a comprehensive software package used to analyze observational data alone or pair observations with gridded model data for chemical transport models. Currently, the observational datasets included are focused within North America, except for AERONET, which is global. MONET performs statistical calculations and creates visualizations to enable researchers to evaluate and improve the scientific understanding of model outputs and observational data. MONET is built entirely on open-source software and while the only model available for use is CMAQ used in the NAQFC, the software is easily able to include different models. To include a new model into MONET, a new object needs to be written that can: (1) read the output and (2) map of the output variables to observation variables needed to pair data; (3) provide an output latitude and longitude for interpolation. Work is ongoing to add the Comprehensive Air Quality Model with Extensions (CAMx) [15] model output to the MONET software package through the use of PseudoNetCDF (https://github.com/barronh/pseudonetcdf).

The MONET software package will continue to grow and improve through internal development at NOAA ARL, and external development through collaborating partners and contributions made through the community. The authors encourage collaborations with intergovernmental agencies, private industry, and academia to improve and further develop MONET. Major software developments are planned for the next release of MONET. One improvement is the inclusion of different models, such as the Weather Research and Forecasting (WRF), CAMx, and the Next Generation Global Prediction System (NGGPS). The next major development planned for MONET is to add different observational sources such as radiosonde, profiler, and radar observations from the Meteorological Assimilation Data Ingest System (MADIS; available online: http://madis.noaa.gov) and AWIPS II (available online: http://www.unidata.ucar.edu/software/awips2/). Another improvement planned is the use of pyresample to re-grid satellite swath data to model grids, as well as calculate comparable variables (e.g., column aggregated aerosol optical depth values from model output).

Analysis enhancements to the MONET software will include implementing non-parametric analysis techniques, such as cumulative distribution frequency plots, and Q-Q plots; adding temporal decomposition capabilities such as Kolmogorov-Zurbenko filtering (Rao and Zurbenko [16]; Wise and Comrie [17]) and spectral density plots; improving the overall speed through the use of parallel processing in pyresample for spatial re-gridding, xarray for temporal averaging, and dask for larger than memory computations. Many improvements to the MONET software are in progress and will be included in the next release.

Acknowledgments: The authors would like to thank the reviewers and editors for their dedication and valuable input in improving this manuscript.

Author Contributions: Barry Baker developed the infrastructure and implementation of MONET while Li Pan assisted in testing and evaluation of the NAQFC simulations.

References

1. George, B.J.; Schultz, B.D.; Palma, T.; Vette, A.F.; Whitaker, D.A.; Williams, R.W. An evaluation of EPA's National-Scale Air Toxics Assessment (NATA): Comparison with benzene measurements in Detroit, Michigan. *Atmos. Environ.* **2011**, *45*, 3301–3308. [CrossRef]

2. Fann, N.; Lamson, A.D.; Anenberg, S.C.; Wesson, K.; Risley, D.; Hubbell, B.J. Estimating the National Public Health Burden Associated with Exposure to Ambient $PM_{2.5}$ and Ozone. *Risk Anal.* **2011**, *32*, 81–95. [CrossRef] [PubMed]
3. Trasande, L.; Malecha, P.; Attina, T.M. Particulate Matter Exposure and Preterm Birth: Estimates of U.S. Attributable Burden and Economic Costs. *Environ. Health Perspect.* **2016**, *124*, 1913–1918. [CrossRef] [PubMed]
4. Ghude, S.D.; Jena, C.; Chate, D.M.; Beig, G.; Pfister, G.G.; Kumar, R.; Ramanathan, V. Reductions in India's crop yield due to ozone. *Geophys. Res. Lett.* **2014**, *41*, 5685–5691. [CrossRef]
5. Tong, D.; Mathur, R.; Schere, K.; Kang, D.; Yu, S. The use of air quality forecasts to assess impacts of air pollution on crops: Methodology and case study. *Atmos. Environ.* **2007**, *41*, 8772–8784. [CrossRef]
6. Dennis, R.; Fox, T.; Fuentes, M.; Gilliland, A.; Hanna, S.; Hogrefe, C.; Irwin, J.; Rao, S.T.; Scheffe, R.; Schere, K.; et al. A framework for evaluating regional-scale numerical photochemical modeling systems. *Environ. Fluid Mech.* **2010**, *10*, 471–489. [CrossRef] [PubMed]
7. Appel, K.W.; Gilliam, R.C.; Davis, N.; Zubrow, A.; Howard, S.C. Overview of the Atmospheric Model Evaluation Tool (AMET) V1.1 for Evaluating Meteorological and Air Quality Models. *Environ. Model. Softw.* **2011**, *26*, 434–443. [CrossRef]
8. Lee, P.; McQueen, J.; Stajner, I.; Huang, J.; Pan, L.; Tong, D.; Kim, H.; Tang, Y.; Kondragunta, S.; Ruminski, M.; et al. NAQFC developmental forecast guidance for fine particulate matter ($PM_{2.5}$). *Weather Forecast.* **2017**. [CrossRef]
9. Stajner, I.; Lee, P.; McQueen, J.; Draxler, R.; Dickerson, P.; Upadhayay, S. Update on NOAA's Operational Air Quality Predictions. In *Air Pollution Modeling and Its Application XXIV*; Steyn, D.G., Chaumerliac, N., Eds.; Springer: Cham, Switzerland, 2016; pp. 593–597.
10. McKinney, W. Data Structures for Statistical Computing in Python. In Proceedings of the 9th Python in Science Conference, Austin, TX, USA, 28 June–3 July 2010; pp. 51–56.
11. McKinney, W. *Python for data analysis: Data wrangling with Pandas, NumPy, and IPython*; O'Reilly Media, Inc.: Sebastopol, CA, USA, 2012.
12. Van der Walt, S.; Colbert, S.C.; Varoquaux, G. The NumPy Array: A Structure for Efficient Numerical Computation. *Comput. Sci. Eng.* **2011**, *13*, 22–30. [CrossRef]
13. Hunter, J.D. Matplotlib: A 2D graphics environment. *Comput. Sci. Eng.* **2007**, *9*, 90–95. [CrossRef]
14. Taylor, K.E. Summarizing multiple aspects of model performance in a single diagram. *J. Geophys. Res. Atmos.* **2001**, *106*, 7183–7192. [CrossRef]
15. ENVIRON. *User's Guide, Comprehensive Air Quality Model with Extensions (CAMx), Version 5.40*; ENVIRON: Arlington, TX, USA, 2011.
16. Rao, S.T.; Zurbenko, I.G. Detecting and Tracking Changes in Ozone Air Quality. *Air Waste* **1994**, *44*, 1089–1092. [CrossRef] [PubMed]
17. Wise, E.K.; Comrie, A.C. Extending the Kolmogorov–Zurbenko Filter: Application to Ozone, Particulate Matter, and Meteorological Trends. *J. Air Waste Manag. Assoc.* **2005**, *55*, 1208–1216. [CrossRef] [PubMed]

9

Pollution Characteristics of Particulate Matter ($PM_{2.5}$ and PM_{10}) and Constituent Carbonaceous Aerosols in a South Asian Future Megacity

Afifa Aslam [1], Muhammad Ibrahim [1], Imran Shahid [2], Abid Mahmood [1], Muhammad Kashif Irshad [1], Muhammad Yamin [3], Ghazala [1], Muhammad Tariq [1] and Redmond R. Shamshiri [4,*]

1 Department of Environmental Sciences & Engineering, Government College University Faisalabad, Faisalabad 38000, Pakistan; afifaaslam22@gcuf.edu.pk (A.A.); mibrahim@gcuf.edu.pk (M.I.); drabid@gcuf.edu.pk (A.M.); kashifirshad@gcuf.edu.pk (M.K.I.); ghazala@gcuf.edu.pk (G.); tariqnazir159@gmail.com (M.T.)
2 Environmental Science Centre, Qatar University, Doha P.O. Box 2713, Qatar; ishahid@qu.edu.qa
3 Department of Farm Machinery & Power, Faculty of Agricultural Engineering & Technology, University of Agriculture, Faisalabad 38040, Pakistan; yamin529@uaf.edu.pk
4 Leibniz Institute for Agricultural Engineering and Bioeconomy, Max-Eyth-Allee 100, 14469 Potsdam-Bornim, Germany
* Correspondence: rshamshiri@atb-potsdam.de.

Abstract: The future megacity of Faisalabad is of prime interest when considering environmental health because of its bulky population and abundant industrial and anthropogenic sources of coarse particles (PM_{10}) and fine airborne particulate matter ($PM_{2.5}$). The current study was aimed to investigate the concentration level of $PM_{2.5}$ and PM_{10}, also the characterization of carbonaceous aerosols including organic carbon (OC), elemental carbon (EC) and total carbon (TC) in $PM_{2.5}$ and PM_{10} samples collected from five different sectors (residential, health, commercial, industrial, and vehicular zone). The data presented here are the first of their kind in this sprawling city having industries and agricultural activities side by side. Results of the study revealed that the mass concentration of $PM_{2.5}$ and PM_{10} is at an elevated level throughout Faisalabad, with ambient $PM_{2.5}$ and PM_{10} points that constantly exceeded the 24-h standards of US-EPA, and National Environment Quality Standards (NEQS) which poses harmful effects on the quality of air and health. The total carbon concentration varied between 21.33 and 206.84 $\mu g/m^3$, and 26.08 and 211.15 $\mu g/m^3$ with an average of 119.16 ± 64.91 $\mu g/m^3$ and 124.71 ± 64.38 $\mu g/m^3$ for $PM_{2.5}$ in summer and winter seasons, respectively. For PM_{10}, the concentration of TC varied from 34.52 to 289.21 $\mu g/m^3$ with an average of 181.50 ± 87.38 $\mu g/m^3$ (for summer season) and it ranged between 44.04 and 300.02 $\mu g/m^3$ with an average of 191.04 ± 87.98 $\mu g/m^3$ (winter season), respectively. No significant difference between particulate concentration and weather parameters was observed. Similarly, results of air quality index (AQI) and pollution index (PI) stated that the air quality of Faisalabad ranges from poor to severely pollute. In terms of AQI, moderate pollution was recorded on sampling sites in the following order; Ittehad Welfare Dispensary > Saleemi Chowk > Kashmir Road > Pepsi Factory, while at Nazria Pakistan Square and Allied Hospital, higher AQI values were recorded. The analysis and results presented in this study can be used by policy-makers to apply rigorous strategies that decrease air pollution and the associated health effects in Faisalabad.

Keywords: particulate matter; aerosols; vehicular exhaust; industrial activity; WHO; NEQS; US-EPA

1. Introduction

In many developing countries, increasing industrialization and overpopulation becomes the reason for escalating air pollution [1]. According to various researches conducted in high-income countries situated in the Asian region, the level of many air pollutants is normally beyond the ambient air quality standards and WHO guidelines. In many developing countries, the use of non-renewable fuel like biomass and diesel is associated with the increasing level of air pollution at the regional level. Airborne particulate matter is abundant in the atmosphere and is the foremost indicator of the quality of air in a specified area. Chemical composition, concentration and size of particulate matter varied widely and are delimited universally under acceptable standards built on size elements ranging from $PM_{2.5}$ to PM_{10} to Total Suspended Particles (TSP), while PM_4 was also identified as the respirable size fraction [2]. Particulate matter instigates from a diversity of anthropogenic (e.g., rapid industrialization, agricultural activities, refineries, waste incineration, biomass burning, motor vehicles, utilities, brick kiln, industrial emissions power plants, factories, large population and heavy traffic) are responsible for bad air quality in the cities due to elevated levels of gaseous and particulate pollutants [3] and natural (e.g., dust storm and sea spray) sources, besides secondary formation processes. However, also mineral dust transport from deserted areas is considered a significant source for regional pollution in Asia [4]. Henceforth, for the air quality management and epidemiological studies, the assessment of the concentration of atmospheric particulate matter (PM) and its associated toxic constituents is a prerequisite [5]. It was consistently confirmed by epidemiological studies that there is a strong association between ambient particulate matter comprising toxic components and cardiovascular- and respiratory-related upsurges in mortality and morbidity, particularly in urban areas [6]. This connection has been revealed to be stronger for $PM_{2.5}$ rather than for PM_{10} or total suspended particles since $PM_{2.5}$ can infiltrate deep into the alveolar areas of the human lungs [6]. The transport and distribution of particulate matter in the atmosphere are distinctly allied with meteorological parameters such as air temperature, relative humidity, atmospheric pressure, wind direction, speed, and rainfall [7]. In various parts of the world, different monitoring programs on atmospheric PM have been directed which exposed varied instabilities and disproportions among the trace element constituents and particulate matter [8].

Components of carbonaceous aerosol, elemental carbon (EC) and organic carbon (OC), account for a large element of atmospheric particulate matter and, on average, subsidize 20–35% of coarse particulate and 20–45% of fine particulate [9]. Carbonaceous aerosols have a chief role in the interactions of light-particles within the atmosphere and are one of the significant components of fine and coarse particulate matter; they are therefore associated with the negative climatic and environmental impacts and the worsening in public health and air quality [10]. Elemental carbon is often used as a substitute for black carbon (BC) and is discharged into the atmosphere mostly through the processes of combustion [11]. Elemental carbon is primarily accountable for the absorption of light in the atmosphere, which sturdily influence the radiative balance of the earth [12]. The six main sources of elemental carbon have been recognized using organic tracers as coal combustion, biomass burning, vehicle exhaust, cigarette smoke, cooking and vegetative detritus [13]. Carbonaceous aerosols were found dominant in $PM_{2.5}$ (which is attained from agricultural waste and wood-fuel burning) and have a strong effect on the decline in visibility and air-quality and also stimulates radiative forcing on a regional scale [14].

In Pakistan, control of air pollution has not yet become a democratic issue because of a lack of suitable information for policy and decision-makers, though some infrequent reports that identify airborne particulate matter as a great health and environmental concern in urban regions of Pakistan are present [15]. Generally, the concentration of particulate matter is many folds higher than the acceptable limits documented by the World Health Organization (WHO), National Environmental Quality Standards (NEQS) and the United States Environmental Protection Agency (US-EPA). According to the World Bank [16], the annual burden of health because of particulate matter was 1% of the GDP and is accountable for 700 deaths among children and 22,000 premature deaths amongst adults in Pakistan. However, due to the absence of air quality management competencies, the country is suffering from the deterioration of air quality. Evidence from many international bodies and governmental organizations has indicated that air pollution is a momentous risk to the health of residents, environment and quality of life [17]. According to a study directed by the World Health Organization, Bombay, Calcutta and Tehran were found to be the most contaminated cities in Asia [18]. Similarly, Faisalabad (the textile city of Pakistan) is also highlighted to be an extremely polluted city in this study. Due to increased industrialization and construction of commercial zones and rapid urbanization, the atmosphere of the city is getting worst day by day [19]. This state of concern stimulated us to conduct a comprehensive study on the status of air pollution in Faisalabad. As a result of the burning issue of air pollution and associated health impacts, a study was planned to examine the quality of air in Faisalabad city for which 12 different sites were selected and categorized as residential, commercial, industrial, and health centers.

Keeping in view the facts discussed above, the present study was conducted with the following objectives: (a) to measure the quality of air with its allied consequences within varying activity zones of Faisalabad city; (b) to compare the ambient air quality of Faisalabad with air pollution indexes such as NEQS -Pakistan, National Ambient Air Quality Standards (NAAQS)-US-EPA and WHO; and (c) to provide an opportunity to conduct additional studies on source identification, impact assessment, and trend analysis for this zone. It is expected that the current study will be supportive for designing and establishing emission regulations and abatement strategies in the future.

2. Methodology

2.1. Study Area and Sampling Sites

Faisalabad is the third-largest city of Pakistan and a major industrial hub (dominated by textile and chemical industries); consequently, the air quality of the city is a major environmental problem. It covers an area of 1230 km^2 and is occupied by more than four million people. The summer season is very hot with a humid climate while a cold winter (falls to 0 °C some days) is experienced by the Faisalabad city. The climate of the city touches extreme hotness and humidity during summer and cold during winter. The sampling sites were banquets around Faisalabad and its vicinities. Twelve sampling sites were nominated based on current anthropogenic activities accountable for atmospheric pollution, and the dominant direction of wind for pollutant dispersion and distribution in the area. The locations were selected based on the zones in the city. The selected locations are comprised of medical units, residential areas, commercial areas, industrial areas, and automobiles rich areas. The average wind speed of 3–6 km in winter and 6–13 km in summer was observed. The map showing the locations on the Faisalabad (Figure 1) represents the coordinates of the location within Faisalabad geography.

2.2. Data Collection

Data of meteorological parameters were obtained from Agromet. The PM concentrations were determined by the first author herself. We took the samples from all the locations and then measurements were made in the Lab. We took sample readings sector-wise and readings for all the residential sites were taken at the same time. A similar trend was followed for commercial, industrial, health, and automobile sites. At Provincial and Federal EPAs, Data Logging systems retrieve the data about the quality of ambient air from air monitoring stations with the help of data processing software. The seasonal average was intended to find out the difference in the mass concentration of $PM_{2.5}$ in summer and winter seasons. For this study, 12 discrete sampling sites under five diverse sectors (residential, health, commercial, industrial, and automobile vehicles) were selected for the evaluation of $PM_{2.5}$ and PM_{10} with the help of high air volume sampler. The interpretations were taken at three diverse times (morning, noon, and evening) daily from November 1 to December 31 for winter and from May 1 to June 30 for summer. It should be noted that wind speed and direction influence the rate of diffusion of pollution. The temperature inversion is also directly linked to solar radiation making the air softer, hence the air converts into fog because pollutants and dust are no longer raised from the surface. This can become a problem in metropolises where numerous pollutants exist. The data of temperature and radiation used in this study was collected from the Agricultural Metrological Cell Agromet Bulletin.

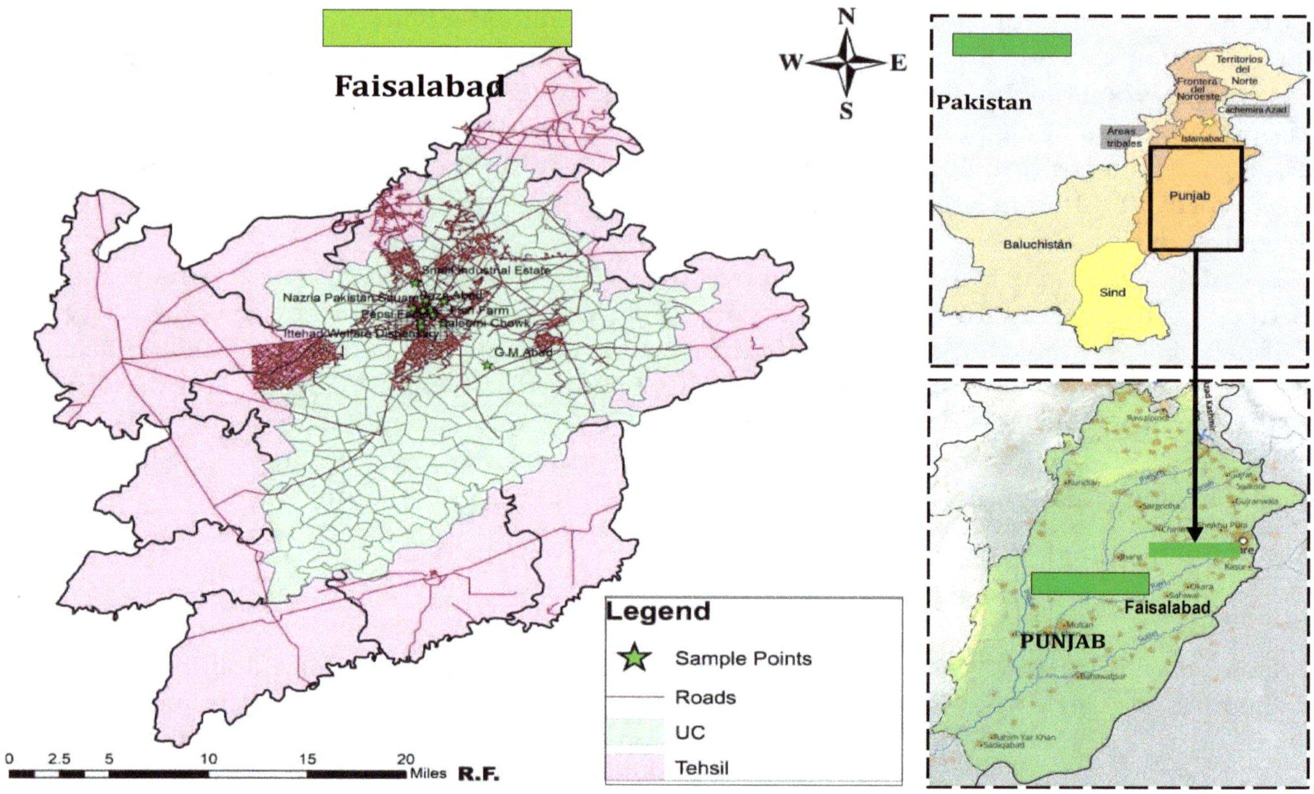

Figure 1. Study sites within the Faisalabad city and location of the city within the political map of Pakistan.

2.3. Sample Analysis (Chemical, Gravimetric, and Carbonaceous Aerosols)

After sample collection, the filter papers were kept in exact environmental conditions at a relative humidity of 30–40% and temperature of 20–23 °C for 24 h as per the US-EPA standard. Before mass analysis, the filter paper with fine and coarse particulates samples was equilibrated for 24 h in silica gel desiccators to abolish the effect of humidity and to attain accurate particulate matter measurements. The $PM_{2.5}$ and PM_{10} masses of each sample were determined gravimetrically by deducting the initial average mass of the blank filter from the final average mass of the sampled filter. Gravimetric analysis is the determination of particulate concentration based on weight difference. Individual filters (Teflon®, 46.2 mm) were weighed on an electronic micro-balance pre and post field sampling. Particulate matter <2.5 μm was collected from ambient air on the filters throughout the sample duration of 24 h. The net variances between pre- and post-sampling filter weights were used to estimate the mass concentration in the ambient air of the city. After post weighing, filters can be stored for a minimum of one year. Using the post-sample and pre-sample filter weights, the total filter mass gain ($PM_{2.5}$) and the concentration of PM_{10} were respectively calculated from Equation (1) and Equation (2):

$$PM_{2.5} = (M_{Post} - M_{Pre})(10^3)/M_{Pre} \tag{1}$$

$$PM10 = (Wf - Wi)(106)/V \tag{2}$$

Here $PM_{2.5}$ is the total mass gain in μg, M_{Post} is the post sample filter weight in mg, M_{Pre} is the pre-sample filter weight in μg, W_f is the filter paper weight, W_i is the initial mass of filter paper, and V is the total air sampled in m^3. In the current study, elemental carbon was determined by a two-step combustion method described by [20]. Filters were heated for 2 h at 340 °C in an oxygen atmosphere to remove organic carbon (OC). The calibration procedure was done using tartaric acid dyed in aluminum foil. While, total carbon was determined by a combustion method, where all material on the filter is combusted in pure oxygen at 1000 °C and the resulting CO_2 is measured by non-dispersive IR photometry (NDIR, Maihak) [21].

2.4. Air Quality and Pollution Index

An Air Quality Index is defined as a complete scheme that converts the weighed values of parameters related to individual air pollution (e.g., the concentration of pollutant) into a sole number or set of numbers [22]. Air Quality Index (AQI) is a tool to detect the present scenario of air quality. AQI was calculated based on the arithmetic mean of the ratio of the concentration of pollutants to the standard value of that pollutant such as PM_{10}, $PM_{2.5}$, NO_2, and SO_2. The average is then multiplied by 100 to arrive at the AQI index. The pollutant AQI and the pollution index (PI) of the potentially noxious element were respectively derived from Equations (3) and (4):

$$AQI = (W * C/Cs) \tag{3}$$

$$PI = Cn/Bn \tag{4}$$

where W is the pollutant weighted, C is the observed value ($PM_{2.5}$, PM_{10}, SO_2 and NO_2), Cs is the CPCB standard for the residential area [23], B_n is the background concentration, and C_n is the measured concentration of the element. It should be noted that in Pakistan, the National Air Quality Index is followed. Moreover, it should be highlighted that the pollution index of the potentially contaminated elements is the ratio between the concentration of toxic elements and the reference background concentration of the consistent elements obtained from a previous published study [24].

3. Results

3.1. Mass Concentration of $PM_{2.5}$ and PM_{10} in Winter Season

The results of the present investigations in Faisalabad city for which 12 different sites were selected and categorized as residential, commercial, industrial and health centers are presented (Figure 2). The concentration of $PM_{2.5}$ had the following decreasing order in the air samples collected near medical centers of Mian Trust Hospital (38.50 ± 0.30 μg/m^3), Ittehad Welfare Dispensary (37.35 + 0.45 μg/m^3), Allied Hospital (36.65 ± 0.27 μg/m^3). While, in residential areas, the highest concentration of fine particulate was found in Ghulam Muhammad Abad (39.1 ± 0.50 μg/m^3) followed by Raza Abad (35.2 ± 0.23 μg/m^3), Saleemi Chowk (33.83 ± 0.74 μg/m^3), Kashmir Road (32.25 ± 0.14 μg/m^3). In the selected commercial areas, Nazria Pakistan (43.63 ± 0.59 μg/m^3) was the most polluted site of the city. The average concentration of $PM_{2.5}$ in small industrial estate was recorded as (37.93 ± 0.19 μg/m^3) followed by Pepsi factory area (37.17 ± 0.62 μg/m^3); while in and near transport station (35.08 ± 0.61 μg/m^3) was analyzed. In contrast, the highest concentration level of PM_{10} was found in the samples collected from Nazria Pakistan (800.85 ± 0.93 μg/m^3) followed by Mian Trust Hospital (586.6 ± 3.88 μg/m^3), Allied Hospital (584.62 ± 3.41 μg/m^3), Small Industrial Estate (469.1 ± 0.57 μg/m^3) and Ghulam Muhammad Abad (440.2 ± 0.10 μg/m^3); while the lowest concentration was analyzed in the ambient air of Saleemi Chowk (280.18 ± 0.12 μg/m^3). The average particulate concentration can be compared with guideline values proposed by WHO, US-EPA and NEQS. During the present study, the concentration of $PM_{2.5}$ was within the guideline value of the US-EPA and NEQS (35 μg/m^3) but still higher than the WHO guideline value of 25μg/m^3 [25] for 24 h average, while coarse particles were drastically exceeding the safe limits of all the quality standards. The composition of particulate matter is strongly reliant with its sources, i.e., anthropogenic or natural (Figure 2).

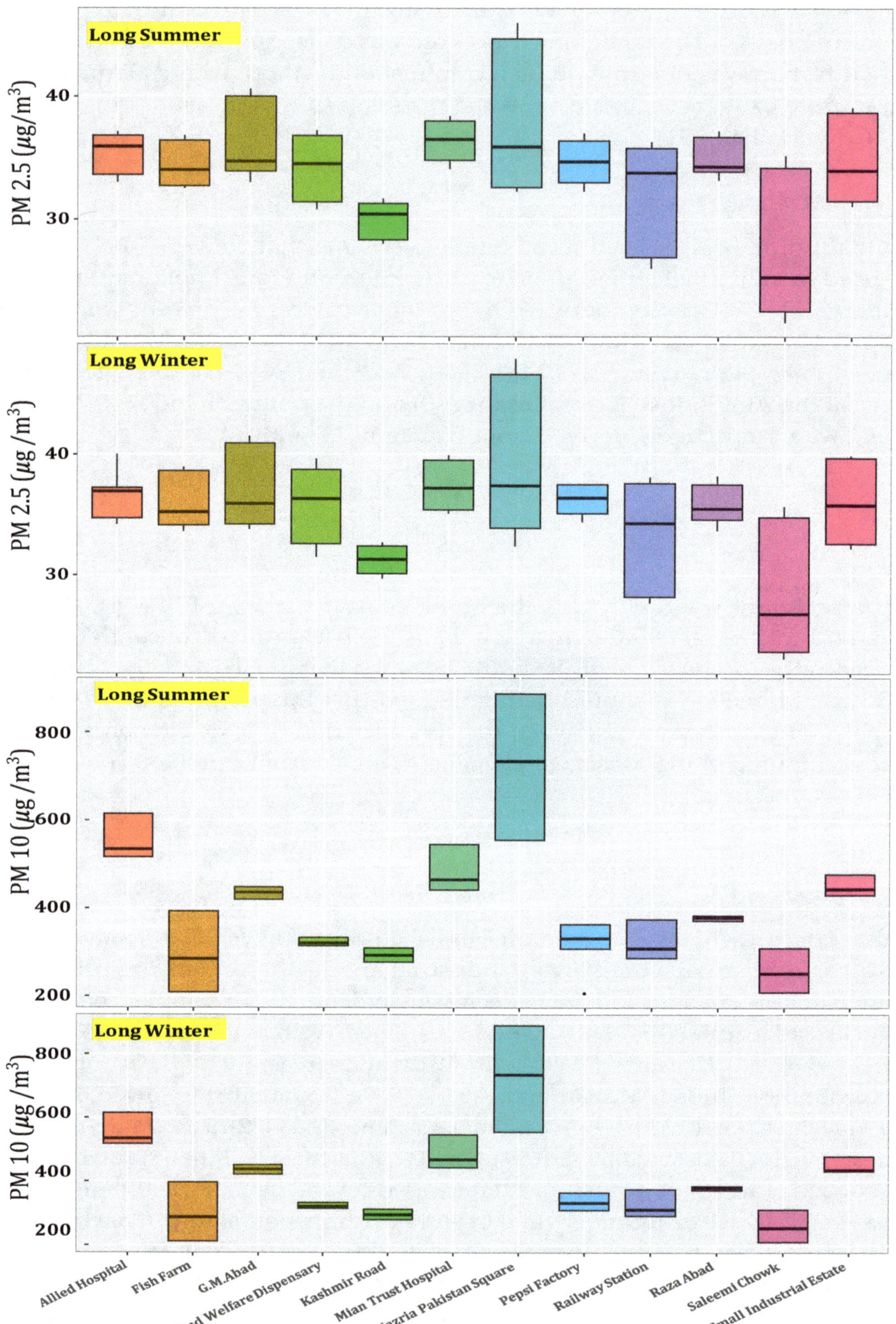

Figure 2. Average mass concentrations of $PM_{2.5}$ and PM_{10} in long summer and long winter in Faisalabad city.

3.2. Mass Concentration of $PM_{2.5}$ and PM_{10} in Summer Season

Figure 2 shows the mass concentration of fine ($PM_{2.5}$) and coarse (PM_{10}) particles calculated from the samples collected during the winter (Dec 2016 to Jan 2017) and summer (May to June 2017) seasons from different selected areas of Faisalabad. It was clear from the results that the concentration levels of both particulates were lower in the summer season as compared to the winter season. According to the results, it was cleared that the contamination level that the concentration level of both particulates was lower in the summer season as compared to the winter season. According to the results, the highest concentration of $PM_{2.5}$ was found in the air samples collected from Nazria Pakistan (42.5 ± 0.57 μg/m^3) followed by a Ghulam Muhammad Abad (38.4 ± 0.34 μg/m^3), Mian Trust Hospital and Small Industrial Estate air samples (36.9 ± 0.34 μg/m^3). While, the mass volume of $PM_{2.5}$ was lower in the air samples of Station Chowk (33.5 ± 0.35 μg/m^3), Raza Abad (34.2 ± 0.28 μg/m^3), Saleemi Chowk (31.8 ± 0.94 μg/m^3) and Kashmir Road (31.2 ± 0.21 μg/m^3) when compared with US-EPA and NEQS guidelines rather than WHO safe limits. The rest of the areas were slightly higher in $PM_{2.5}$ concentrations than US-EPA and NEQS safe limits but still highly polluted if compared with WHO guidelines. Table 1 represents the concentration values of coarse particles obtained after analysis. The decreasing order was followed as Nazria Pakistan (800.6 ± 2.16 μg/m^3), Allied Hospital (477.4 ± 3.7 μg/m^3), Mian Trust Hospital (477.3 ± 0.82 μg/m^3), Small Industrial Estate (465.4 ± 1.16 μg/m^3), Ghulam Muhammad Abad (440.1 ± 0.13 μg/m^3), Raza Abad (380.1 ± 0.08 μg/m^3) with the lowest value obtained at Kashmir Road (297.3 ± 0.50 μg/m^3). According to the results, the mass concentration of PM_{10} is exceeding the safe guidelines of all the selected air quality standards throughout the study area.

Table 1. (National Air Quality Index, CPCB, October 2014).

Category	Range
Good	0–50
Satisfactory	51–100
Moderately Polluted	101–200
Poor	201–300
Very Poor	301–400
Severe	401–500

3.3. Seasonal Impact on $PM_{2.5}$ and PM_{10} Concentration

Figure 2 represents the average concentration of $PM_{2.5}$ and PM_{10} in the long summer and winter seasons during the study period. Figure 2 shows the highest fine particulates concentration in a commercial area on the average 39.18 ± 4.70 μg/m^3 and 40.73 ± 2.9 μg/m^3 with the lowest obtained concentration in the residential sector ranging from 33.91 ± 3.27 μg/m^3 and 35.1 ± 2.93 μg/m^3 for summer and winter season, respectively. The WHO safe limit for $PM_{2.5}$ is 25 μg/m^3 and for PM_{10} is 50 μg/m^3 (WHO, 2005). Similarly, the US-EPA and NEQS safe limit for $PM_{2.5}$ is 35 μg/m^3 and PM_{10} is 150 μg/m^3 [26,27]. Coarse particles (PM_{10}) were similar in trend as shown by $PM_{2.5}$ with decreasing trend as commercial areas, hospital areas, industrial areas, automobile station and residential areas were in the range of 575.19 ± 66.26 μg/m^3, 499.81 ± 148.62 μg/m^3, 409.63 ± 59.44 μg/m^3, 379.63 ± 0.81 μg/m^3 and 350.98 ± 74.29 μg/m^3 in winter and 573.14 ± 321.64 μg/m^3, 427.52 ± 86.30 μg/m^3, 405.63 ± 84.49 μg/m^3, 349.41 ± 74.48 μg/m^3 and 349.46 ± 74.49 μg/m^3 in the summer season, respectively. Correspondingly, Figure 2 showed a strong positive correlation between $PM_{2.5}$ and PM_{10} in both winter and summer seasons on average.

3.4. Analysis of Carbonaceous Aerosols in fine ($PM_{2.5}$) and Coarse Particulate (PM_{10}) Samples

The concentration level of carbonaceous aerosols is presented in Figures 3–5 for summer and winter seasons. It was clear that concentration of EC and OC was higher in winter (Figures 3 and 4) which was quite similar with previous studies. According to the results, EC was found in low concentration on average in the samples of fine particles collected from Kashmir Road (8.56 ± 1.86 μg/m^3) in summer season while highest EC contamination was found in the ambient air of Nazria Pakistan (89.67 ± 1.52 μg/m^3). A similar trend was found for OC with the lowest concentration in the $PM_{2.5}$ samples collected from Kashmir Road (19.93 ± 0.42 μg/m^3) categorized as one of the residential sites, while the highest values were obtained in the ambient air of Nazria Pakistan (178.4 ± 3.51 μg/m^3) nominated as the busiest commercial zone of Faisalabad with a variety of businesses (Figure 4).

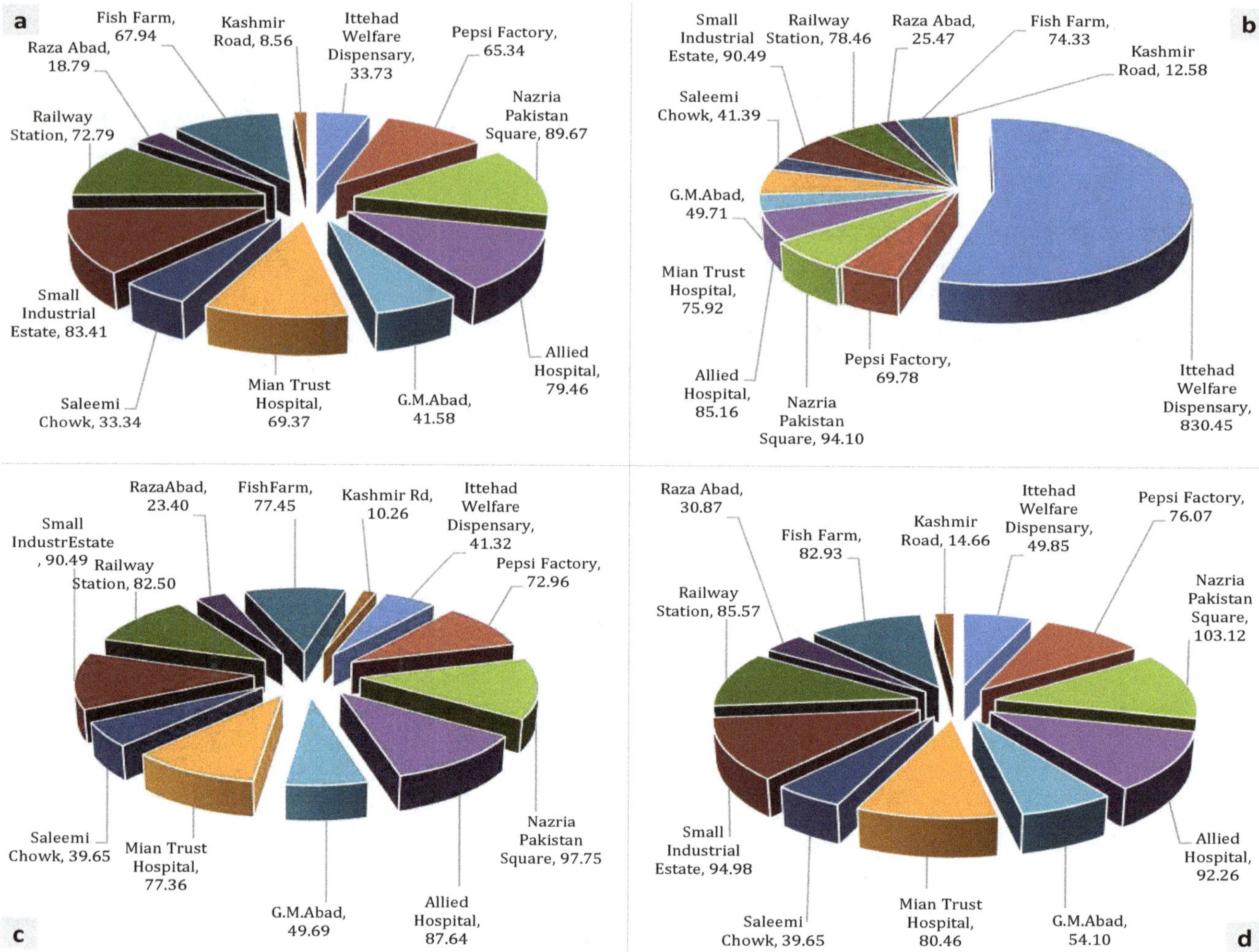

Figure 3. Concentration of carbonaceous species—elemental carbon (EC) in $PM_{2.5}$ in summer (**a**); $PM_{2.5}$ in winter (**b**); PM_{10} in summer (**c**); PM_{10} in winter (**d**). All units are expressed in μg/m^3.

When considering TC for the investigated sites (Figure 5), it was observed that Nazria Pakistan was heavily contaminated (268.08 ± 5.03 μg/m^3) followed by Small Industrial Estate (248.23 ± 5.79 μg/m^3), Fish Farm (235.68 ± 5.02 μg/m^3), Allied Hospital (224.34 ± 3.62 μg/m^3), Vehicular station (223.21 ± 3.82 μg/m^3), Mian Trust Hospital (202.77 ± 8.42 μg/m^3) and Pepsi Factory (184.84 ± 4.27 μg/m^3) with positive OC/EC correlation which indicates the common source of emission of TC in these zones. While, residential areas (G.M Abad, Saleemi Chowk, Raza Abad and Kashmir Road) were less contaminated (114.71 ± 2.76 μg/m^3; 64.97 ± 3.58 μg/m^3; 51.16 ± 3.14 μg/m^3; and 28.49 ± 2.27 μg/m^3, respectively), as compared to the other sites indicating negative OC/EC correlation. While samples of fine particulates collected in the winter season from the same investigating sites were analyzed for carbonaceous aerosol concentration. The levels of EC and OC were higher in winter as compared to the summer season. This may be due to the more wood and fossil fuel burning to warm up the surroundings as well as extra consumption of diesel and petrol by vehicles to warm up the engines in sizzling cold weather. The concentration of TC was much higher in the ambient air of Nazria Pakistan (277.5 ± 4.9 μg/m^3) followed by Small Industrial Estate (262.01 ± 3.68 μg/m^3), Fish Farm (248.07 ± 5.28 μg/m^3), Allied Hospital (237.13 ± 2.89 μg/m^3), Vehicular Station (234.4 ± 4.08 μg/m^3), Mian Trust Hospital (215.97 ± 8.82 μg/m^3), Pepsi Factory (194.27 ± 4.30 μg/m^3) and Ittehad Welfare Dispensary (142.58 ± 7.69 μg/m^3). On the other side, mix community of domestic zones showed less concentration of both EC and OC but still falls in contamination categories that are not safe for human health. Kashmir Road was detected with the least concentration of TC (38 ± 3.43 μg/m^3) while Ghulam Muhammad Abad was higher in TC (130.04 ± 2.6 μg/m^3) concentration in the residential zone. A positive OC/EC correlation was observed in the maximum of the investigating sites in the winter season. Samples of coarse particulate were also analyzed to evaluate the concentration level of carbonaceous aerosols collected from the investigating sites of Faisalabad city for summer and winter seasons, respectively.

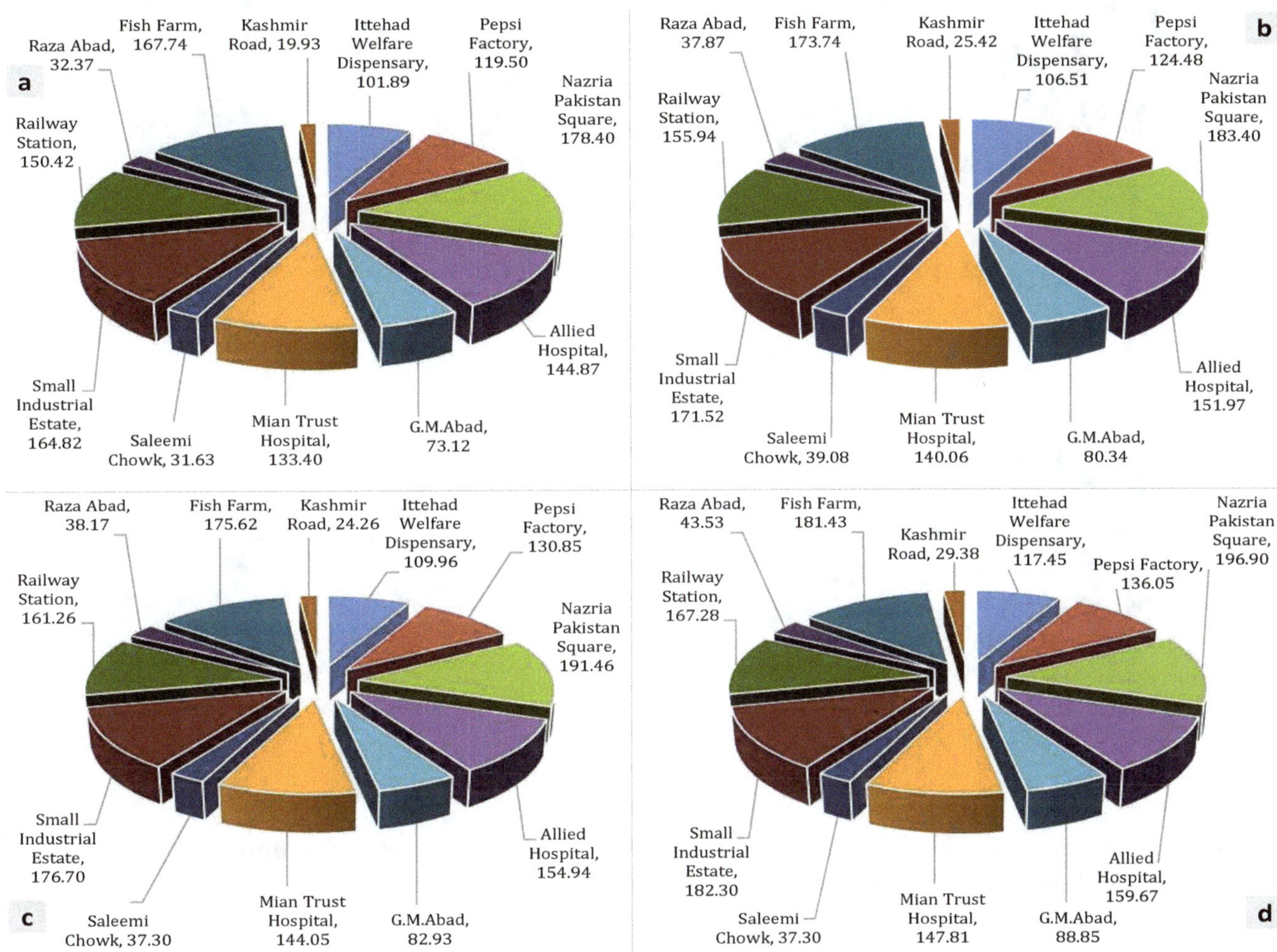

Figure 4. Concentration of carbonaceous species—organic carbon (OC) in $PM_{2.5}$ in summer (**a**); $PM_{2.5}$ in winter (**b**); PM_{10} in summer (**c**); PM_{10} in winter (**d**). All units are expressed in μg/m^3.

We found similar trends of concentration level as experienced with aerosols available in fine particulates but higher in concentration than observed in $PM_{2.5}$ samples. Commercial areas of Faisalabad were enriched with TC (289.21 ± 2.75 μg/m^3 and 300.02 ± 3.25 μg/m^3 for Nazria Pakistan and 253.06 ± 5.59 μg/m^3 and 264.36 ± 4.16 μg/m^3 for Fish Farm) at an elevated level among all the sites. Coarse particulate samples collected from the Small Industrial Estate were also found to be extremely high (267.19 ± 4.28 μg/m^3 and 277.28 ± 3.16 μg/m^3) after Nazria Pakistan followed by Vehicular Station (243.75 ± 3.66 μg/m^3 and 252.85 ± 2.9 μg/m^3), Allied Hospital (242.58 ± 5.24 μg/m^3 and 251.93 ± 4.44 μg/m^3), Mian Trust Hospital (221.4 ± 6.51 μg/m^3 and 228.27 ± 5.53 μg/m^3) and Ittehad Welfare Dispensary (151.28 ± 12.09 μg/m^3 and 167.31 ± 4.57 μg/m^3) for summer and winter seasons, respectively. Concentration level of carbonaceous aerosol was higher in Ghulam Muhammad Abad (132.62 ± 3.4 μg/m^3 and 142.95 ± 3.65 μg/m^3) while considering the residential zone of the city followed by Saleemi Chowk (76.95 ± 4.87 μg/m^3 and 76.95 ± 3.9 μg/m^3), Raza Abad (61.57 ± 3 μg/m^3

and 74.4 ± 2.32 μg/m^3) and Kashmir Road (34.52 ± 2.02 μg/m^3 and 44.04 ± 1.97 μg/m^3) for both the seasons accordingly.

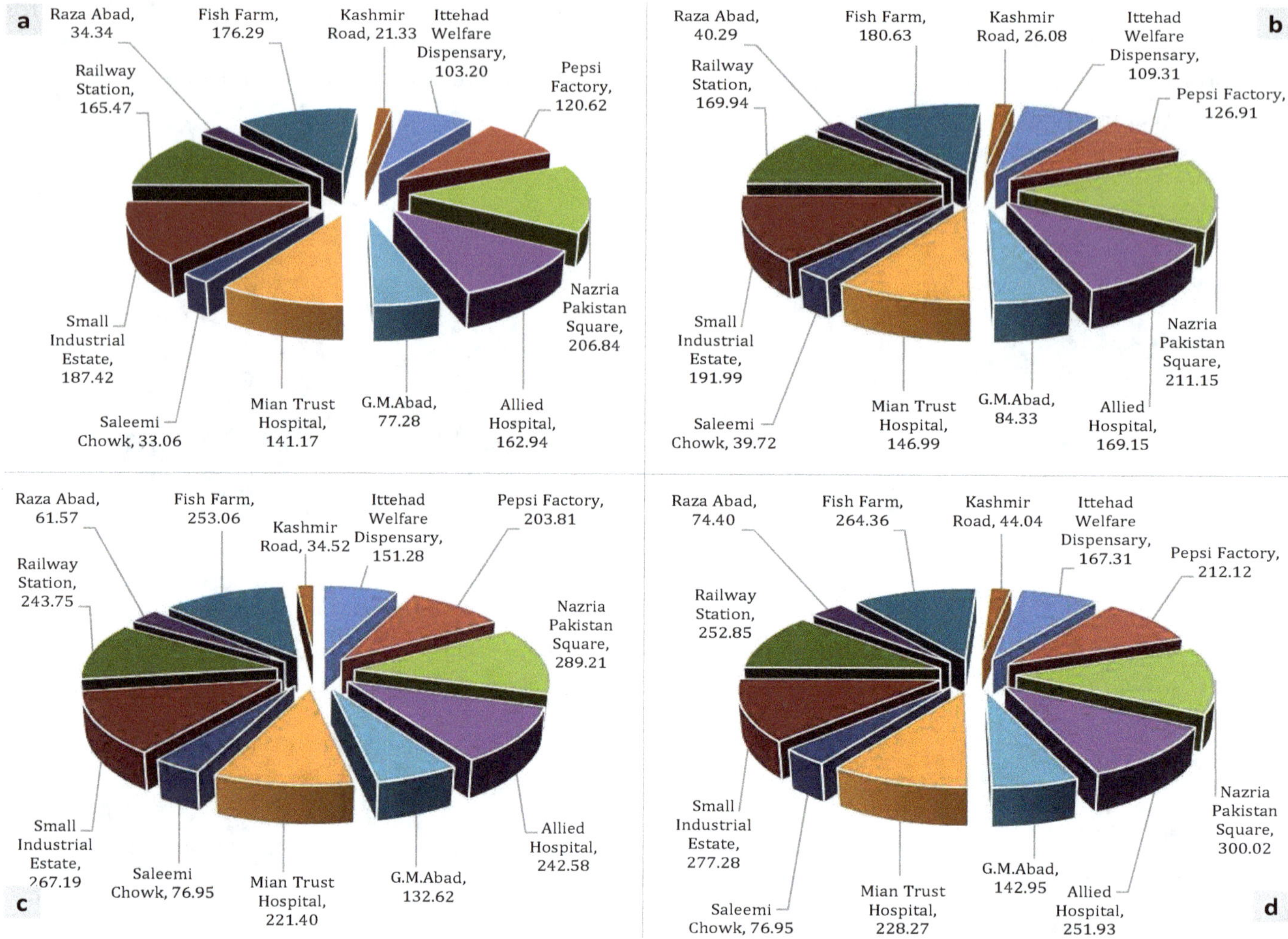

Figure 5. Concentration of carbonaceous species—total carbon (TC) in $PM_{2.5}$ in summer (**a**); $PM_{2.5}$ in winter (**b**); PM_{10} in summer (**c**); PM_{10} in winter (**d**). All units are expressed in μg/m^3.

3.5. Air Quality and Pollution Index

The air temperature of the study area fluctuated between 18–25 °C in winter which is considered a typical range while the trend in May–June 2017, as shown in Figure 6, showed an increase to 37.2 °C on average. Figure 6 also demonstrates the trend of relative humidity with a mean value of that varies from 39.45% in summer to 60.4% in winter. Table 1 presented the categories of air quality according to the AQI while Figure 7 illustrates the AQI index of selected sites of Faisalabad city with detrimental outcomes. It was found that the ambient air of Faisalabad city ranges from moderately polluted with the sequence of Ittehad Welfare Dispensary > Saleemi Chowk > Kashmir Road > Pepsi Factory, while severely polluted air was found in the vicinity of Nazria Pakistan Square, followed by Allied Hospital. The air quality of Mian Trust Hospital and the Small Industrial Estate was categorized as very poor besides GM Abad, Raza Abad and Station Chowk where the air quality was poor. The overall air quality of Faisalabad city was not good for health and other activities that require urgent attention from Government institutes and ministries involved in making and implementing policies to safeguard the environment.

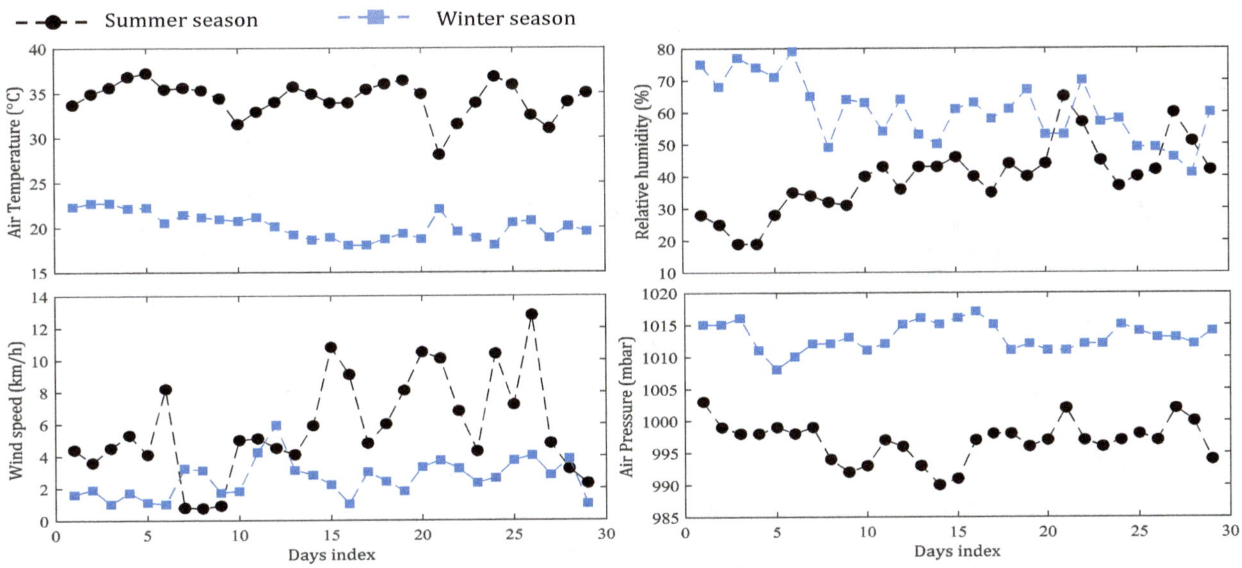

Figure 6. Relation of weather parameters in summer and winter seasons.

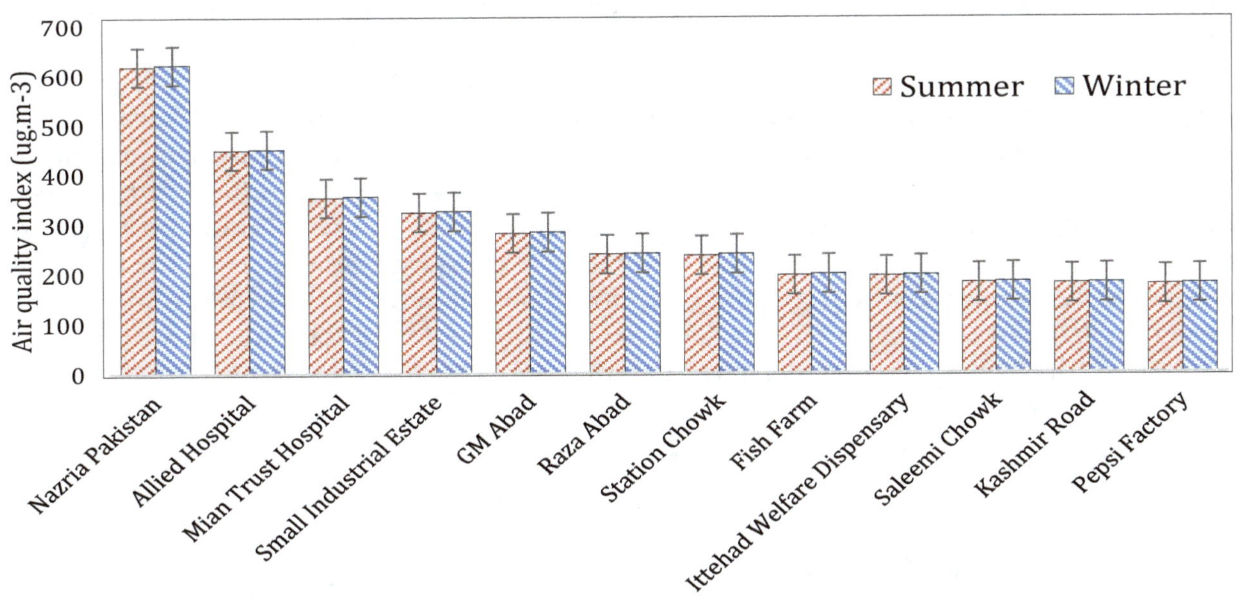

Figure 7. Air Quality Index (AQI) of Faisalabad city after analysis.

The average values of pollution index (PI) for each potential toxic element at selected sites of Faisalabad city for both seasons have been shown in Figure 8. In some residential areas, PI of $PM_{2.5}$ was found in the average level of pollution. While the PI of PM_{10} was estimated for the same areas showed a middle level of pollution $1 < PI \leq 2$, and sample site which is located near Nazria Pakistan Square suggested a high level of environmental pollution $PI > 4$. The $PM_{2.5}$ concentrations for almost all the sampling sites also showed a low level of environmental pollution of $PI \geq 1$. At Saleemi Chowk, Fish Farm, and Kashmir Road, the PI of PM_{10} showed a low level of environmental pollution $PI \leq 1$, while samples collected in the vicinities of Ittehad Welfare Dispensary, Pepsi Factory, Station Chowk, Raza Abad, GM Abad, and Mian Trust Hospital showed the middle level of pollution $(2 < PI \leq 3)$ in the environment. It should be noted that the samples of Allied Hospital and Nazria Pakistan showed the highest environmental pollution level of $3 < PI < 4$. This can be due to the toxic elements in urban dust which accumulate and originate mainly from traffic, paint, and many other nonspecific urban sources in the megacity.

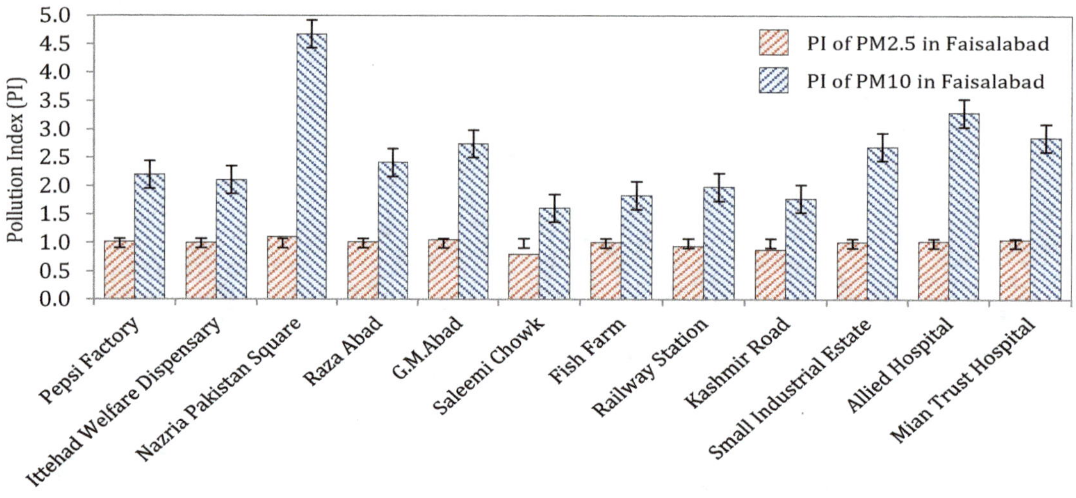

Figure 8. Pollution index flow diagram of $PM_{2.5}$ and PM_{10} in Faisalabad.

In order to show the difference between the mass concentrations of $PM_{2.5}$ and PM_{10} were statistically significant between the 12 locations, four sets of null hypotheses with H_0: $\mu(Location1) = \mu(Location2) = \ldots = \mu(Location12)$ were tested against the alternative hypotheses that the means of mass concentrations in the 12 locations were not equal. Based on the very small p-value that resulted from the one-way analysis of variance tests ($p < 0.0001$), all null hypotheses were rejected at any significant level and we concluded that the difference between locations was statistically significant. This result has been also shown in Figure 9 by means of the four scatter plots that demonstrate a visual comparison between the mean values of $PM_{2.5}$ and PM_{10} data in the 12 locations with respect to the daytime (morning, afternoon, evening) and season (summer or winter). Taking plot labeled (c) of Figure 9 as an example (mean of PM_{10} data in summer), the difference between morning concentrations of $PM_{2.5}$ in the 12 locations is clearly visible. It should be noted that for 61 days of winter the total number of measurements was 549. That is 61 days multiplied by 3 daytime (morning, afternoon, and evening) multiplied by 3 replications for each time.

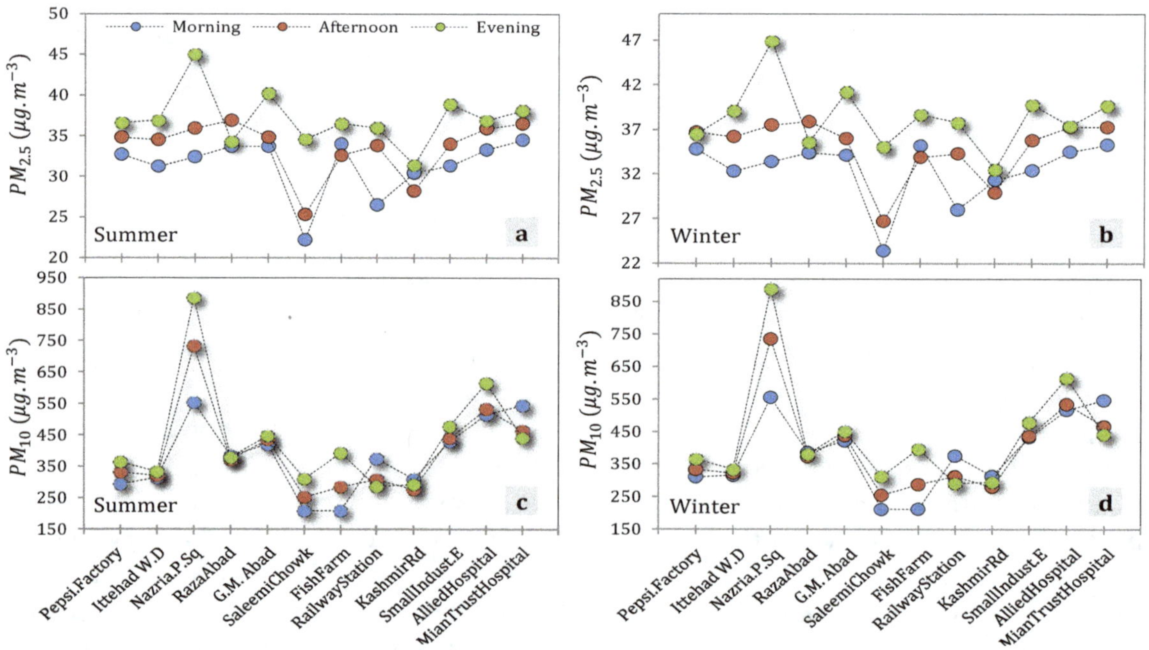

Figure 9. A comparison between mean values of $PM_{2.5}$ and PM_{10} data with respect to the daytime (morning, afternoon, evening) and season (summer or winter) in the 12 locations for (**a**): $PM_{2.5}$ in summer, (**b**): $PM_{2.5}$ in winter, (**c**) PM_{10} in summer, (**d**) PM_{10} in winter.

4. Discussion

Usually, the sources of gaseous pollutants are measured in three categories: natural emission mechanism, combustion sources and industrial manufacturing processes. Industrial sources of particulates, like steel, heavy traffic loads, indiscriminate burning of solid wastes and cement factories are the main sources of PM [2,28] besides gaseous and noise pollutants. The increasing air pollution is, after losses of properties, crops and increased health care costs. Airborne particulate matter is abundant in the atmosphere and varies extensively (temporally and spatially) in size, chemical composition and concentration. Emissions of particulate matter towards air are the focal environmental challenges for the transport and industrial sectors [4]. According to the WHO report, South Asia has developed as one of the most polluted zones in the globe because of its increasing population and rapid industrialization. Acquaintance to PM leads to more appointments to the emergency room or doctor. Health effects include premature deaths with existing lung and heart diseases, lung damage, coughing, aggravated asthma, wheezing and shortness of breath. Specifically, in the dry and cold season in major urban areas of Pakistan, people of almost all ages suffer from throat infections as reported [2,4,17]. In Pakistan, like the other developing countries, the emissions from vehicles have been conquered by emissions from poorly maintained and old vehicles that subsidize to heightened mass concentrations of carbon monoxide and fine particulates [17,28].

The current study focused on particulate volume has reported the highest concentration of $PM_{2.5}$ and PM_{10} in a commercial area on the average 39.18 ± 4.70 μg/m^3, 573.14 ± 321.64 μg/m^3 and 40.73 ± 2.9 μg/m^3, 575.19 ± 225.66 μg/m^3 for summer and winter, respectively (Figure 2). While, the lowest concentration of $PM_{2.5}$ and PM_{10} was obtained in the residential sector ranging (33.91 ± 3.27 μg/m^3), (35.1 ± 2.93 μg/m^3) and (349.46.75 ± 74.49 μg/m^3), (350.98 ± 74.29 μg/m^3) for summer and winter season correspondingly with a strong positive correlation between $PM_{2.5}$ and PM_{10} in both seasons on average. According to the guidelines, the reference value for $PM_{2.5}$ and PM_{10} are WHO (25 μg/m^3 and 50 μg/m^3) [17,25], NEQS and US-EPA (35 μg/m^3, 150 μg/m^3) [26,27] and most samples examined in the present study had values higher than the reference values. Elemental carbon is discharged from a variety of ignition procedures, categorized as a short-lived climate forcer that put up to atmospheric warming and also allied with human mortality and morbidity [28]. Common sources of atmospheric primary and secondary organic carbon antecedents are biomass burning, vehicular exhaust, biogenic emission and industrial emissions [29]. During the winter season, a higher level of pollutants especially the mass concentration of $PM_{2.5}$ persists in the ambient air of Faisalabad, owing to reduced atmospheric dispersion due to high relative humidity. Similarly, it was observed that $PM_{2.5}$ and PM_{10} sources were frequently localized as depicted by high concentrations at low wind speeds, mostly by the emissions from road vehicles [2,29]. This demonstrates the fact that $PM_{2.5}$ and PM_{10} concentrations were lower in summer than in winter (Figure 2) due to an increase in wind speed and temperature.

Prior studies conducted in the carbonaceous aerosols were assessed to account for about 50–60% of the total mass of $PM_{2.5}$ in metropolises in Jordan, Israel and Palestine [30]. Not unexpectedly, since production and processing of oil was prevalent transversely in the Middle East, substantial oil burning was valued to contribute 18% to total mass of PM_{10} and 69% to the total mass of $PM_{2.5}$ in Jeddah, Saudi Arabia [31]. Likewise, in Faisalabad, Pakistan, the quality of air not only reflects the impact of regional and local dust but also momentous local sources which include numerous industries and a heavy traffic weight. In municipal areas, the higher concentrations of $PM_{2.5}$ and PM_{10} are symbolic of the higher density of traffic as presented in the current study (Figures 2 and 4). Additionally, the burden of particulates is higher in the daytime than nighttime one, demonstrating more urban activities throughout day time. In Faisalabad, the textile industry, the topographical configuration and the geographical location make the problem of air pollution so perilous that it is very crucial to study it (Figure 5). The current study aimed at finding out whether or not the situation of air pollution in Faisalabad was previously seriously abundant to warrant the establishment of a regular air quality management system through which intercession measures can be premeditated and executed.

The analysis result of ambient air samples of selected sites of Faisalabad city displays that the level of particulate matter in most of the areas of the city is above the indorsed levels of the WHO, NEQS and US-EPA. Most of the city's commercial and residential areas are within the sensitive zone with the maximum concentrations of PM, which is constant with their proximity to the city's industrial areas.

When compared with the other studies conducted in other cities, it was found that the $PM_{2.5}$ level at Industrial Estate I-10 and IJP Road has reached the critical level (>35 $\mu g/m^3$) whereas at Industrial Estate I-9 it was moderate to the high level (31.9 $\mu g/m^3$ to 41.1 $\mu g/m^3$) [27]. While, the mean concentration of $PM_{2.5}$ and PM_{10} for Peshawar city during the study period has been calculated to be respectively 172 $\mu g/m^3$ and 480 $\mu g/m^3$ [32]. A similar high mass concentration of particulate matter was observed by [33] at Lahore, Pakistan and documented that the average $PM_{2.5}$ mass was 190 $\mu g/m^3$, and ranged from 89 $\mu g/m^3$ to 476 $\mu g/m^3$, far over US-EPA standards. Much higher PM_{10} mass concentration was experienced in Faisalabad when compared with other megacities [34], In addition, the PM_{10} concentrations were quite higher than the annual mean PM_{10} concentrations in Eastern Mediterranean and Africa [35,36] (WHO. Ambient (outdoor) 2014), Malaysia [34] and Bogota, Egypt, Los Angeles and Mexico [34]. It was also identified that PM_{10} is the dominant pollutant in the index value [37]. While, according to the results obtained after the analysis of particulate matters samples, the highest concentration of elemental carbon was 103.12 ± 1.46 $\mu g/m^3$ and the highest concentration of organic carbon was 196.9 ± 1.79 $\mu g/m^3$. While, 300.02 ± 3.25 $\mu g/m^3$ was the highest TC concentration found in the samples of coarse particulate matter collected in the vicinity of Nazria Pakistan (Figure 3). When compared, it was found that these concentrations are comparatively higher than in other metropolises in the areas like Punjab, India (116 $\mu g/m^3$), Hangzhou, China (119 $\mu g/m^3$), Kolkatta, India (197 $\mu g/m^3$), New Delhi, India (219 $\mu g/m^3$) and Lahore, Pakistan (233 $\mu g/m^3$) [38–40]. For elemental carbon, a large number of sources are identified, e.g., biomass and coal-fired power plant, two-stroke vehicles, fossil fuel burning, diesel engines and low burning efficiency. Elemental carbon is also utilized as a tracer for vehicular emission [17,40]. It was stated by [41] that diesel and gasoline motor vehicles and traffic exhaust are key sources of elemental carbon, followed by biomass burning. Organic carbon can be released straight from sources identified as primary carbon as a result of biomass and fossil combustion or can be formed as a result of a chemical reaction recognized as secondary organic carbon [42]. Temperature means are also under the normal limit but the increasing trend shows the alarming state of affairs and the same case is with radiations. Relative humidity has a value that is normal and considered healthy but a decreasing trend precedes the deterioration of ambient air quality. AQI and PI indicated that the ambient air quality of Faisalabad city falls from poor to severely polluted categories which are not safe to breathe and perform our daily activities.

5. Conclusions

Studying particle matters with aerodynamic diameters below 10 μm and 2.5 μm have received research attention for atmospheric pollution characteristics due to their severe effects on the human health issue. In this paper, we studied PM_{10} and $PM_{2.5}$ and highlighted that atmospheric pollution has become a significant issue as a result of growing industries in the megacity of Faisalabad, leading to the increased risk factors for chronic respiratory diseases in elderly and accelerated loss of lung function in newborns. To determine the pollution characteristics of particular matter, as well as the source and factors affecting them, we concentrated our study on 12 different sites that were selected and categorized as residential, commercial, industrial and health centers. Results of our study showed that the PM concentrations measured during current study periods (Dec 2016–Jan 2017) at various zones of Faisalabad were surprisingly higher than summer (May–June 2017). The enormous difference between fine ($PM_{2.5}$) and coarse (PM_{10}) particulate specifies that Faisalabad is inclined by a high loading of "coarse" particulate dust. Commercial areas are heavily polluted with fine and coarse particulate pollution. The average levels of pollution for fine and particulate matter were recorded as 39.18 ± 4.70, 573.14 ± 321.64 and 40.73 ± 2.9, 575.19 ± 225.66 during summer and winter, respectively (values in $\mu g/m^3$). The average $PM_{2.5}$ and PM_{10} concentrations were higher as compared to other major cities like

Islamabad, Lahore, and Peshawar. The quality of ambient air of Faisalabad has deteriorated beyond the safe limits set by WHO, US-EPA and NEQS. We also concluded that carbonaceous aerosols are in higher concentration in the air of the study sites. The air quality of Faisalabad city ranges from poor to severely polluted category which is highly unsafe for human health. These demands for an effort to introduce appropriate pollution control and management plans such as plantation and green belts for the betterment of civic life. A sustainable solution to improve air quality in Faisalabad would be to reduce emission by replacing high-energy consuming industries with renewable and clean energy sources, besides other strategies that reduce the use of fossil energy. Future studies may involve the use of wavelet analysis to explore the temporal characteristics of $PM_{2.5}$ and PM_{10}, or to investigate the relationship between meteorological factors and PM_{10}.

Author Contributions: A.A., M.I., I.S. and A.M. conceived and designed the experiments; A.A., M.T. and A.M. performed the experiments. M.K.I., G., M.Y., M.T. used software; R.R.S. worked on the final review and editing, figures and analysis; A.A., M.T., R.R.S. performed formal analysis and writing the draft; M.I., A.M. and G. did project funding and finding acquisition; M.I., I.S., A.M. and G. performed supervision of the experiments. All authors have read and agreed to the published version of the manuscript.

Acknowledgments: The authors acknowledge the financial support by the Open Access Publication Fund of the Leibniz Association, Germany, the partial research funding and editorial supports from Adaptive AgroTech Consultancy International and the administrative supports from Benjamin Mahns at the Leibniz Institute for Agricultural Engineering and Bioeconomy in Potsdam, Germany.

References

1. Hamid, A.; Akhtar, S.; Atique, S.A.; Huma, Z.; Mohay Uddin, S.G.; Asghar, S. Ambient air quality & noise level monitoring of different areas of Lahore (Pakistan) and its health impacts. *Pol. J. Environ. Stu.* **2019**, *28*, 623–629.
2. Javed, W.; Wexler, A.S.; Murtaza, G.; Iqbal, M.M.; Zhao, Y.; Naz, T. Chemical characterization and source apportionment of atmospheric particles across multiple sampling locations in Faisalabad, Pakistan. *Clean Soil Air Water* **2016**, *44*, 753–765. [CrossRef]
3. Gurjar, B.R.; Butler, T.M.; Lawrence, M.G.; Lelieveld, J. Evaluation of emissions and air quality in megacities. *Atmos. Environ.* **2008**, *42*, 1593–1606. [CrossRef]
4. Shahid, I.; Kistler, M.; Mukhtar, A.; Ghauri, B.M.; Cruz, C.R.S.; Bauer, H.; Puxbaum, H. Chemical characterization and mass closure of PM10 and PM2.5 at an urban site in Karachi, Pakistan. *Atmos. Environ.* **2016**, *128*, 114–123. [CrossRef]
5. Cheng, F.J.; Lee, K.H.; Lee, C.W.; Hsu, P.C. Association between particulate matter air pollution and hospital emergency room visits for pneumonia with septicemia: A retrospective analysis. *Aerosol Air Qual. Res.* **2019**, *19*, 345–354. [CrossRef]
6. Correia, A.W.; Pope III, C.A.; Dockery, D.W.; Wang, Y.; Ezzati, M.; Dominici, F. The effect of air pollution control on life expectancy in the United States: An analysis of 545 US counties for the period 2000 to 2007. *Epidemiology* **2013**, *24*, 23. [CrossRef]
7. Pakbin, P.; Hudda, N.; Cheung, K.L.; Moore, K.F.; Sioutas, C. Spatial and temporal variability of coarse (PM10−2.5) particulate matter concentrations in the Los Angeles area. *Aerosol. Sci. Technol.* **2010**, *44*, 514–525. [CrossRef]
8. Leghari, S.K.; Zaidi, M.A.; Ahmed, M.; Sarangzai, A.M. Assessment of suspended particulate matters level and role of vegetation in ambient air of North-East Balochistan, Pakistan. *Fuuast J. Biol.* **2013**, *3*, 37–43.
9. Dinoi, A.; Cesari, D.; Marinoni, A.; Bonasoni, P.; Riccio, A.; Chianese, E.; Tirimberio, G.; Naccarato, A.; Sprovieri, F.; Andreoli, V.; et al. Inter-comparison of carbon content in $PM_{2.5}$ and PM_{10} collected at five measurement sites in southern Italy. *Atmosphere* **2017**, *8*, 243. [CrossRef]
10. Choomanee, P.; Bualert, S.; Thongyen, T.; Salao, S.; Szymanski, W.W.; Rungratanaubon, T. Vertical variation of carbonaceous aerosols with in the $PM_{2.5}$ fraction in Bangkok, Thailand. *Aerosol. Air Qual. Res.* **2020**, *20*, 43–52. [CrossRef]
11. Ji, D.; Zhang, J.; He, J.; Wang, X.; Pang, B.; Liu, Z.; Wang, L.; Wang, Y. Characteristics of atmospheric organic and elemental carbon aerosols in urban Beijing, China. *Atmos. Environ.* **2016**, *125*, 293–306. [CrossRef]

12. Li, C.; Chen, P.; Kang, S.; Yan, F.; Hu, Z.; Qu, B.; Sillanpää, M. Concentrations and light absorption characteristics of carbonaceous aerosol in $PM_{2.5}$ and PM_{10} of Lhasa city, the Tibetan Plateau. *Atmos. Environ.* **2016**, *127*, 340–346. [CrossRef]
13. Li, Y.C.; Yu, J.Z.; Ho, S.S.; Schauer, J.J.; Yuan, Z.; Lau, A.K.; Louie, P.K. Chemical characteristics and source apportionment of fine particulate organic carbon in Hong Kong during high particulate matter episodes in winter 2003. *Atmos. Res.* **2013**, *120*, 88–98. [CrossRef]
14. Bisht, D.S.; Tiwari, S.; Dumka, U.C.; Srivastava, A.K.; Safai, P.D.; Ghude, S.D.; Chate, D.M.; Rao, P.S.; Ali, K.; Prabhakaran, T.; et al. Tethered balloon-born and ground-based measurements of black carbon and particulate profiles within the lower troposphere during the foggy period in Delhi, India. *Sci. Total Environ.* **2016**, *573*, 894–905. [CrossRef]
15. Colbeck, I.; Nasir, Z.A.; Ali, Z. The state of ambient air quality in Pakistan-a review. *Environ. Sci. Poll. Res.* **2010**, *17*, 49–63. [CrossRef]
16. World Bank and Institute for Health Metrics and Evaluation. *The Cost of Air Pollution: Strengthening the Economic Case for Action*; World Bank Group: Washington, DC, USA, 2016.
17. Niaz, Y.; Zhou, J.; Iqbal, M.; Nasir, A.; Dong, B. Ambient air quality evaluation: A comparative study in China and Pakistan. *Pol. J. Environ. Stu.* **2015**, *24*, 1723–1732. [CrossRef]
18. Gunawardana, C.; Goonetilleke, A.; Egodawatta, P.; Dawes, L.; Kokot, S. Source characterization of road dust based on chemical and mineralogical composition. *Chemosphere* **2012**, *87*, 163–170. [CrossRef]
19. Shahid, M.A.K.; Hussain, K.; Awan, M.S. Characterization of Solid Aerosols related to Faisalabad Environment and their probable sources. *Coden Jnsmac.* **2012**, *52*, 09–29.
20. Cachier, H.; Bremond, M.P.; Buat Ménard, P. Determination of atmospheric soot carbon with a simple thermal method. *Tellus B* **1989**, *41*, 379–390. [CrossRef]
21. Shahid, I.; Kistler, M.; Shahid, M.Z.; Puxbaum, H. Aerosol Chemical Characterization and Contribution of Biomass Burning to Particulate Matter at a Residential Site in Islamabad, Pakistan. *Aeros. Air Qual. Res.* **2019**, *19*, 148–162. [CrossRef]
22. Ott, W.R. *Environmental Indices: Theory and Practices*; Ann Arbor Science Publishers Inc.: Ann Arbor, MI, USA, 1978.
23. Central Pollution Control Board (CPCB). *Guidelines for National Ambient Air Quality Monitoring*; Series: NAAQM/25/2003- 04; Central Pollution Control Board: Delhi, India, 2009.
24. Chen, T.B.; Zheng, Y.M.; Lei, M.; Huang, Z.C.; Wu, H.T.; Chen, H.; Fan, K.K.; Yu, K.; Wu, X.; Tian, Q.Z. Assessment of heavy metal pollution in surface soils of urban parks in Beijing, China. *Chemosphere* **2005**, *60*, 542–551. [CrossRef]
25. World Health Organization. *Mental Health: Facing the Challenges, Building Solutions: Report from the WHO European Ministerial Conference*; WHO Regional Office Europe: Geneva, Switzerland, 2005.
26. Abbas, M.; Tahira, A.; Jamil, S. Air quality monitoring of particulate matter (PM2.5 & PM10) at Niazi and Daewoo bus station, Lahore. *FUUAST J. Biol.* **2017**, *7*, 13–18.
27. Hassan, M.; Malik, A.H.; Waseem, A.; Abbas, M. Air pollution monitoring in urban areas due to heavy transportation and industries: A case study of Rawalpindi and Islamabad. *J. Chem. Soc. Pak.* **2013**, *35*, 1623.
28. Huang, X.H.; Bian, Q.J.; Louie, P.K.K.; Yu, J.Z. Contributions of vehicular carbonaceous aerosols to $PM_{2.5}$ in a roadside environment in Hong Kong. *Atmos. Chem. Physics.* **2014**, *14*, 9279–9293. [CrossRef]
29. Weinhold, B. Global bang for the buck: Cutting black carbon and methane benefits both health and climate. *Environ. Health Perspect.* **2012**, *120*, A245. [CrossRef]
30. Abdeen, Z.; Qasrawi, R.; Heo, J.; Wu, B.; Shpund, J.; Vanger, A.; Sharf, G.; Moise, T.; Brenner, S.; Nassar, K.; et al. Spatial and temporal variation in fine particulate matter mass and chemical composition: The Middle East consortium for aerosol research study. *Sci. World J.* **2014**, 878704. [CrossRef]
31. Khodeir, M.; Shamy, M.; Alghamdi, M.; Zhong, M.; Sun, H.; Costa, M.; Chen, L.-C.; Maciejczyk, P. Source apportionment and elemental composition of $PM_{2.5}$ and PM_{10} in Jeddah City, Saudi Arabia. *Atmos. Pollut. Res.* **2012**, *3*, 331–340. [CrossRef]
32. Alam, K.; Rahman, N.; Khan, H.U.; Haq, B.S.; Rahman, S. Particulate matter and its source apportionment in Peshawar, Northern Pakistan. *Aerosol Air Qual. Res.* **2015**, *15*, 634–647. [CrossRef]
33. Husain, L.; Dutkiewicz, V.A.; Khan, A.J.; Ghauri, B.M. Characterization of carbonaceous aerosols in urban air. *Atmos. Environ.* **2007**, *41*, 6872–6883. [CrossRef]

34. Safar, Z.S.; Labib, M.W. Assessment of particulate matter and lead levels in the Greater Cairo area for the period 1998–2007. *J. Advan. Res.* **2010**, *1*, 53–63. [CrossRef]
35. World Health Organization. *Ambient (Outdoor) Air Quality and Health*; Fact sheet No. 313; 2014; Available online: https://www.who.int/news-room/fact-sheets/detail/ambient-(outdoor)-air-quality-and-health (accessed on 25 April 2020).
36. Ul-Saufie, A.; Yahya, A.; Ramli, N.; Hamid, H. Future PM_{10} concentration prediction using quantile regression models. In *International Conference on Environmental and Agriculture Engineering*; IACSIT Press: Singapore, 2012; Volume 37.
37. Nigam, S.; Rao, B.P.S.; Kumar, N.; Mhaisalkar, V.A. Air quality index-A comparative study for assessing the status of air quality. *Res. J. Eng. Technol.* **2015**, *6*, 267–274. [CrossRef]
38. Cao, J.; Shen, Z.; Chow, J.C.; Qi, G.; Watson, J.G. Seasonal variations and sources of mass and chemical composition for PM10 aerosol in Hangzhou, China. *Particuology* **2009**, *7*, 161–168. [CrossRef]
39. Awasthi, A.; Agarwal, R.; Mittal, S.K.; Singh, N.; Singh, K.; Gupta, P.K. Study of size and mass distribution of particulate matter due to crop residue burning with seasonal variation in rural area of Punjab, India. *J. Environ. Monit.* **2011**, *13*, 1073–1081. [CrossRef] [PubMed]
40. Alam, K.; Mukhtar, A.; Shahid, I.; Blaschke, T.; Majid, H.; Rahman, S.; Khan, R.; Rahman, N. Source apportionment and characterization of particulate matter (PM10) in urban environment of Lahore. *Aerosol. Air Qual. Res.* **2014**, *14*, 1851–1861. [CrossRef]
41. Ghauri, B.; Lodhi, A.; Mansha, M. Development of baseline (air quality) data in Pakistan. *Environ. Monit. Assess.* **2007**, *127*, 237–252. [CrossRef]
42. Seinfeld, J.H.; Pandis, S.N. *Atmospheric Chemistry and Physics: From Air Pollution to Climate Change*; John Wiley & Sons: Hoboken, NJ, USA, 2016.

10

Characterization of Particulate Matter ($PM_{2.5}$ and PM_{10}) Relating to a Coal Power Plant in the Boroughs of Springdale and Cheswick, PA

Casey D. Bray *, William Battye, Pornpan Uttamang, Priya Pillai and Viney P. Aneja

Department of Marine, Earth, and Atmospheric Sciences, North Carolina State University, Raleigh, NC 27695, USA; whbattye@ncsu.edu (W.B.); puttama@ncsu.edu (P.U.); prpillai@ncsu.edu (P.P.); vpaneja@ncsu.edu (V.P.A.)

* Corresponding: cdbray@ncsu.edu.

Abstract: Ambient concentrations of both fine particulate matter ($PM_{2.5}$) and particulate matter with an aerodynamic diameter less than 10 micron (PM_{10}) were measured from 10 June 2015 to 13 July 2015 at three locations surrounding the Cheswick Power Plant, which is located between the boroughs of Springdale and Cheswick, Pennsylvania. The average concentrations of PM_{10} observed during the periods were 20.5 ± 10.2 µg m^{-3} (Station 1), 16.1 ± 4.9 µg m^{-3} (Station 2) and 16.5 ± 7.1 µg m^{-3} (Station 3). The average concentrations of $PM_{2.5}$ observed at the stations were 9.1 ± 5.1 µg m^{-3} (Station 1), 0.2 ± 0.4 µg m^{-3} (Station 2) and 11.6 ± 4.8 µg m^{-3} (Station 3). In addition, concentrations of $PM_{2.5}$ measured by four Pennsylvania Department of Environmental Protection air quality monitors (all within a radius of 40 miles) were also analyzed. The observed average concentrations at these sites were 12.7 ± 6.9 µg m^{-3} (Beaver Falls), 11.2 ± 4.7 µg m^{-3} (Florence), 12.2 ± 5.3 µg m^{-3} (Greensburg) and 12.2 ± 5.5 µg m^{-3} (Washington). Elemental analysis for samples (blank – corrected) revealed the presence of metals that are present in coal (i.e., antimony, arsenic, beryllium, cadmium, chromium, cobalt, lead, manganese, mercury, nickel and selenium).

Keywords: $PM_{2.5}$; PM_{10}; coal-fired power plant; particulate matter emissions

1. Introduction

With over 600 active coal-fired power plants in the US, 39% of the country's total electricity generation is attributed to coal (Available online: http://www.eia.gov). While these plants may bring jobs and prosperity to their surrounding regions, they also emit dangerous pollutants into the atmosphere. According to the US Environmental Protection Agency (EPA) National Emissions Inventory, coal-fired power plants emit sulfur dioxide, oxides of nitrogen and particulate matter into the atmosphere. They also emit 84 of the 187 Hazardous Air Pollutants (HAP) regulated by the US EPA [1]. In addition to the stack emissions from the coal-fired power plant, coal handling may also emit pollutants into the atmosphere and thus degrade the air quality in the vicinity near the power plant [2]. Several studies have been conducted on the relationship between particulate matter and emissions from coal-fired power plants [2–14]. Particulate matter measured near a coal-fired power plant is known to contain a number of harmful chemicals that are present in coal and are also known to have carcinogenic properties, such as antimony, arsenic, beryllium, cadmium, chromium, cobalt, lead, manganese, mercury, nickel, selenium and polycyclic organic matter (POM) [2–4]. All of these chemicals are known to be hazardous and are thus regulated by the US EPA as HAPs under the Clean Air Act, as amended in 1990 [15]. The emission of these pollutants into the atmosphere is known to be dangerous to both human health and welfare. Exposure to high concentrations of HAPs can lead to a number of adverse health effects such as damage to the eyes, skin, lungs, kidneys and the nervous system, and can even cause cancer, pulmonary disease and cardiovascular disease [14,16]. Particulate matter ($PM_{2.5}$), which is particulate matter with an aerodynamic diameter of 2.5 microns or less, is also a dangerous atmospheric pollutant due to its small size, which can travel deep into people's lungs and lead to a number of severe health effects. Elevated concentrations of $PM_{2.5}$ are known to be associated with cardiovascular issues (heart disease, heart attacks, etc.) as well as respiratory issues, reproductive issues and even cancer [16,17]. In addition to harming human health, coal-fired power plants can also lead to a number of environmental impacts as well, such as acidification of the environment, bioaccumulation of toxic metals, the contamination of water sources, reduced visibility due to haze as well as degradation of buildings and monuments.

The Cheswick Power Plant is located in the southwestern part of Pennsylvania along the Allegheny River right between the boroughs of Springdale (40.5414° N, 79.7821° W) and Cheswick (40.5416° N, 79.8002° W) (Figure 1). While the Cheswick Power Plant has brought economic benefits to the small boroughs of Springdale and Cheswick, PA, the plant also brought with it a multitude of harmful effects on both human health and welfare. According to the Allegheny County Health Department's Point Source Emissions Inventory Report, the Cheswick Power Station is the largest point source emitter of both criteria pollutants and hazardous air pollutants (HAP) in Allegheny County, PA [18]. In 2006, the plant was listed as the 17th Dirtiest Power Plant for sulfur dioxide [1]. In 2010, the plant once again made headlines, ranking 41st in the Top Power Plant Hydrochloric Acid Emitters and 91st in the Top US Power Plant Lead Emitters [19]. Since then, significant efforts have been made to reduce emissions [18]. Overall, emissions of carbon monoxide and sulfur dioxide did decrease in 2011, while emissions of nitrogen oxides, particulate matter, NO_x, and volatile organic compounds (VOC) increased [18]. However, the plant remains Allegheny County's largest point source emitter of both criteria pollutants and hazardous air pollutants. The objective of this study was to measure the air quality at three different locations within the two boroughs surrounding the power plant in order to determine concentrations of both $PM_{2.5}$ and PM_{10} due to emissions from the power plant. Particulate matter measurements measured by the PA Department of Environmental Protection (DEP) were also analyzed for the study period and compared against the measurements taken for this study.

2. Experiments

2.1. NCSU Monitoring Stations

Three experimental sites were set up within Springdale and Cheswick, PA, each within a mile of the Cheswick power plant (Figure 1). There were two sampling periods in this study. Two of the three sites took samples from 10 June 2015 to 27 June 2015 and then from 30 June 2015 to 13 July 2015. Station 1 was located at 244 Center St., Springdale, PA (40.5402° N, 79.7884° W), which is less than half of a mile from the power plant. Samples were taken at this location during both sampling periods. Station 2 was located at 200 Hill Avenue, Cheswick, PA (40.542° N, 79.80° W), which is approximately one mile from the power plant. The sampling period for this study was from 10 June 2015 to 27 June 2015. The third station was located at 1212 Fairmont St, Cheswick, PA (40.5465° N, 79.8031° W). The sampling period for this site was from 30 June 2015 to 13 July 2013. All three of these sites are in residential areas, very close to the power plant. The monitoring sites were chosen considering the prevailing wind direction (west southwesterly) such that during both sampling periods, one site was upwind of the power plant (those sites in Cheswick) and one site was downwind of the power plant (the site in Springdale). However, it must be noted that the wind during the sampling period was highly variable.

Each experimental site was equipped to measure $PM_{2.5}$ and PM_{10} using Reference Ambient Air Sampler (RAAS) high volume air samplers by Anderson Instruments as well as meteorological data measured by Met One Model Automet portable weather stations equipped with an onboard data logger. The sampling period for each instrument was set to 24 h, and the air volumetric flow on both types of instruments was set at the standard flow rate of 16.67 Liters/minute. The sampling filters used to collect both $PM_{2.5}$ and PM_{10} were 47 mm diameter Teflon filters. For $PM_{2.5}$ sampling, there was an additional smaller filter used (~39 mm diameter) in the WINS impactor, which was wetted with impactor oil, in order to restrict particles larger than 2.5 micron from reaching the 47 mm Teflon filter. All the filters were pre-tared and numbered.

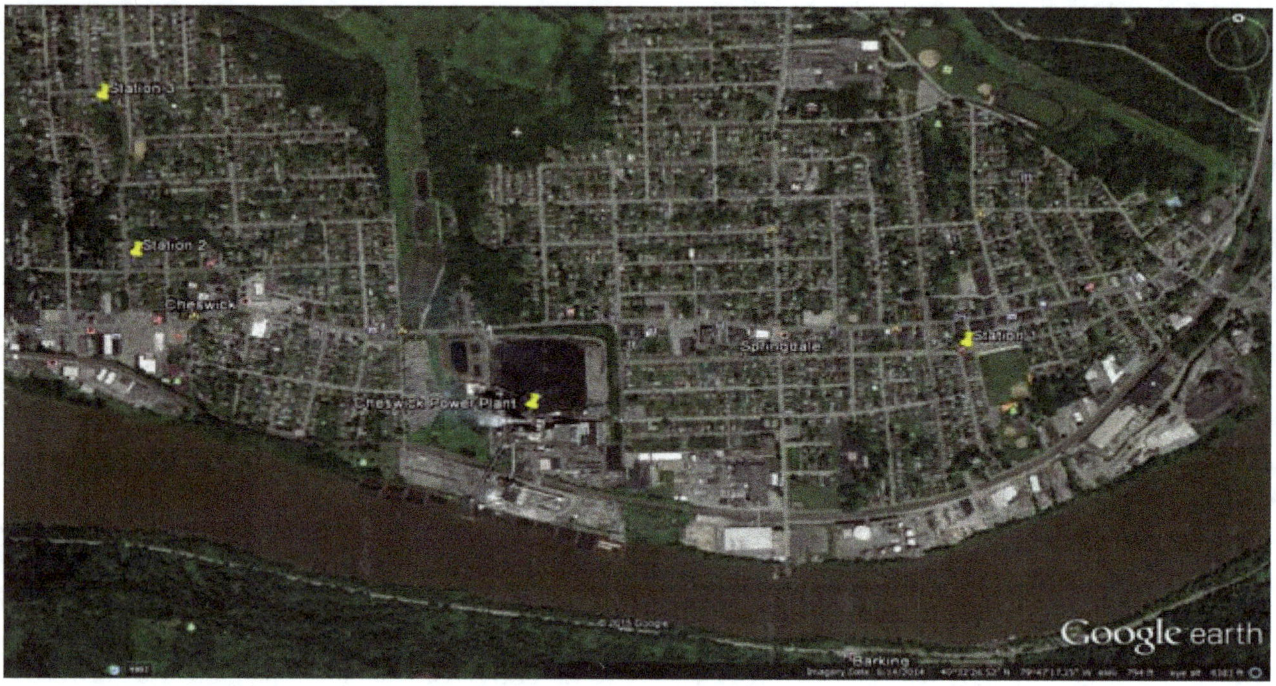

Figure 1. Location of monitoring stations in relation to the Cheswick Power Plant and the two boroughs of Springdale and Cheswick.

High Volume (Hi Vol) 47 mm MTL Teflon filters were used to analyze mass concentration of PM_{10} while Hi Vol 39 mm Teflon filters were used to analyze mass concentration of $PM_{2.5}$. Filter weighing measurements took place inside a temperature/relative humidity-controlled ISO Class 6 (<1000 $PM_{0.5}$/cf) clean room employing a draft-shielded microbalance with anti-static wand. Temperature and relative humidity were controlled to 21 °C and 35%, respectively. The filter mass was determined through five weightings of the filter, with each weight bracketed by a reading of the internal zero of the balance. A buoyancy correction was applied to the mass and the average of the zeros bracketing the mass was then subtracted from the result. This was repeated four additional times, with the average of these five results being reported as the final mass. The gravitational analysis was performed by the Research Triangle Park's (RTP) office of Applied Research Associates, Inc. (ARA).

Ten samples were chosen from this study to undergo an inorganic analysis, i.e., five for $PM_{2.5}$ and five for PM_{10}. In addition, two sample blanks were analyzed for calibration. Samples were analyzed from all three monitoring locations. The samples were selected to represent a mix of particulate matter concentrations collected during times when the plant was running. The inorganic analysis was then performed by X-Ray Fluorescence (XRF) and then Ion Chromatography (IC) analysis.

2.2. PA $PM_{2.5}$ Monitoring Sites

In addition to the measurement sites used in the study, data from four monitoring stations run by the state of PA (PA Department of Environmental Protection) was also used to compare against measurements taken in the field campaign. The four monitoring sites used in this study (Figure 2) were Beaver Falls (40.7478° N, 80.3157° W), Florence (40.4454° N, 80.4212° W), Greensburg (40.3043° N, 79.5060° W) and Washington (40.1706° N, 80.2617° W). These four sites contain continuous Met One BAM 1020 $PM_{2.5}$ monitors [20]. It is important to note that while the NCSU monitoring sites were located fairly close to the Cheswick Power Plant (i.e., less than 5 miles), the PA DEP monitoring sites were located 20–40 miles from the power plant.

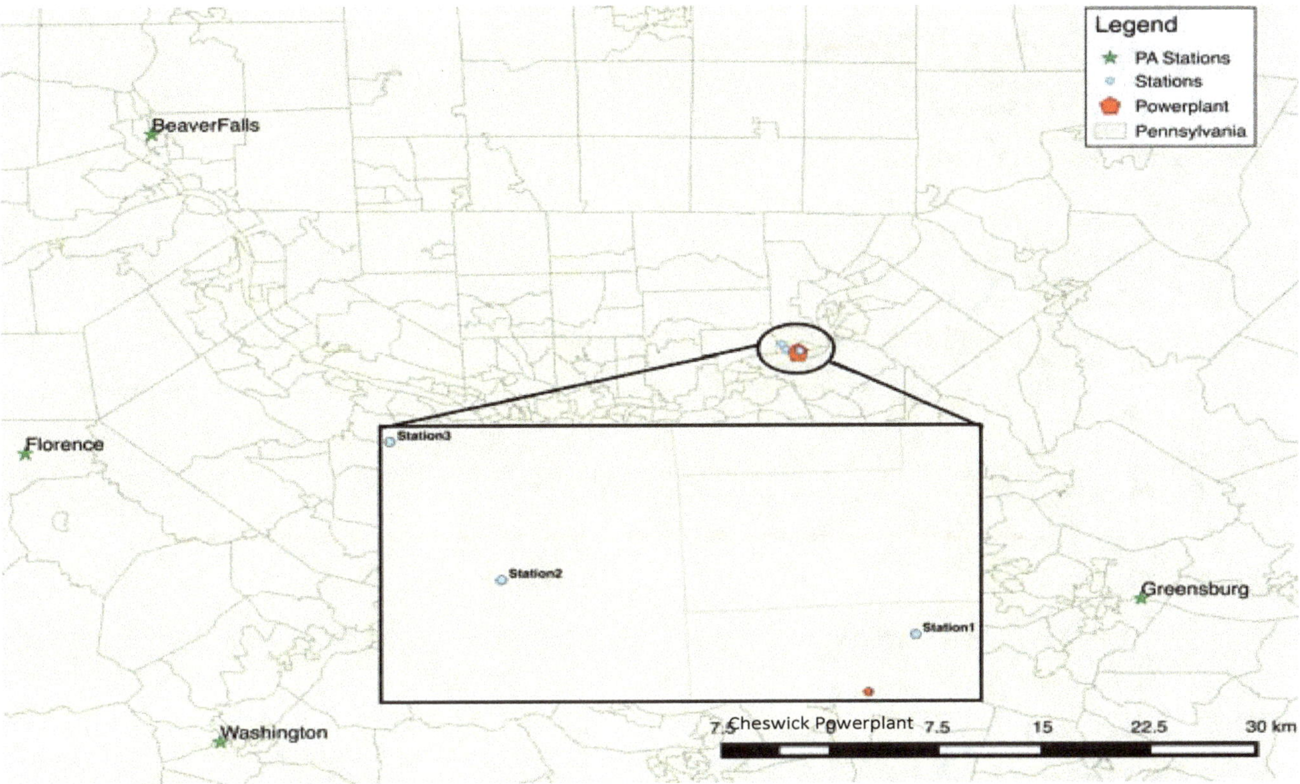

Figure 2. Location of the PA DEP monitoring sites in relation to the NCSU monitoring stations. The PA DEP monitoring sites are represented by the green stars while the NCSU monitoring sites are represented by the blue dots.

3. Results and Discussion

The observed average 24-hour concentrations of both $PM_{2.5}$ (0.02–22.76 μg m^{-3}) and PM_{10} (0.9−44.15 μg m^{-3}) from the NCSU monitoring sites were found to be lower than the US EPA National Ambient Air Quality Standard of 35 μg m^{-3} and 150 μg m^{-3}, respectively (Figure 3).

The average concentrations of PM_{10} observed during the periods were 20.5 ± 10.2 μg m^{-3} (Station 1), 16.1 ± 4.9 μg m^{-3} (Station 2) and 16.5 ± 7.1 μg m^{-3} (Station 3). The average concentrations of $PM_{2.5}$ observed at the stations were 9.1 ± 5.1 μg m^{-3} (Station 1), 0.2 ± 0.4 μg m^{-3} (Station 2) and 11.6 ± 4.8 μg m^{-3} (Station 3). Station 1 observed the maximum concentrations of both $PM_{2.5}$ (22.76 μg m^{-3}) and PM_{10} (44.15 μg m^{-3} μg m^{-3}). While Station 3 also observed higher concentrations of particulate matter (both $PM_{2.5}$ and PM_{10}), the results of a t test (at $\alpha = 0.05$) indicated that the concentrations at Station 1 were indeed higher than those observed at Stations 2 and 3. There are several plausible causes for this: the location of Station 1 versus Stations 2 and 3 in relation to the meteorological conditions as well as additional potential sources of particulate matter into the atmosphere (i.e., traffic emissions, roadway dust, transport). Concentrations of particulate matter tended to be higher when conditions were warm and sunny. Conversely, the lowest concentrations of particulate matter were primarily observed during rainy conditions. This was expected due to the removal of the particulates from the atmosphere through wet deposition. However, there were a few days where elevated concentrations were observed when the plant was running during rainy conditions. This may have resulted from the stability of the atmosphere in these conditions, which allows for the atmospheric accumulation of particulate matter. Nevertheless, comparisons of particulate matter concentrations with meteorological conditions (Figures 4 and 5) failed to show any strong correlations.

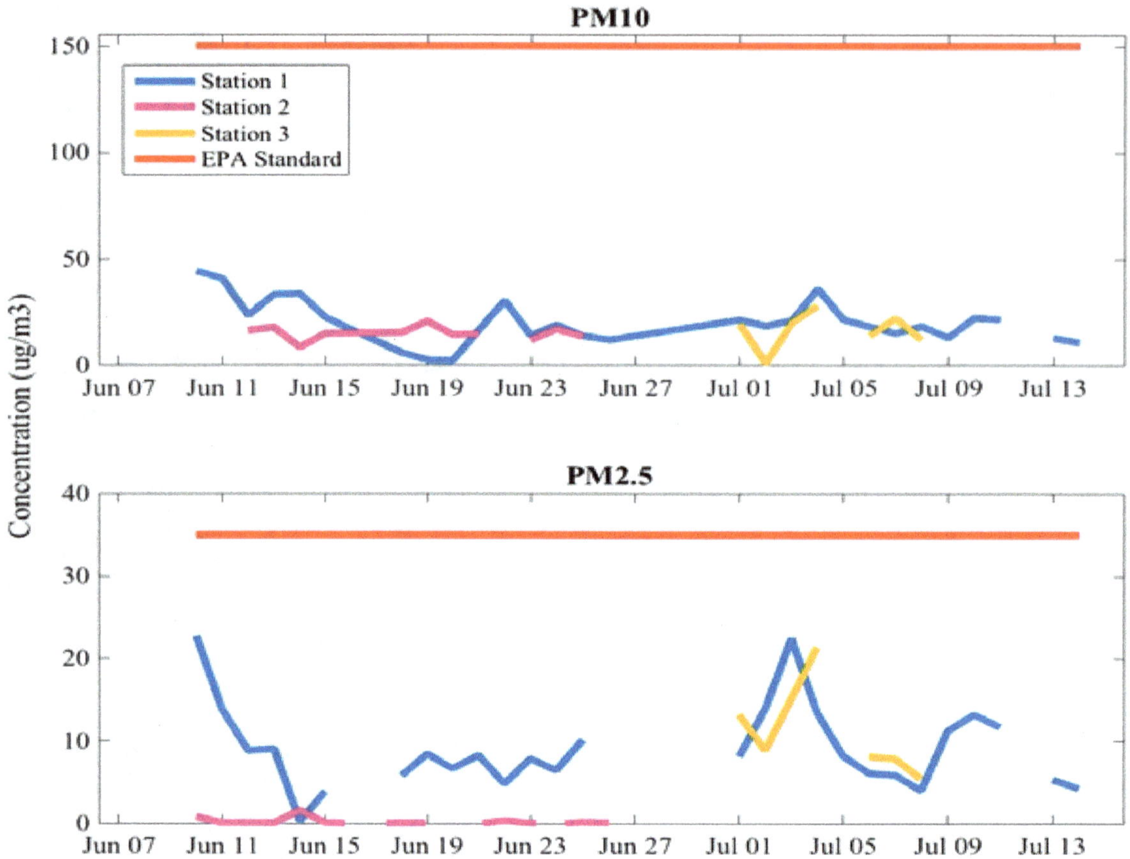

Figure 3. The daily 24-hour concentration of PM_{10} and $PM_{2.5}$ at Station 1 (blue), Station 2 (magenta) and Station 3 (gold) plotted against the US EPA National Ambient Air Quality Standards value for 24-hour average PM_{10} concentration (red).

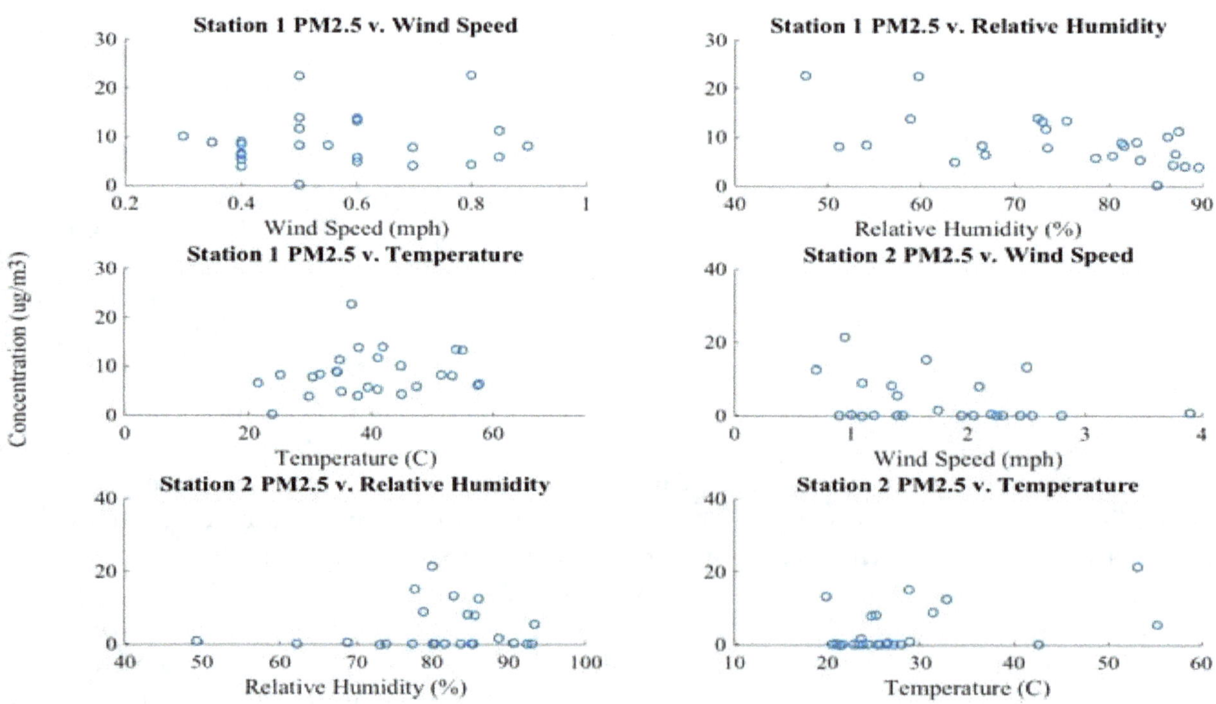

Figure 4. Comparing concentrations of $PM_{2.5}$ with meteorological variables for both Station 1 and Stations 2 and 3, denoted by Station 2 in this figure.

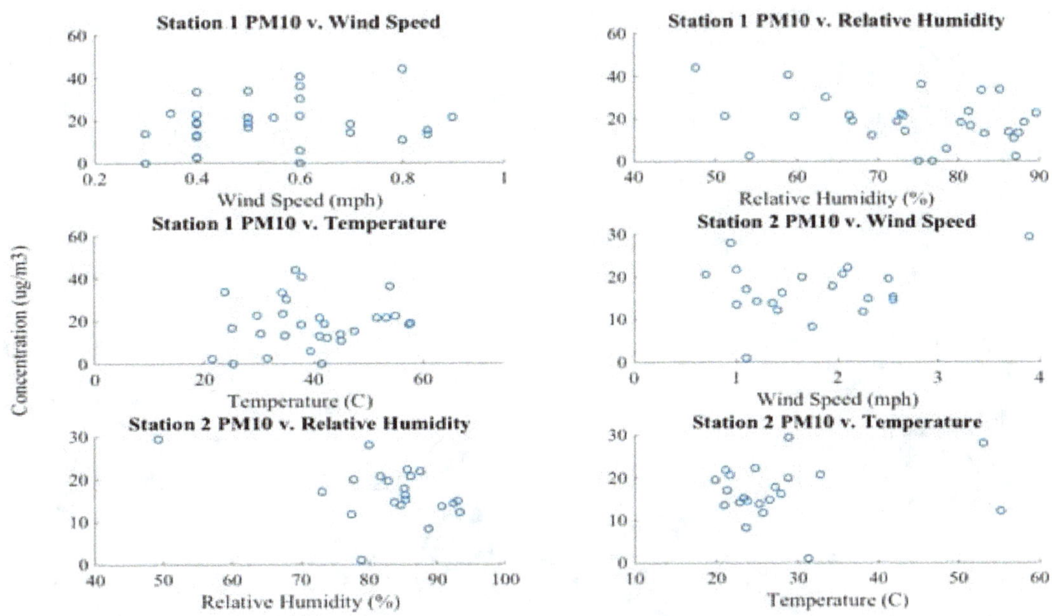

Figure 5. Comparing concentrations of PM_{10} with meteorological variables for both Station 1 and Stations 2 and 3, denoted by Station 2 in this figure.

When comparing the median wind speed versus the concentration of PM_{10} at Stations 1 and 2, positive correlations of 0.2 and 0.27, respectively, were observed. When comparing the median relative humidity with the concentration of PM_{10}, negative correlations of 0.3 and 0.44 were observed, respectively, while minor positive correlations of 0.08 and 0.12 were observed for temperature and concentration at Stations 1 and 2, respectively. When comparing meteorological conditions with concentrations of $PM_{2.5}$ at Stations 1 and 2, a negative correlation of 0.32 between wind speed and concentration of $PM_{2.5}$ was observed at Station 2 while Station 1 observed a positive correlation of 0.12. The strongest meteorological correlation was a negative correlation of -0.56 between relative humidity and the concentration of $PM_{2.5}$ at Stations 1, which is nominally downwind from the plant. In contrast, the correlation between relative humidity and $PM_{2.5}$ was weakly positive, at 0.09, for Station 2. Positive correlations of 0.54 and 0.46 were observed when comparing temperature versus concentrations of $PM_{2.5}$ at Stations 1 and 2, respectively. However, when considering these correlations, it is important to acknowledge the short sample period and thus small sample.

The median wind direction at upwind versus downwind stations was compared with the average 24-hour concentrations of particulate matter, where the categorizations of upwind and downwind were denoted based on the daily median wind direction. The results of a t test provide $p = 0.5$ and $p = 0.4$, for concentrations of PM_{10} and $PM_{2.5}$, respectively, which suggests that the relationship is not statistically significant.

Despite the fact that the median wind direction was not significantly correlated with the concentrations of particulate matter, the local hourly wind direction likely did impact concentrations of particulate matter. Furthermore, it is likely that some of the relatively high concentrations of both particulate matter observed at Station 1 were found to have come from the direction of the coal site on the power plant's property (Figures 6 and 7). In addition to observing elevated concentrations coming directly from the plant, elevated concentrations of both $PM_{2.5}$ and PM_{10} were also observed coming from other directions. This can likely be attributed to a number of sources, such as emissions from vehicles, road dust. In addition, the elevated concentrations can also be attributed to transport due to local and regional meteorological phenomena. Furthermore, fairly elevated concentrations of PM_{10} are observed at faster wind speeds. While wind speeds would generally reduce ambient concentrations of particulate matter in the atmosphere, the observed elevated concentrations could simply be due to an emission source located extremely close to the monitoring site (i.e., a stalled truck on the side of the road).

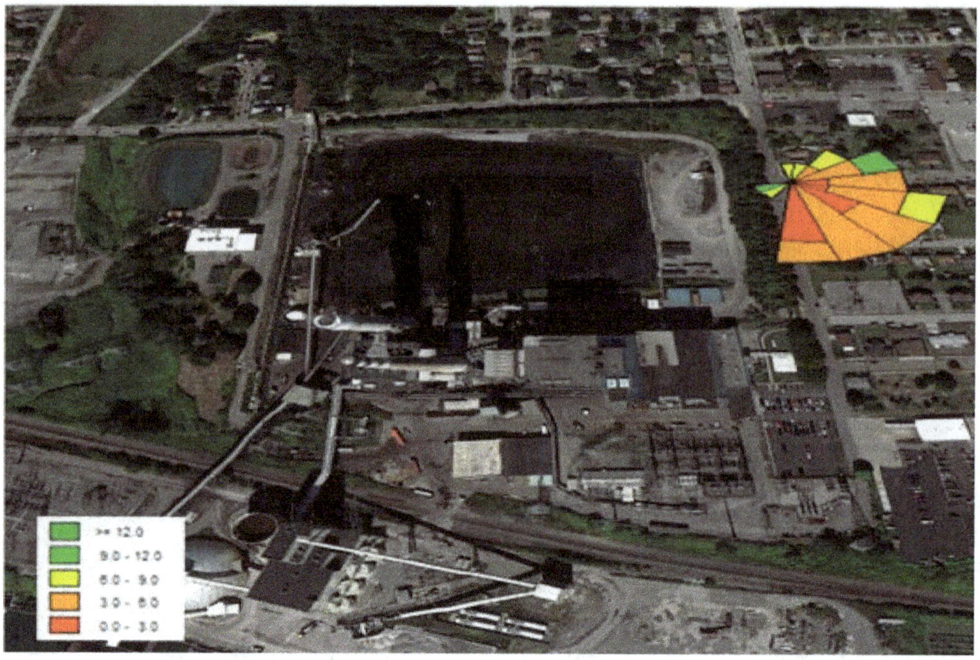

Figure 6. Wind rose created for $PM_{2.5}$ concentrations measured at Station 1. Green colors represent the highest concentrations of particulate matter while the orange and red colors represent lower concentrations of particulate matter, as shown in the legend. Units of particulate matter concentration are $\mu g\ m^{-3}$. The upper side of the wind rose represents North. The length of the wind rose vectors represents average wind speed, where the longer vectors represent a higher average wind speed.

Figure 7. Wind rose created for PM_{10} concentrations measured at Station 1. Green colors represent the highest concentrations of particulate matter while the orange and red colors represent lower concentrations of particulate matter, as shown in the legend. Units of particulate matter concentration are $\mu g\ m^{-3}$. The upper side of the wind rose represents North. The length of the wind rose vectors represents average wind speed, where the longer vectors represent a higher average wind speed.

In contrast to this, neither Stations 2 nor 3 (not pictured) showed extremely high concentrations coming from the power plant. However, this is not entirely surprising for two reasons: the first reason is that the winds were primarily such that those stations were upwind of the power plant and that both of these stations were twice as far as Station 1 for the plant, thus allowing for some dispersion of pollutants emitted from the plant and the coal site to occur.

In addition to this, concentrations of both $PM_{2.5}$ and PM_{10} were compared at calm wind conditions (median wind speeds less than 1 kt) and non-calm wind conditions (Table 1). When comparing average concentrations of $PM_{2.5}$ during calm winds (9.29 ± 5.59 μg m^{-3}) with average concentrations during not calm winds (3.53 ± 5.18 μg m^{-3}), it is clear that the average concentrations during calm winds are higher than the average concentrations during not calm winds. Similarly, average concentrations of PM_{10} observed during calm winds (19.83 ± 10.29 μg m^{-3}) was higher than the average concentrations observed when the winds were not calm (16.25 ± 4.87 μg m^{-3}). Two *t* tests were also conducted in order to determine whether or not the comparison between concentrations of $PM_{2.5}$ and PM_{10} during calm and not calm winds was statistically significant. The results of the t tests provide $p = 0.052$ and $p = 0.001$, for concentrations of PM_{10} and $PM_{2.5}$, respectively. This suggests that while the comparison between concentrations of $PM_{2.5}$ during calm and not calm winds is statistically significant, the comparison for concentrations of PM_{10} is not quite statistically significant.

Concentrations of particulate matter were also compared with the plant's gross load during the period (power plant gross load obtained via: EPA. Available online: http://www.ampd.epa.gov/ampd). Figure 8 compares the Cheswick Power Plant's daily gross load with the daily 24-hour average concentration of PM_{10} and $PM_{2.5}$. When comparing the daily gross load with the particulate matter concentrations, the correlation coefficients (ranging between −0.23 and 0.15) suggest that the concentrations of particulate matter at each station are not statistically correlated with the plant gross load.

Table 1. Comparison of concentrations of $PM_{2.5}$ and PM_{10} (μg m^{-3}) during both calm and not calm wind conditions. Calm wind conditions are considered to be conditions where the median wind speed is less than 1 kt.

	Calm Winds (μg m^{-3})		Not Calm Winds (μg m^{-3})	
	$PM_{2.5}$	PM_{10}	$PM_{2.5}$	PM_{10}
Maximum	22.76	44.15	15.05	29.31
Minimum	0.16	0.91	0.02	8.30
Mean	9.29	19.83	3.53	16.25
Median	8.26	20.55	0.35	15.03
Standard Deviation	5.59	10.29	5.18	4.87
Number of Samples	31	33	15	16

Five PM_{10} samples and five $PM_{2.5}$ samples were subjected to XRF and IC analysis. The XRF analysis represents the PM samples as elements, while the IC analysis represents the PM samples as ions. When comparing the results of the analyses, it is evident that they are consistent. The results of the XRF analysis (Figure 9) showed that the primary constituents of PM_{10} are sulfur (0.66–1.77 μg m^{-3}), silicon (0.14–2.47 μg m^{-3}), aluminum (0.03–1.06 μg m^{-3}) and iron (0.08–0.95 μg m^{-3}). Similarly, the primary constituents of $PM_{2.5}$ were found to be sulfur (0.41–1.17 μg m^{-3}), silicon (0.02–0.23 μg m^{-3}) and iron (0.03–0.06 μg m^{-3}). Particulate matter is typically composed of a complex mixture of chemicals that are strongly dependent on source characteristics. Inorganic analysis of the particulate matter revealed the presence of antimony, arsenic, beryllium, cadmium, chromium, cobalt, lead, manganese, mercury, nickel and selenium (Figure 9). All of these metals present in the PM samples are known to be present in coal [21]. While these elements are all found within coal and coal ash, they are also present in crustal material. Based on the small number of samples analyzed, it was not possible to discern differences between the upwind and downwind samples. The highest concentration of sulfur observed in a sample during the period was observed at Station 1. In this case, it is possible that this can be attributed to the coal pit, which is located less than half of a mile away from the station. However, it is also important to note that there are other coal-fired power plants located within 50 miles of the Cheswick Power plant, which could also contribute to the elevated sulfur concentrations.

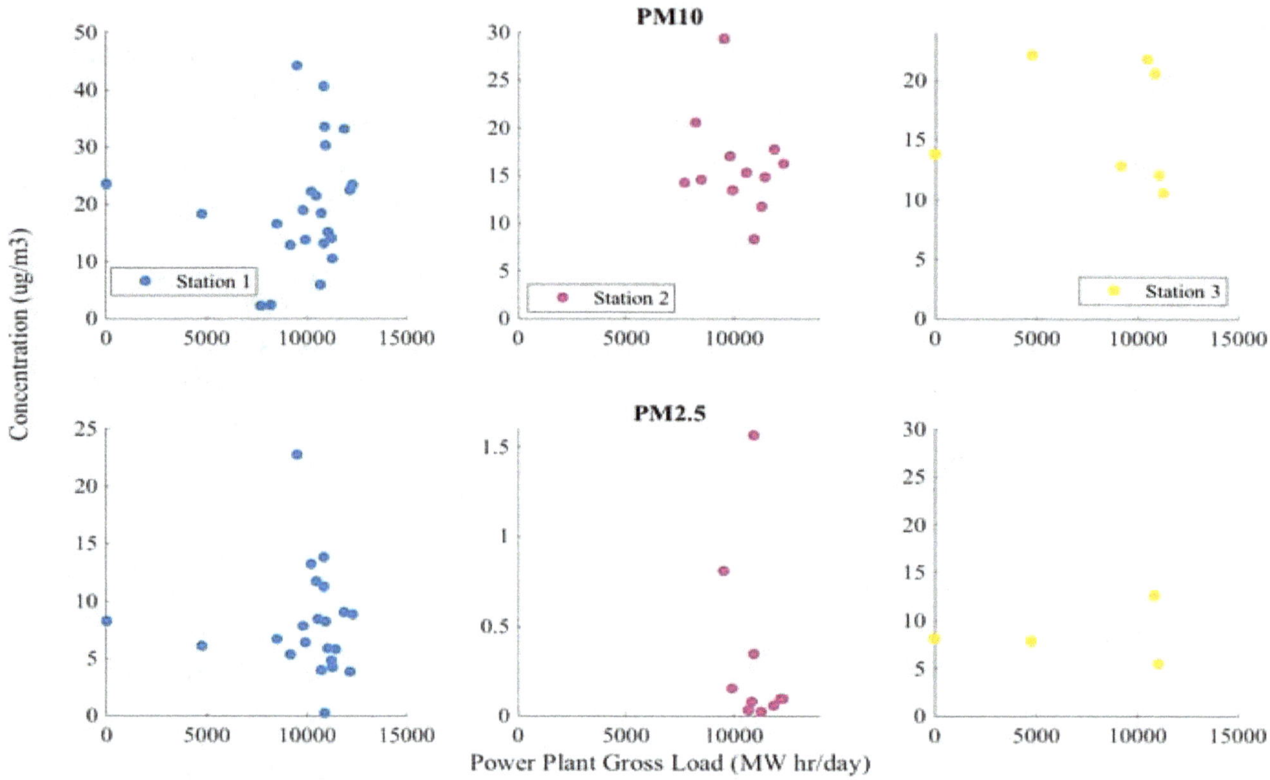

Figure 8. Comparing concentrations of PM_{10} and $PM_{2.5}$ at each station with the power plant gross load. The blue dots represent the concentrations at Station 1, the magenta dots represent the concentrations at Station 2 and the yellow/gold dots represent the concentrations at Station 3.

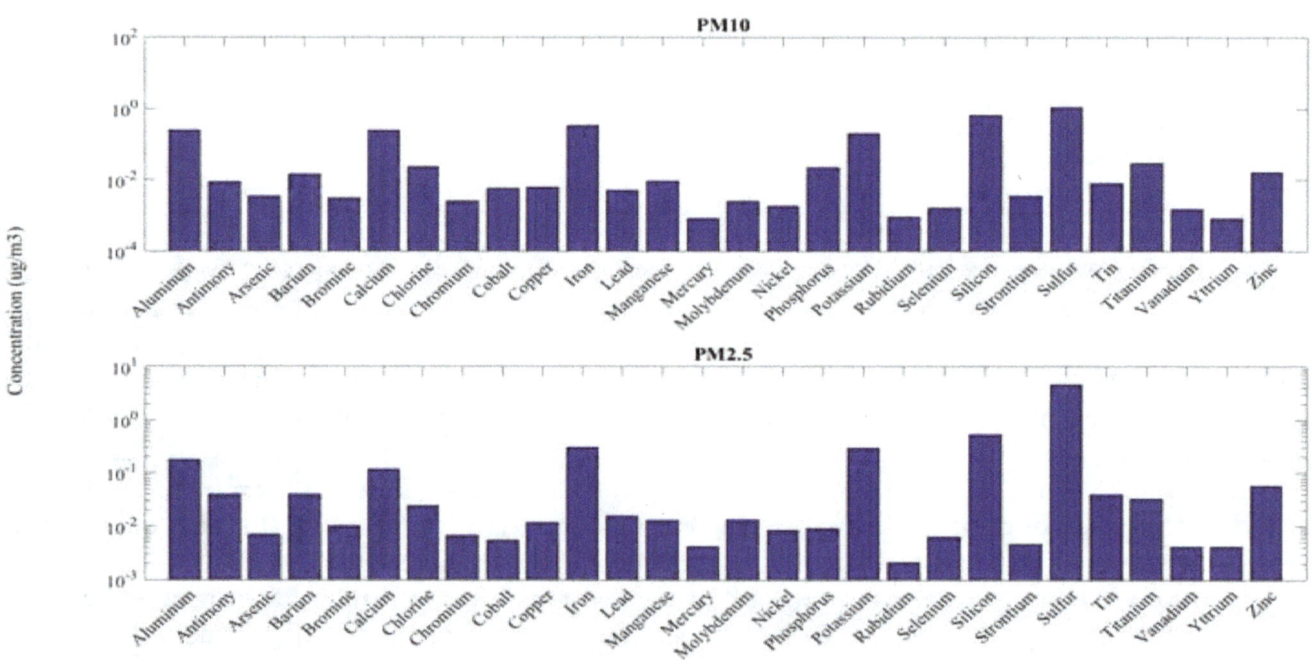

Figure 9. The results of the XRF analysis plotted on a log scale.

The results of the IC analysis (Figure 10) showed that the dominant species of particulate matter in this region are primarily sulfate, nitrate and ammonium. Within the $PM_{2.5}$ samples analyzed, concentrations of sulfate ranged from 2.31 μg m^{-3} to 3.08 μg m^{-3}, nitrate concentrations ranged from 0.03 μg m^{-3} to 0.58 μg m^{-3}, and concentrations of ammonium ranged from 0.82 μg m^{-3} to 1.15 μg m^{-3}. Within the analyzed PM_{10} samples, concentrations of sulfate ranged from 1.05 μg m^{-3} to 4.73 μg m^{-3}, concentrations of nitrate ranged from 0.01 μg m^{-3} to 0.22 μg m^{-3} and concentrations of ammonium ranged from 0.38 μg m^{-3} to 1.69 μg m^{-3}. As described above, it was not possible to discern differences between the upwind and downwind samples based on the small sample size. It is also important to note that sulfate, nitrate and ammonium are also common constituents of particulate matter and thus cannot be attributed entirely to the power plant.

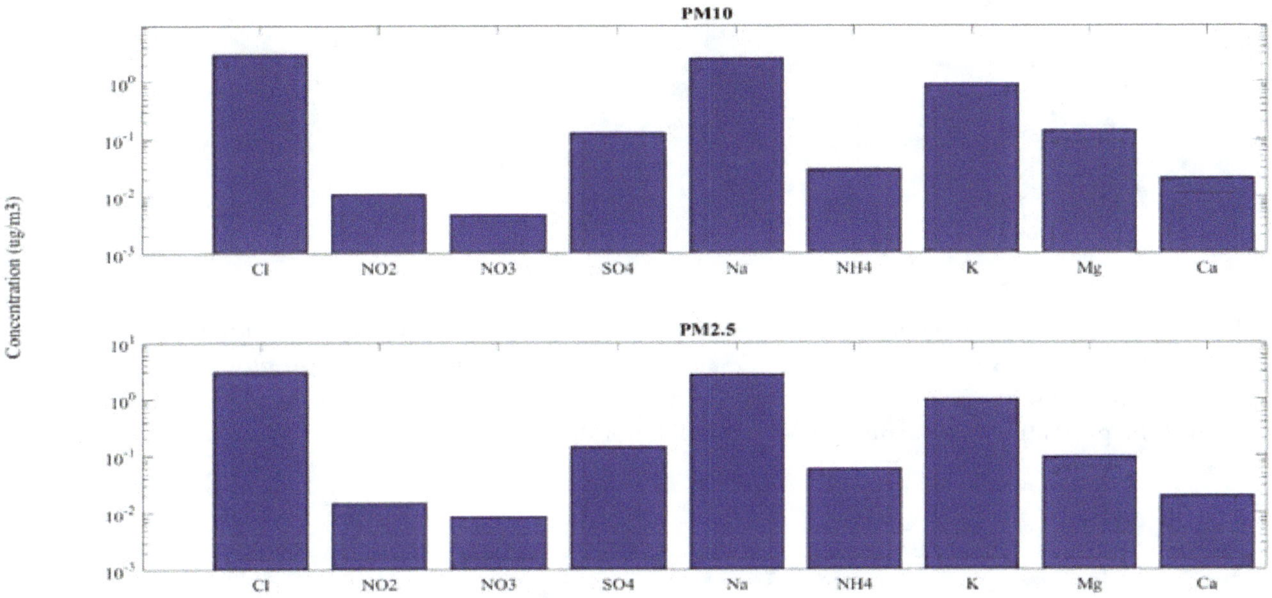

Figure 10. The results of the IC analysis, plotted on a log scale.

The concentrations of $PM_{2.5}$ measured by four PA Department of Environmental Protection air quality monitors were also analyzed in this study. The observed average 24-hour concentrations at these sites were 12.7 ± 6.9 μg m^{-3} (Beaver Falls), 11.2 ± 4.7 μg m^{-3} (Florence), 12.2 ± 5.3 μg m^{-3} (Greensburg) and 12.2 ± 5.5 μg m^{-3} (Washington). Similar to what was observed at the NCSU monitoring sites, the PA DEP monitors also observed 24-hour average concentrations of fine particulate (2–28.2 μg m^{-3}) matter below the US EPA 24-hour NAAQS (35 μg m^{-3}), with the exception of the Beaver Falls monitoring stations, which observed a concentration of 42.6 μg m^{-3} on 4 July 2015 (Figure 11). However, this elevated concentration can likely be explained by the presence of fireworks and other combustion processes due to the holiday.

When comparing the average PA DEP monitoring station $PM_{2.5}$ concentrations [20] with the average NCSU monitor $PM_{2.5}$ concentrations (Figure 12), it is evident that the monitors used in this study observed lower concentrations than the concentrations observed at the PA DEP monitoring sites. However, the general trend in concentrations for both monitoring networks is the same. There are several possible explanations for this discrepancy. One potential cause for the differences in observations could be due to differences in the instrumentation used to measure the particulate matter. In addition, another potential cause for the observed differences could be due to location. Since the monitors are farther away, it is possible that the pollutants were transported aloft before mixing down to the surface level. Furthermore, it is also possible that there are other potential sources of particulate matter located near the sites.

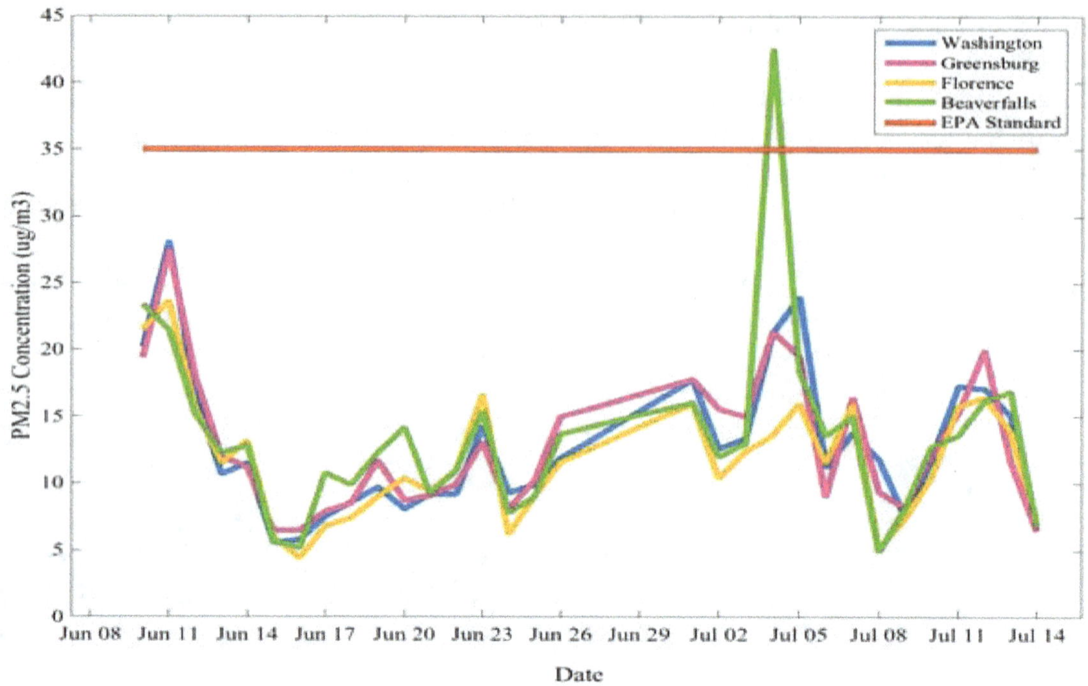

Figure 11. $PM_{2.5}$ measurements taken by the PA DEP monitoring sites during the study period. The red line represents the US EPA 24-hour NAAQS, the blue line represents the Washington monitoring site, the magenta represents the Greensburg monitoring site, the gold line represents the Florence monitoring site and the green line represents the Beaver Falls monitoring site.

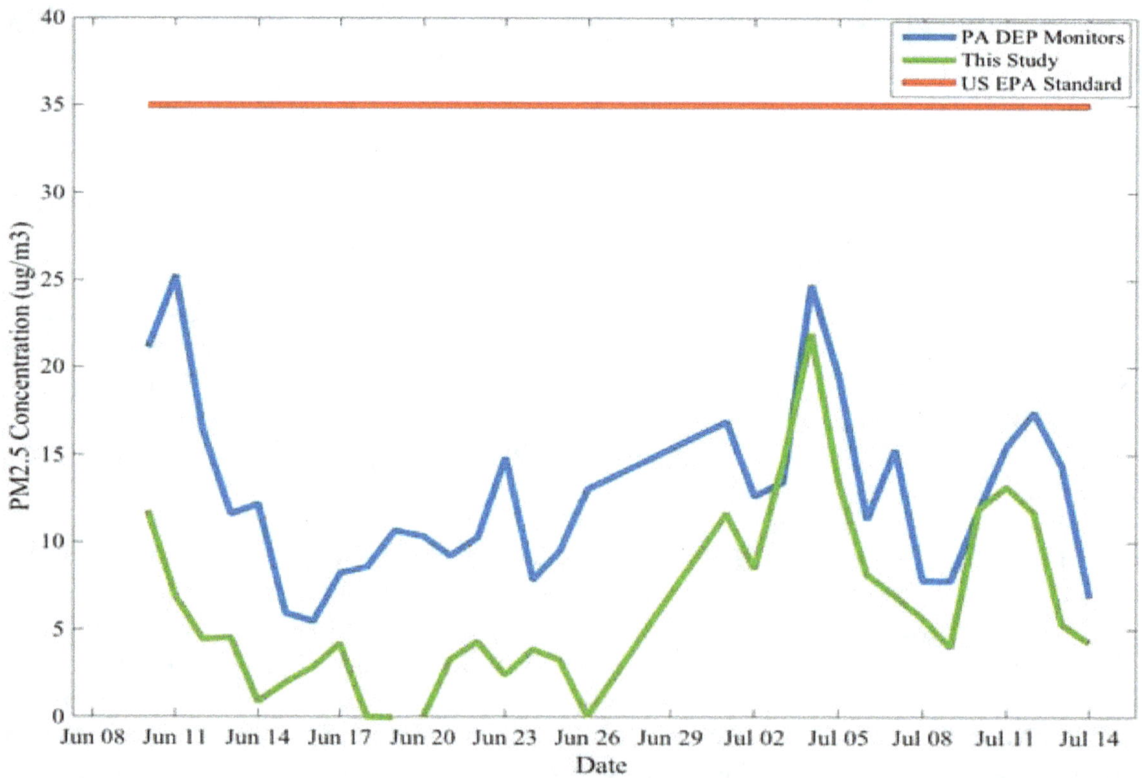

Figure 12. The average concentration measured by the NCSU monitors (this study, green line) compared with the average concentration measured by the PA DEP monitors (blue line line) and the US EPA 24-hour NAAQS (red line) for $PM_{2.5}$.

4. Conclusions

The ambient 24-hour average concentrations of both $PM_{2.5}$ (measured in this study as well as by the PA DEP) and PM_{10} did not reach levels higher than what is permitted by the US EPA 24-hour NAAQS. The average concentrations of PM_{10} observed during the periods were 20.5 ± 10.2 μg m^{-3} for Station 1, 16.1 ± 4.9 μg m^{-3} for Station 2 and 16.5 ± 7.1 μg m^{-3} for Station 3. The average concentrations of $PM_{2.5}$ observed at the stations were 9.1 ± 5.1 μg m^{-3} for Station 1, 0.2 ± 0.4 μg m^{-3} for Station 2 and 11.6 ± 4.8 μg m^{-3} for Station 3. The highest average concentration was observed at Station 1 for PM_{10} and at Station 3 for $PM_{2.5}$. However, Station 1 observed the highest daily average concentration for both $PM_{2.5}$ and PM_{10}. Not only was Station 1 the closest to the power plant, but the wind rose analysis showed that some of the elevated concentrations potentially came directly from the power plant's coal pit, in addition to other PM sources. However, other elevated concentrations were observed coming from other directions, suggesting that there are several sources for particulate matter located in the region.

The IC analysis showed that the dominant species of particulate matter were primarily sulfate, nitrate and ammonium. The results of the XRF analysis showed that the primary constituents of PM_{10} and $PM_{2.5}$ were sulfur, silicon, aluminum and iron. While these are all constituents that are observed from coal combustion emissions, these constituents are also prominent in most particulate matter speciation and therefore cannot be contributed directly to emissions from the Cheswick Power Plant.

The low particulate matter emissions and results of the speciation analyses could be attributed to a number of factors. Because the study period was so short, it is likely that the local scale meteorological conditions led to bias in the results. In addition, it is also possible that there were errors in the instrumentation that led to such low concentrations, particularly for $PM_{2.5}$ concentrations observed at Station 2. Furthermore, the low concentrations of particulate matter observed at the sites could also be attributed to the emission control technology that is added to the Cheswick power plant, which includes wet lime flue gas desulfurization, low NO_x burner technology with separated over fire air selective catalytic reduction and an electrostatic precipitator (Available online: http://www.ampd.epa.gov/ampd), which would explain the reduced concentrations of pollutants observed in this study.

Based on the results of this study, is not possible to determine a concrete conclusion on the role of the Cheswick Power Plant in concentrations of particulate matter in the Cheswick and Springdale boroughs. These inconclusive results can be attributed to a number of factors, including unfavorable meteorological conditions, potential issues with the measurement equipment as well as an extremely short sampling period. It must be emphasized that this study was conducted over a limited time period. Therefore, it is recommended that further work be done on this matter, with longer sampling periods occurring in each season in order to capture a seasonal profile of concentrations of particulate matter in this region.

Acknowledgments: The authors acknowledge the support from the Sierra Club. We want to thank Marty Blake, Karen Petito, and Natalie Bick for their help with the monitoring stations. We also want to thank Karsten Baumann for help with the sample mass analysis, John Walker for help with the inorganic analysis of PM samples, and Dennis Mikel for his help with the X-Ray Fluorescence analysis of the PM samples. We would also like to thank Sean Nolan and the PA Department of Environmental Protection, Bureau of Air Quality, for their assistance in obtaining the data for the PA DEP monitoring sites. In addition, we would also like to thank Zachary Fabish for his assistance with this project.

Author Contributions: Viney Aneja was the PI for this project. Priya Pillai led the measurement process. Pornpan Uttamang helped take measurements for the project and helped in research for the manuscript. William Battye helped in the data analysis and in the revision and improvement of the manuscript. Casey Bray assisted in the measurement process, analyzed the data and wrote the manuscript.

References

1. Dirty Kilowatts. America's most Polluting Power Plants 2006. Available online: http://www.dirtykilowatts.org/dirty_kilowatts.pdf (accessed on 15 September 2015).

2. Aneja, V.P.; Isherwood, A.; Morgan, P. Characterization of Particulate Matter (PM_{10}) Related to Surface Coal Mining Operations in Appalachia. *Atmos. Environ.* **2012**, *54*, 496–501. [CrossRef]
3. Nielsen, M.T.; Livbjerg, H.; Fogh, C.L.; Jensen, J.N.; Simonsen, P.; Lund, C.; Poulsen, K.; Sander, B. Formation and Emission of Fine Particles from Two Coal-Fired Power Plants. *Combust. Sci. Technol.* **2002**, *174*, 79–113. [CrossRef]
4. Tian, H.; Wang, Y.; Xue, Z.; Qu, Y.; Chai, F.; Hao, J. Atmospheric emissions estimation of Hg, As, and Se from coal-fired power plants in China. *Sci. Total Environ.* **2007**, *409*, 3078–3081. [CrossRef] [PubMed]
5. Contini, D.; Cesari, D.; Conte, M.; Donateo, A. Application of PMF and CMB Receptor Models for the Evaluation of the Contribution of a Large Coal-Fired Power Plant to PM_{10} Concentrations. *Sci. Total Environ.* **2016**, *560*, 131–140. [CrossRef] [PubMed]
6. Chow, J.C.; Watson, J.G. Review of $PM_{2.5}$ and PM_{10} Apportionment for Fossil Fuel Combustion and Other Sources by the Chemical Mass Balance Receptor Model. *Energy Fuels* **2002**, *16*, 222–260. [CrossRef]
7. Cohen, D.D.; Crawford, J.; Stelcer, E.; Atanacio, A.J. Application of Positive Matrix Factorization, Multi-Linear Engine and Back Trajectory Techniques to the Quantification of Coal–Fired Power Station Pollution in Metropolitan Sydney. *Atmos. Environ.* **2012**, *61*, 204–211. [CrossRef]
8. Gladney, E.S.; Small, J.A.; Gordon, G.E.; Zoller, W.H. Composition and Size Distribution of in-Stack Particulate Material at a Coal-Fired Power Plant. *Atmos. Environ.* **1976**, *10*, 1071–1077. [CrossRef]
9. Lee, S.W. Source Profiles of Particulate Matter Emissions from a Pilot-Scale Boiler Burning North American Coal Blends. *J. Air Waste Manag. Assoc.* **2001**, *51*, 1568–1578. [CrossRef]
10. Liu, X.; Xu, M.; Yao, H.; Yu, D.; Zhang, Z.; Lü, D. Characteristics and Composition of Particulate Matter from Coal-Fired Power Plants. *Sci. China Ser. E* **2009**, *52*, 1521–1526. [CrossRef]
11. Wu, H.; Pedersen, A.J.; Glarborg, P.; Frandsen, F.; Dam-Johansen, K.; Sander, B. Formation of Fine Particles in Co-Combustion of Coal and Solid Recovered Fuel in a Pulverized Coal-Fired Power Station. *Proc. Combust. Inst.* **2011**, *33*, 2845–2852. [CrossRef]
12. Young, G.S.; Fox, M.A.; Trush, M.; Kanarek, N.; Glass, T.A.; Curriero, F.C. Differential Exposure to Hazardous Air Pollution in the United States: A Multilevel Analysis of Urbanization and Neighborhood Socioeconomic Deprivation. *Int. J. Environ. Res. Public Health* **2012**, *9*, 2204–2225. [CrossRef] [PubMed]
13. Zhou, W.; Cohan, D.; Pinder, R.; Neuman, J.; Holloway, J.; Peischl, J.; Ryerson, T.; Nowak, J.; Flocke, F.; Zheng, W. Observation and Modeling of the Evolution of Texas Power Plant Plumes. *Atmos. Chem. Phys.* **2012**, *12*, 455–468. [CrossRef]
14. Anderson, H.R.; Atkinson, R.W.; Bremner, S.A.; Marston, L. Particulate air pollution and hospital admissions for cardiorespiratory diseases: Are the elderly at greater risk? *Eur. Pespir. J.* **2003**, *21*, 39–46. [CrossRef]
15. Congress.gov. Clean Air Act Amendments of 1990. Available online: https://www.congress.gov/bill/101st-congress/senate-bill/1630/amendments (accessed on 24 July 2017).
16. MacIntosh, D.; Spengler, J. Emissions of Hazardous Air Pollutants from Coal Fired Power Plants. 2011. Available online: http://www.lung.org/assets/documents/healthy-air/coal-fired-plant-hazards.pdf (accessed on 15 September 2015).
17. Pope, C.A., III; Dockery, D.W. Health Effects of Fine Particulate Air Pollution: Lines that Connect. *J. Air Waste Manag. Assoc.* **2006**, *56*, 709–742. [CrossRef]
18. Kelly, M.; Fischman, G. Point Source Emission Inventory Report. 2011. Available online: http://www.achd.net/air/pubs/pdf/2011_emissions_inventory_report.pdf (accessed on 15 September 2015).
19. Levin, I. America's Top Power Plant Toxic Polluters 2011. Available online: http://www.environmentalintegrity.org/documents/report-topuspowerplanttoxicairpolluters.pdf (accessed on 15 September 2015).
20. Nolan, S.; PA DEP. PADEP $PM_{2.5}$ Air Quality Monitoring Data. Personal Communication, 2017.
21. Finkelman, R.B. Modes of occurrence of environmentally-sensitive trace elements in coal. In *Environmental Aspects of Trace Elements in Coal; Anonymous*; Springer: Dordrecht, The Netherlands, 1995; pp. 24–50.

11

Multi-Year (2013–2016) $PM_{2.5}$ Wildfire Pollution Exposure over North America as Determined from Operational Air Quality Forecasts

Rodrigo Munoz-Alpizar [1,*], Radenko Pavlovic [1], Michael D. Moran [2] , Jack Chen [2], Sylvie Gravel [2], Sarah B. Henderson [3], Sylvain Ménard [1], Jacinthe Racine [1], Annie Duhamel [1], Samuel Gilbert [1], Paul-André Beaulieu [1], Hugo Landry [1], Didier Davignon [1], Sophie Cousineau [1] and Véronique Bouchet [1]

1 Canadian Meteorological Centre Operations, Environment and Climate Change Canada, 2121 Rte Transcanadienne, Montreal, QC H9P 1J3, Canada; radenko.pavlovic@canada.ca (R.P.); sylvain.menard@canada.ca (S.M.); jacinthe.racine@canada.ca (J.R.); annie.duhamel@canada.ca (A.D.); samuel.gilbert@canada.ca (S.G.); paul-andre.beaulieu@canada.ca (P.-A.B.); hugo.landry@canada.ca (H.L.); didier.davignon@canada.ca (D.D.); sophie.cousineau@cafnada.ca (S.C.); veronique.bouchet@canada.ca (V.B.)

2 Air Quality Research Division, Environment and Climate Change Canada, Toronto, ON M3H 5T4, Canada; mike.moran@canada.ca (M.D.M.); jack.chen@canada.ca (J.C.); sylvie.gravel@canada.ca (S.G.)

3 Environmental Health Services, British Columbia Centre for Disease Control, Vancouver, BC V5Z 4R4, Canada; sarah.henderson@bccdc.ca

* Correspondence: rodrigo.munoz-alpizar@canada.ca.

Abstract: FireWork is an on-line, one-way coupled meteorology–chemistry model based on near-real-time wildfire emissions. It was developed by Environment and Climate Change Canada to deliver operational real-time forecasts of biomass-burning pollutants, in particular fine particulate matter ($PM_{2.5}$), over North America. Such forecasts provide guidance for early air quality alerts that could reduce air pollution exposure and protect human health. A multi-year (2013–2016) analysis of FireWork forecasts over a five-month period (May to September) was conducted. This work used an archive of FireWork outputs to quantify wildfire contributions to total $PM_{2.5}$ surface concentrations across North America. Different concentration thresholds (0.2 to 28 $\mu g/m^3$) and averaging periods (24 h to five months) were considered. Analysis suggested that, on average over the fire season, 76% of Canadians and 69% of Americans were affected by seasonal wildfire-related $PM_{2.5}$ concentrations above 0.2 $\mu g/m^3$. These effects were particularly pronounced in July and August. Futhermore, the analysis showed that fire emissions contributed more than 1 $\mu g/m^3$ of daily average $PM_{2.5}$ concentrations on more than 30% of days in the western USA and northwestern Canada during the fire season.

Keywords: air quality modeling; wildfire smoke; fine particulate matter; wildfire pollution exposure

1. Introduction

Wildfires are large, uncontrolled vegetation fires that result from natural processes or anthropogenic activities. In North America they are a major natural hazard, with high interannual variability in both the number of fires and the total burned area. Every year, wildfires consume millions of hectares of forest in North America, resulting in several community evacuations due to the direct threat of fire or the indirect threat of heavy smoke [1]. According to the 2016 report of the Canadian Interagency Forest Fire Centre (CIFFC), during the last decade an average of 7000 wildfires occurred each year in Canada and burned an average of 2.6 million hectares per year [2]. Annual costs of wildfire suppression in Canada have ranged from about \$0.5 billion to \$1 billion in the last decade [3]. As an extreme example, one wildfire in western Canada from 1 May to 4 July 2016 burned an area of 590,000 ha, roughly the size of the Canadian province of Prince Edward Island, and forced the evacuation of nearly 90,000 people from the city of Fort McMurray in northeastern Alberta. The damages caused by this fire were estimated to be on the order of $9.5 billion [4].

In addition to economic impacts, wildfires can adversely affect both air quality (AQ) and human health. The AQ impacts depend on the amount and chemical composition of the emissions from these fires, the smoke plume dynamics, and the meteorological conditions that drive the transport and diffusion of wildfire smoke. Biomass burning from wildfires can release significant amounts of pollutants into the atmosphere, including particulate matter (PM), ammonia (NH_3), and ozone (O_3) precursors such as nitrogen oxides (NOx), volatile organic compounds (VOCs), and carbon monoxide (CO) [5]. Although many of these species are harmful to human health, the population health impacts of wildfire smoke have been attributed mainly to short-term concentrations of PM less than 2.5 μm in aerodynamic diameter ($PM_{2.5}$). Two recent systematic reviews found that short-term smoke exposure is strongly associated with small increases in daily mortality from all causes, and with acute respiratory outcomes ranging in severity from increased reporting of symptoms through to increased risk of hospital admissions. Associations were weaker for acute cardiovascular morbidity and birth outcomes, but suggestive of effects in both cases [6,7]. In addition, new evidence about wildfire smoke is emerging rapidly given the severity of fires across North America over the past decade. Furthermore, wildfire smoke is playing an increasingly important role in long-term air pollution as fires get larger and other sources such as motor vehicles and industry come under increasingly strict regulation [8]. Long-term exposure to $PM_{2.5}$ is associated with the development of a wide and growing range of chronic diseases [9].

During wildfire events, $PM_{2.5}$ concentration at the ground level may be significantly increased, such that it exceeds the levels established by regulatory agencies to protect the environment and human health. In Canada, for example, the established Canadian Ambient Air Quality Standards (CAAQS) for $PM_{2.5}$ concentration are an annual mean of 10 μg/m^3 and a daily mean of 28 μg/m^3 [10]. The CAAQS metric for annual concentration of $PM_{2.5}$ is based on the three-year average of the annual average concentrations, and on the three-year average of the annual 98th percentile of the daily 24-h average concentrations. Furthermore, the AQ impacts of wildfire emissions are not limited to the local or regional scales. Under some meteorological conditions, wildfire smoke plumes can disperse widely and travel thousands of kilometers, affecting people living far away from the fire location [11]. Observational evidence indicates that the long-range transport of wildfire smoke can episodically increase PM and O_3 ground-level concentrations at regional and continental scales. For example, smoke from Canadian wildfires was associated with high concentrations of PM in areas great distances from the fire source, such as Baltimore and Washington, D.C. in the eastern USA and as far away as Europe [11–14]. Moreover, Canadian wildfires have also been linked to increased O_3 concentrations in Houston, TX and the northeastern USA, as well as Europe [15–19]. On the other hand, long-range transport of Siberian wildfire smoke has also contributed to exceedances in O_3 and $PM_{2.5}$ on the west coast of Canada [20,21].

The assessment of human exposure to smoke from wildfires is challenging because such smoke episodes are typically sporadic and short-lived, with highly variable concentrations in both space and time [22,23]. Furthermore, the spatial variability of population exposure to wildfire smoke cannot be correctly represented based solely on regulatory monitoring data because these data provide limited spatial coverage. For example, impacts are often observed in populated non-urban areas where regulatory monitoring networks are sparse or not available. On the other hand, remote sensing data from satellites can be used over very large areas, covering locations where the monitoring networks are missing. However, these measurements provide information about the total atmospheric column of air pollutants rather than the ground-level concentrations. They are also generally not available at night and can be masked by the occurrence of clouds. Additionally, satellite overpasses may occur only once a day or every few days, resulting in large amounts of missing information. Therefore, deterministic AQ forecast models have become a useful tool to fill the temporal and spatial gaps in available measurements and provide guidance about AQ over the coming hours and days [24–31].

AQ forecasting systems consisting of 3D numerical weather prediction (NWP) models with on-line or off-line chemical transport models (CTMs) have become a valuable tool in the past 15 years. They can provide guidance in the production of AQ forecasts and assist public health authorities in understanding pollutant exposures and developing public actions to protect populations against those exposures. The accuracy of pollutant exposure estimates using modeled AQ data depends on the ability of the forecast systems to reproduce observed concentrations of air pollutants. The differences among the current AQ forecast systems that consider anthropogenic emissions of pollutants have been reviewed recently [24,25], but to date only a few AQ forecast systems have been developed that combine information from wildfires and meteorology to retrospectively or prospectively estimate the emissions, transport, and diffusion of wildfire smoke [26,27].

In order to provide guidance to regional AQ forecasters, first responders, and public health decision-makers about the dispersion of smoke from large wildfires, Environment and Climate Change Canada (ECCC) has developed FireWork [27], an on-line, one-way coupled meteorology–chemistry model based on near-real-time wildfire emissions. FireWork was built on the existing ECCC operational AQ forecast system, and was first deployed in 2013 during the Canadian wildfire season to deliver real-time forecasts of wildfire smoke plumes over North America. Previous studies have shown the ability of FireWork to forecast $PM_{2.5}$ in terms of statistical scores and spatial distributions, as well as public health impacts [27–29]. Observed trends and seasonal variability of $PM_{2.5}$ are well captured by the model. However, FireWork's ability to simulate the emission and dispersion of wildfire smoke is currently limited by factors such as the accuracy of wildfire emission factors, the treatment of fire behavior, and the suitability of plume-rise algorithms.

Here we conduct a multi-year (2013–2016) analysis of FireWork forecasted $PM_{2.5}$ concentrations from biomass burning over North America to provide estimates of the population exposure to $PM_{2.5}$ from wildfires for several concentration thresholds. The number of days that exceed these thresholds as well as the magnitude of the area in exceedance was estimated. The goal of this work is to help public health professionals, policymakers, and the general public better understand the human health impacts of wildfire-related $PM_{2.5}$ pollution.

2. Methodology

2.1. North American Wildfire AQ Forecasting System

The FireWork system was first run in an experimental mode beginning in 2013 at ECCC's Canadian Centre for Meteorological and Environmental Prediction (CCMEP). The system became operational in April 2016. The FireWork system is identical to the ECCC operational Regional Air Quality Deterministic Prediction System (RAQDPS) [30–32], except for the inclusion of satellite-derived, near-real-time biomass burning emissions from natural, prescribed, and agricultural fires [27]. The on-line RAQDPS modeling system relies on the GEM-MACH (Global Environmental Multi-scale-Modelling Air quality and Chemistry) model, an on-line, one-way coupled CTM (i.e., meteorology affects chemistry, but chemistry does not affect meteorology), embedded within the Global Environmental Multi-scale (GEM) model. Both the RAQDPS and FireWork systems input the same hourly anthropogenic gridded emissions fields based on processing the 2010 Canadian national Air Pollutant Emission Inventory (APEI), the 2011 U.S. National Emissions Inventory (NEI), and the 1999 Mexican emissions inventory, as well as biogenic and sea-salt emissions from natural sources [31]. Each of the three national anthropogenic inventories accounts for emissions of at least seven criteria air pollutants: SO_2, NO_x, VOC, CO, NH_3, $PM_{2.5}$, and PM_{10}.

The calculation of the near-real-time biomass burning emissions required by FireWork starts with the Canadian Forest Service's operational Canadian Wildland Fire Information System (CWFIS), which provides fire activity and fire danger conditions across Canada and the continental United

States during the active wildfire season [33]. The primary data used by the CWFIS to capture fire activity come from satellite-based detection systems: NASA's Moderate Resolution Imaging Spectroradiometer (MODIS) instrument, NOAA's Advanced Very High Resolution Radiometer (NOAA/AVHRR), and Visible Infrared Imaging Radiometer Suite (VIIRS) imagery through NASA and the U.S. Forest Service Remote Sensing Applications Center [34]. During the fire season, fire activity is updated six times daily in the CWFIS, corresponding to the frequency of available satellite-based retrievals. Relevant fire information estimated by CWFIS for each fire hotspot includes fuel type, surface fuel consumption, crown fuel consumption, total fuel consumption, and forest floor fuel consumption. Estimates of daily biomass-burning emissions for individual hotspots are then obtained using fuel consumption values from the CWFIS and emission factors from the Fire Emission Production Simulator (FEPS), a component of the BlueSky Modeling Framework [26]. More detailed information about the FireWork modelling system framework and its data flow is provided in other recent publications [27,28].

In the current operational setup, the seasonal FireWork system runs twice per day at 00 UTC and 12 UTC during the North American fire season from 1 April to 31 October. FireWork simulation results provide numerical AQ forecast guidance over North America with a 48-h lead time. In 2013, 2014, and 2015, when FireWork was run as an experimental model version at CCMEP, the period from April to October was only partially covered (see Table 1). In April 2016, however, when the FireWork System became operational [35], FireWork forecasts were extended to cover the full wildfire season.

The seasonal peak for wildfire events in Canada occurs in the months of June, July, and August [2], and initially our analysis focused on this three-month period. However, due to the extreme wildfires that occurred in northern Alberta in May 2016, we decided to extend our analysis to a five-month period from May to September. In order to backfill this five-month period for the years 2013–2015 (Table 1), FireWork was rerun using the operational forecasting approach [27] with the same FireWork version that had been used each year. This required three older versions of FireWork to be run because new, updated versions of the RAQDPS had been introduced before each fire season [31,35].

Table 1. 2013–2016 experimental and operational FireWork start/end forecast periods together with additional periods for which FireWork was rerun retrospectively.

Year	Experimental/Operational FireWork		Added Periods
	Start	End	
2013	June 1	August 31	May 1–31; Sep. 1–31
2014	June 9	October 1	May 1–June 8
2015	May 21	October 31	May 1–20
2016	April 1	October 31	-

Seasonal fire emissions for May 1 through September 30 were estimated for North America using the FireWork emissions system [27]. The total fire emissions of key trace gases and particulate species for each year from 2013–2016 show a significant variation in the seasonal totals (Table S1; see Supplementary Materials). Maximum emissions were observed for the 2014 season, and there was substantial spatial variability in the regional estimates (data not shown). The mean seasonal FireWork primary $PM_{2.5}$ emissions for North America for 2013–2016 of 1.4 Tg/season are lower than but comparable to previous estimates of North American annual $PM_{2.5}$ emissions from wildfires (1.9 and 2.2 Tg/y) [36]. For context, total U.S. anthropogenic $PM_{2.5}$ emissions in 2010 were estimated to be 4.1 Tg/y [37], so fire emissions are an important source of $PM_{2.5}$.

The FireWork domain covers most of Canada and the USA (including Alaska), as well as northern Mexico, with a 10 km × 10 km grid (Figure 1). A new operational version of FireWork with a new domain and new 10 km × 10 km grid (Figure 1) was introduced during the 2016 wildfire season [35]. Results presented in Section 3 are calculated on the original grid used by FireWork prior to 7 September 2016. FireWork results for the period after the grid change (approximately three weeks) were interpolated to the original grid to allow for consistent analysis. The area covered by the original FireWork domain was 33,085,648 km^2.

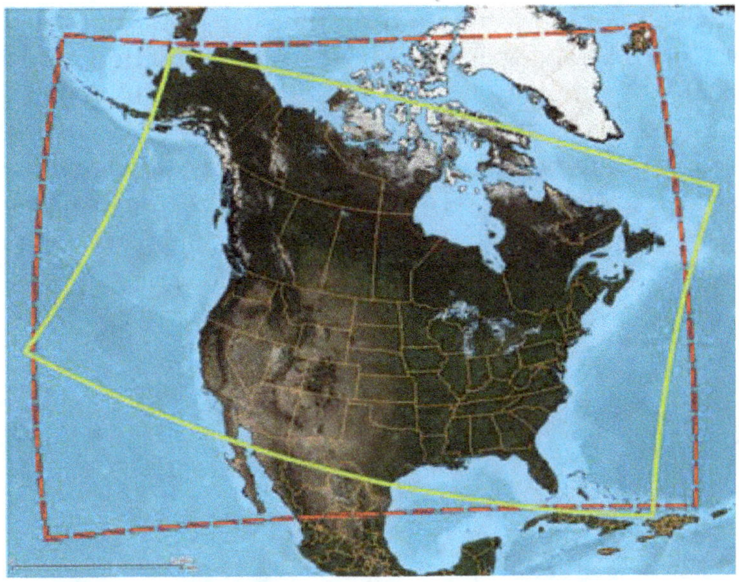

Figure 1. FireWork domain boundaries before (green) and after (red) 7 September 2016. The 10 km × 10 km grid is not shown.

2.2. *Wildfire Emissions' Contribution to $PM_{2.5}$ Pollution*

We used FireWork forecasts to analyze the contribution of fire-originated fine particulate matter (fire-$PM_{2.5}$) to $PM_{2.5}$ pollution over North America. In order to estimate the direct contribution of fire-$PM_{2.5}$ to the total $PM_{2.5}$ concentration forecasted by FireWork, the RAQDPS forecast $PM_{2.5}$ concentration field valid at the same hour was subtracted from the FireWork field. This simple strategy removes the contribution of the anthropogenic sources and other natural sources considered by both the RAQDPS and FireWork, and makes it possible to isolate wildfire smoke plume locations and follow their evolution over time [27]. The analysis of forecasted wildfire smoke presented in this paper is based on the set of hourly $PM_{2.5}$ concentration fields generated by this subtraction. Note that fire-$PM_{2.5}$ includes contributions from both primary $PM_{2.5}$ emissions and secondary aerosol formation from primary gas-phase emissions.

An essential part of characterizing the impacts of exposure to wildfire pollution is to understand both long-term (monthly to yearly) and short-term (hourly to daily) exposures. We used multi-year FireWork simulations (2013–2016) to characterize both long- and short-term wildfire pollution exposure over North America by calculating averages based on multi-year, seasonal, monthly, daily, and hourly concentrations and assessing areas affected by different concentration thresholds. Furthermore, we compared these averages with the $PM_{2.5}$ CAAQS of 10 μg/m^3 (annual standard) and 28 μg/m^3 (daily standard) [10], and with lower thresholds of 0.2, 1, and 5 μg/m^3. The 0.2 μg/m^3 threshold was based on the U.S. Environmental Protection Agency (EPA) Significant Impact Level (SIL) guidance document, which defines 0.2 μg/m^3 as the threshold below which any annual $PM_{2.5}$ change is considered negligible [38]. Also, from our own experience with FireWork, 0.2 μg/m^3 is the lowest value not susceptible to numerical noise that can be considered when analyzing the contributions of fire-$PM_{2.5}$ to total forecasted $PM_{2.5}$ concentrations. The 1 and 5 μg/m^3 thresholds were considered to transition between the minimal 0.2 μg/m^3 threshold and the 10 μg/m^3 threshold.

2.3. Population Exposure Estimation

Statistics on population exposure to wildfire smoke for Canada and the USA can be calculated by combining FireWork fields of direct contributions of fire-$PM_{2.5}$ to total surface $PM_{2.5}$ concentration with population data. We used population data from the 2016 Canadian census [39,40] and from the 2010 U.S. census [41]. For the 2016 Canadian census, we used population reported at the dissemination area (DA) level, where each DA typically has a population of 200 to 1000 people. For the 2010 U.S. census, we used population reported at the block-group level (Figure 2). Although U.S. population projections are available for 2016, they are only available at the coarser census-tract level rather than the more finely resolved block-group level (see Figure S1 in the Supplementary Materials).

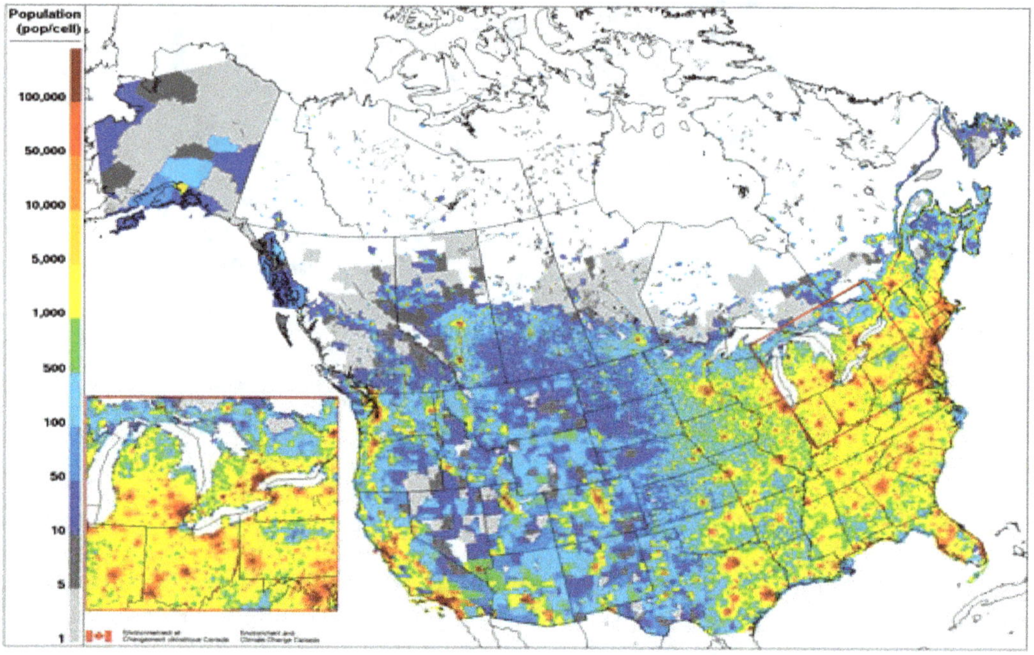

Figure 2. Population count per FireWork grid cell (10 km × 10 km) based on the 2016 Canadian census and the 2010 U.S. census. The red box over the four southern Great Lakes marks the location of the inset.

A number of steps were required to estimate the population affected at different wildfire $PM_{2.5}$ concentration thresholds. The first step was to determine the population for each 10 km × 10 km grid cell on the FireWork domain. To do so, 2016 Canadian population data reported by DA [39] were incorporated into a shapefile containing DA polygons [40]. The same step was performed at the sub-county level for 2010 U.S. population data [41]. The two population shapefiles were then interpolated separately to the 10 km by 10 km FireWork grid using a normalized conservative approach that preserved population within polygons. This interpolation approach divided the population value of each polygon between the grid cells wholly or partly contained within the polygon based on fractional area and assuming uniform population density within a DA or sub-county (Figure 2). The total populations contained in the older FireWork domain (Figure 1) for these two censuses were 35,148,512 in Canada and 305,744,285 in the USA.

The next step was to identify the aggregate population for the set of FireWork grid cells above a $PM_{2.5}$ concentration threshold. The Canadian and U.S. populations were processed separately. For grid cells along the Canada-USA border, it was necessary to determine in which country each cell was mainly located. This was done using a mask indicating the country associated with each grid cell. The population values for the set of FireWork grid cells associated with each $PM_{2.5}$ threshold were then summed together. The final step was to assess population exposure to different $PM_{2.5}$ concentrations.

To do this, five $PM_{2.5}$ thresholds (0.2, 1, 5, 10, and 28 $\mu g/m^3$) were considered and applied to monthly and seasonal (May to September) contribution of fire-$PM_{2.5}$ to total surface $PM_{2.5}$ concentrations.

2.4. Exposure Frequency Estimation

The temporal frequency of $PM_{2.5}$ pollution exposure is another critical factor in determining health impacts from fire-$PM_{2.5}$. One way to characterize this factor is to examine the number of occurrences of hourly or daily fire-$PM_{2.5}$ concentrations above specific thresholds for individual grid cells. This is done by counting the number of hours or days in a wildfire season with wildfire-related $PM_{2.5}$ concentrations above four different levels: 1, 5, 10 and 28 $\mu g/m^3$. In this case the lowest threshold of 0.2 $\mu g/m^3$ was not considered because it is less meaningful over shorter averaging periods.

3. Results

3.1. Area Affected by Wildfire Smoke

In 2013–2016, almost all areas of Canada and the USA included in the FireWork domain were affected by wildfire smoke based on a seasonal fire-$PM_{2.5}$ > 0.2 $\mu g/m^3$ exceedance at least once per grid cell (Figures 3–6). Western North America was more affected by wildfire smoke than eastern North America in all four years. From a continental perspective, 2014 had the most intense wildfire season, based on the seasonal values of the area affected by wildfire pollution for $PM_{2.5}$ concentration thresholds from 1 $\mu g/m^3$ to 28 $\mu g/m^3$ (Table 2). This year was followed by 2015 and 2013, while 2016 was the year least affected by wildfire smoke. Based on the average seasonal concentrations, the percentage areas of the FireWork domain (including land and water areas) above the 0.2 $\mu g/m^3$ threshold were 52%, 49%, 44%, and 22% for the years 2013–2016, respectively (Table 2). Above the 1 $\mu g/m^3$ threshold the corresponding percentage areas were 14%, 17%, 13%, and 6%, and above the 5 $\mu g/m^3$ threshold the values were 0.8%, 1.9%, 1.1%, and 0.4%.

Table 2. FireWork domain area affected (km^2 and percentages) by wildfire pollution above five $PM_{2.5}$ concentration thresholds based on average monthly and seasonal fire-$PM_{2.5}$ contributions to total average monthly and seasonal surface $PM_{2.5}$ concentrations. For reference, the total area of North America is 24.71 million km^2 and for the FireWork domain is 33.09 million km^2. The numbers in parentheses correspond to the percentage of the area affected.

$\mu g/m^3$	>0.2		>1		>5		>10		>28	
					2013					
May	754,020	(2.3%)	30,662	(0.1%)	1377	(0.0%)	196	(0.0%)	0	(0.0%)
June	12,478,900	(37.7%)	2,532,748	(7.7%)	184,058	(0.6%)	48,991	(0.1%)	7378	(0.0%)
July	17,235,170	(52.1%)	7,279,618	(22.0%)	861,986	(2.6%)	194,966	(0.6%)	13,326	(0.0%)
August	20,155,062	(60.9%)	9,873,087	(29.8%)	1,783,453	(5.4%)	604,476	(1.8%)	71,844	(0.2%)
September	9,351,929	(28.3%)	2,072,752	(6.3%)	84,693	(0.3%)	39,022	(0.1%)	10,611	(0.0%)
					2014					
May	1,746,428	(5.3%)	61,310	(0.2%)	6507	(0.0%)	2137	(0.0%)	200	(0.0%)
June	3,411,109	(10.3%)	872,761	(2.6%)	57,926	(0.2%)	13,461	(0.0%)	1686	(0.0%)
July	16,196,537	(49.0%)	7,260,984	(21.9%)	1,562,298	(4.7%)	541,468	(1.6%)	146,616	(0.4%)
August	19,536,859	(59.0%)	8,436,851	(25.5%)	2,560,671	(7.7%)	934,005	(2.8%)	157,462	(0.5%)
September	13,305,055	(40.2%)	4,531,984	(13.7%)	444,316	(1.3%)	137,840	(0.4%)	24,680	(0.1%)
					2015					
May	2,837,325	(8.6%)	252,707	(0.8%)	12,768	(0.0%)	2950	(0.0%)	197	(0.0%)
June	9,375,291	(28.3%)	2,604,739	(7.9%)	291,593	(0.9%)	42,184	(0.1%)	2552	(0.0%)
July	14,245,102	(43.1%)	5,173,825	(15.6%)	1,033,576	(3.1%)	306,687	(0.9%)	30,776	(0.1%)
August	15,581,769	(47.1%)	5,918,077	(17.9%)	1,236,416	(3.7%)	636,671	(1.9%)	143,120	(0.4%)
September	11,999,087	(36.3%)	1,629,073	(4.9%)	102,071	(0.3%)	45,570	(0.1%)	14,173	(0.0%)
					2016					
May	3,829,030	(11.6%)	972,851	(2.9%)	167,935	(0.5%)	58,168	(0.2%)	11,955	(0.0%)
June	2,532,347	(7.7%)	334,211	(1.0%)	35,392	(0.1%)	13,575	(0.0%)	3599	(0.0%)
July	7,866,765	(23.8%)	2,709,907	(8.2%)	223,084	(0.7%)	57,209	(0.2%)	8215	(0.0%)
August	7,807,827	(23.6%)	2,839,017	(8.6%)	477,835	(1.4%)	146,340	(0.4%)	31,658	(0.1%)
September	4,914,590	(14.9%)	1,715,135	(5.2%)	204,326	(0.6%)	71,075	(0.2%)	15,620	(0.0%)

Table 2. *Cont.*

μg/m³	>0.2		>1		>5		>10		>28	
	Seasonal Exceedances (Area)									
2013	17,151,407	(51.8%)	4,534,747	(13.7%)	269,933	(0.8%)	62,226	(0.2%)	9267	(0.0%)
2014	16,335,869	(49.4%)	5,616,429	(17.0%)	625,489	(1.9%)	216,897	(0.7%)	28,417	(0.1%)
2015	14,507,744	(43.8%)	4,203,396	(12.7%)	373,565	(1.1%)	90,825	(0.3%)	16,248	(0.0%)
2016	7,229,006	(21.8%)	1,931,583	(5.8%)	121,897	(0.4%)	40,841	(0.1%)	9388	(0.0%)

In terms of total area burned, in 2013 Canada had its seventh most intense wildfire season of the past 34 years [2], whereas the USA was 37% below its 2006–2016 average [42]. The wildfire season in 2013 effectively started in June, with maximum intensity reached in July and August. During these two months, the area for which the monthly average of fire-$PM_{2.5}$ exceeded the 0.2 μg/m³ threshold covered most of North America (Figure 3). In July, an area of 194,966 km² (the size of South Dakota) had average monthly fire-$PM_{2.5}$ above 10 μg/m³ and 13,326 km² were above 28 μg/m³ (Table 2). In August, using the same thresholds, these areas were 604,476 km² (larger than California; close to the size of Manitoba and Texas) and 71,844 km², respectively. August 2013 was the most active month of the year, with intense activity in northwestern and western Canada and the northwestern USA.

In 2014, Canada had its fifth most intense wildfire season of the past 34 years in terms of total area burned, while in the USA the value was 48% below its 2006–2016 average [2,42]. The extreme wildfire event in the Northwest Territories near the city of Yellowknife started in June and peaked in July (Figure 4). In July, intense wildfires began burning in British Columbia, Alberta, Washington, Oregon, California, and Idaho, bringing the total area with average monthly fire-$PM_{2.5}$ > 10 μg/m³ to 541,468 km², and > 28 μg/m³ to 146,616 km². These fires persisted into August, making August 2014 the most extreme month of the 2013–2016 period in terms of area affected by wildfire smoke. In this month, average monthly fire-$PM_{2.5}$ values above the 5, 10, and 28 μg/m³ thresholds covered 2,560,671 km² (larger than Alaska or Nunavut), 934,005 km² (the size of British Columbia), and 157,462 km² (the size of Georgia), respectively (Table 2).

The 2015 season was the sixth most intense of the past 34 years for Canada in terms of area burned [2]. In the USA, it was the peak year of the 2006–2016 period, with more than 10 million acres (40,500 km²) burned, 45% above the period average [42]. The 2015 fire season was marked by two intense wildfire periods: the first from 15 June to 15 July and the second from 1 to 15 August [27] (Figure 5). In the first period, most of the wildfires occurred in northwestern Canada (Alberta and Saskatchewan), whereas in the second period most of the wildfires occurred in the western USA (Washington, Oregon, Idaho, and California). In July and August areas of 306,687 km² and 636,671 km² had average monthly fire-$PM_{2.5}$ > 10 μg/m³, respectively, and areas of 30,776 km² and 143,120 km² were >28 μg/m³ (Table 2).

The 2016 fire season included unprecedented impacts in Canada on both people and the national economy. The entire city of Fort McMurray, Alberta, with a population of nearly 90,000, was evacuated in May when it was overrun by a large, fast-moving wildfire (Figure 6). Estimated insured fire damages to Fort McMurray were 9.6 billion dollars, the costliest insured natural disaster in Canadian history [4]. Despite this disaster, 2016 was the least intense wildfire season among the four years analyzed in terms of area burned [2] and area affected by wildfire smoke across Canada (Table 2). In the USA, the 2016 area burned was 21% below its 2006–2016 average [42]. On the other hand, the early start to the wildfire season in Canada made the month of May 2016 the most affected by wildfire pollution among the four Mays assessed. For May 2016 the area with average monthly fire-$PM_{2.5}$ > 10 μg/m³ and > 28 μg/m³ was 58,168 km² and 11,955 km² respectively (Table 2).

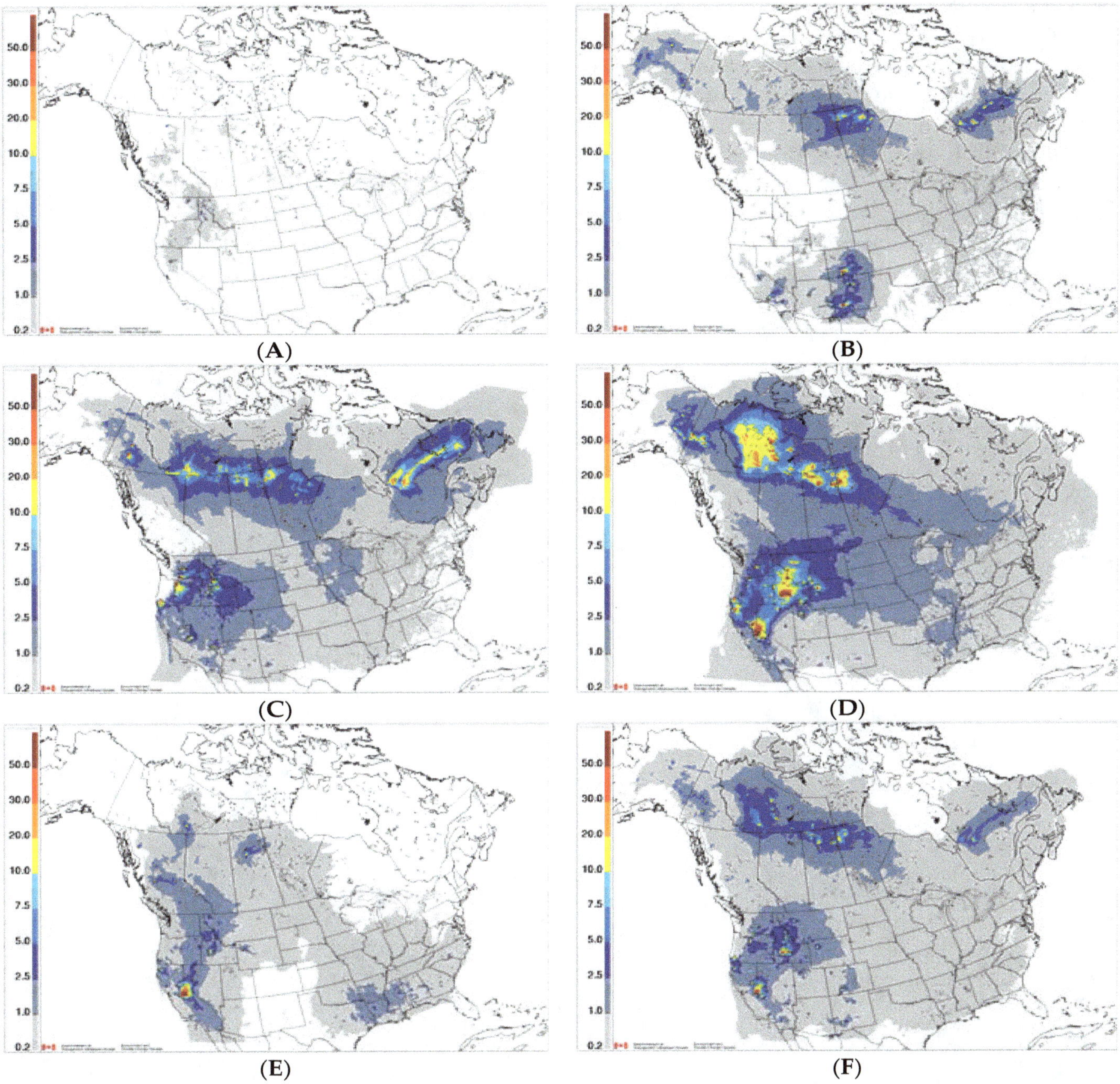

Figure 3. 2013 average monthly fire-$PM_{2.5}$ contribution to total forecasted surface $PM_{2.5}$ concentrations ($\mu g/m^3$) for (**A**) May, (**B**) June, (**C**) July, (**D**) August, (**E**) September, and (**F**) seasonal (May–September average). Note that the color scale is non-linear and white areas indicate values < 0.2 $\mu g/m^3$.

It is also of interest to examine the importance of fire-$PM_{2.5}$ relative to other sources of $PM_{2.5}$. Figure S2 shows the seasonal fire-$PM_{2.5}$ contribution to total $PM_{2.5}$ as a percentage for each of the four years. For 2013 to 2015, the seasonal contribution of fire-$PM_{2.5}$ to total $PM_{2.5}$ was 50% or more over much of northwestern North America and parts of the U.S. mountain west. In 2014 the seasonal wildfire contribution was 90% or greater for a large part of the Northwest Territories and parts of the interior of British Columbia. These results are not surprising considering that these areas have relatively few inhabitants and low anthropogenic emissions.

It is difficult to compare the monthly model forecasted fire-$PM_{2.5}$ directly with $PM_{2.5}$ measurements, as measurements are influenced by all $PM_{2.5}$ sources, not just biomass burning emissions. As an indirect comparison, however, we note that archived near-real-time measurement from the AirNow data feed (www.airnow.gov) include at least one U.S. or Canadian AQ station located close to wildfires having monthly $PM_{2.5}$ concentrations above 30, 50, and even 150 μg/m^3 for each of the four years analyzed here. The most extreme month was August 2015, when 13 $PM_{2.5}$ measurement stations reported mean monthly $PM_{2.5}$ concentrations of 30 μg/m^3 or above, including six stations in Idaho, two stations each in Oregon, Washington State, and British Columbia, and one station in Montana (see Table S2). FireWork also forecasted mean monthly fire-$PM_{2.5}$ concentrations above 30 μg/m^3 for two regions of California in August 2015; although no stations with available measurements were located in these regions, one nearby station in Shasta county in northern California had a mean monthly $PM_{2.5}$ concentration of 24.5 μg/m^3 and 100% data completeness.

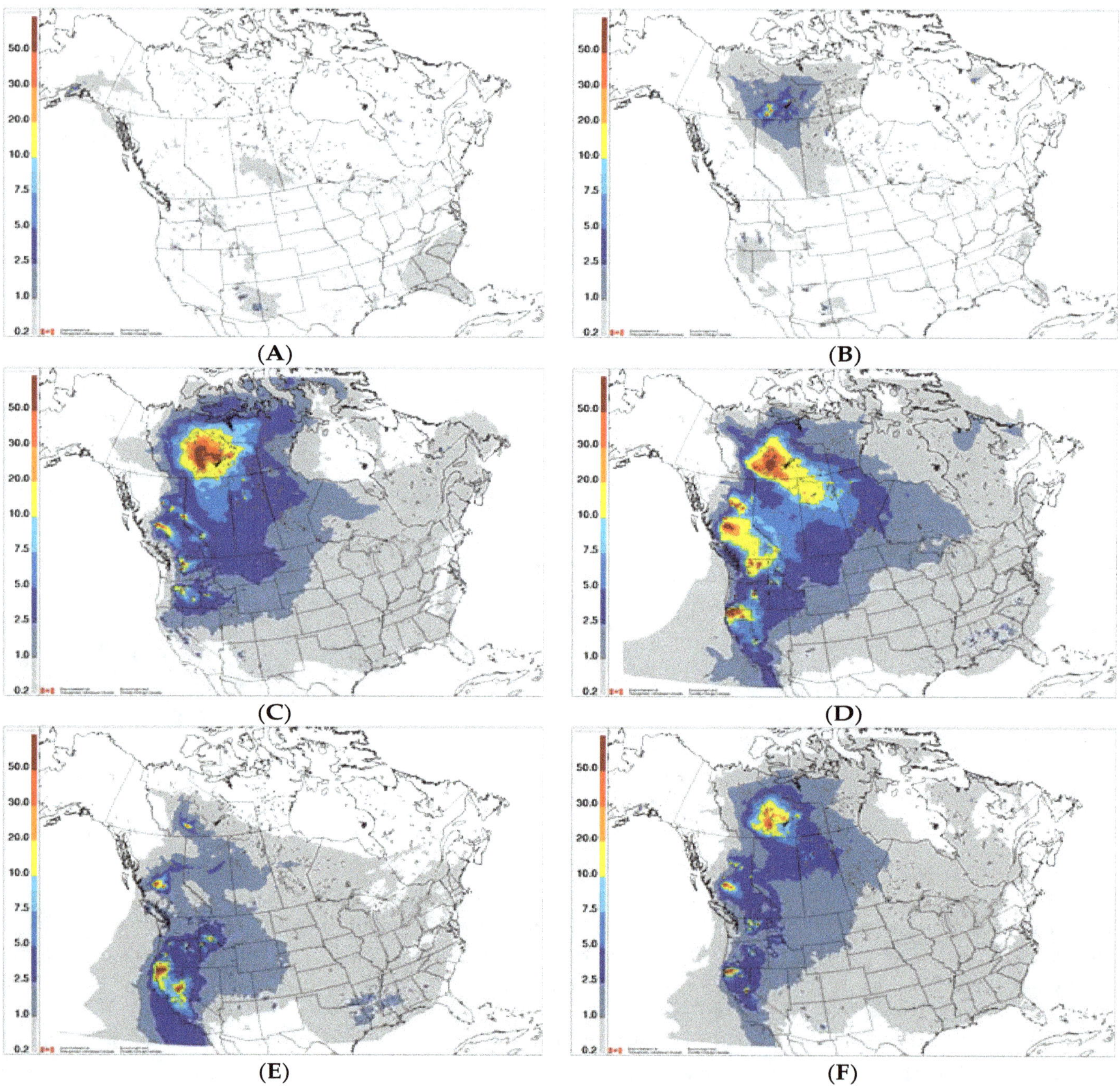

Figure 4. Same as Figure 3 but for 2014.

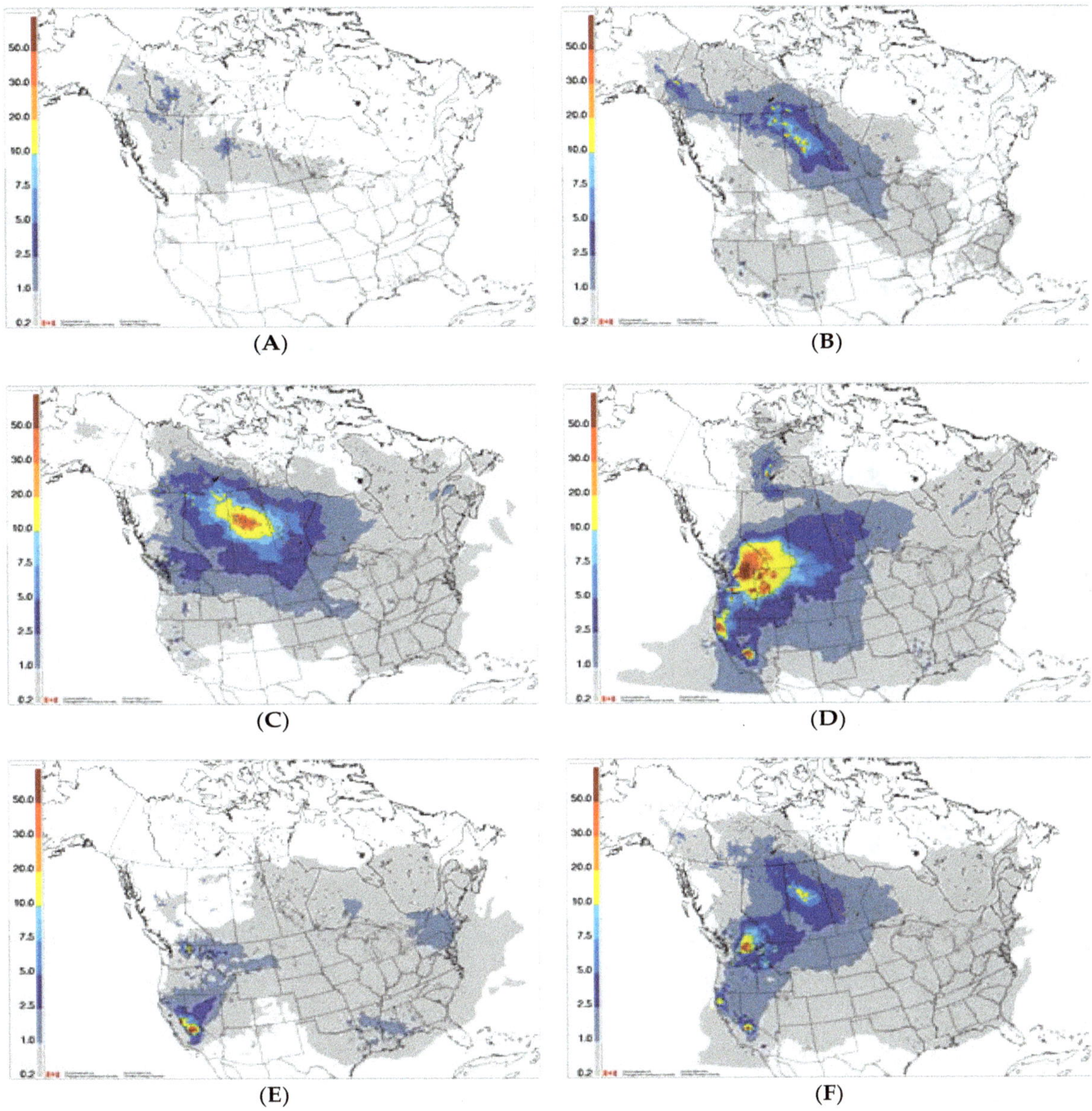

Figure 5. Same as Figure 3 but for 2015.

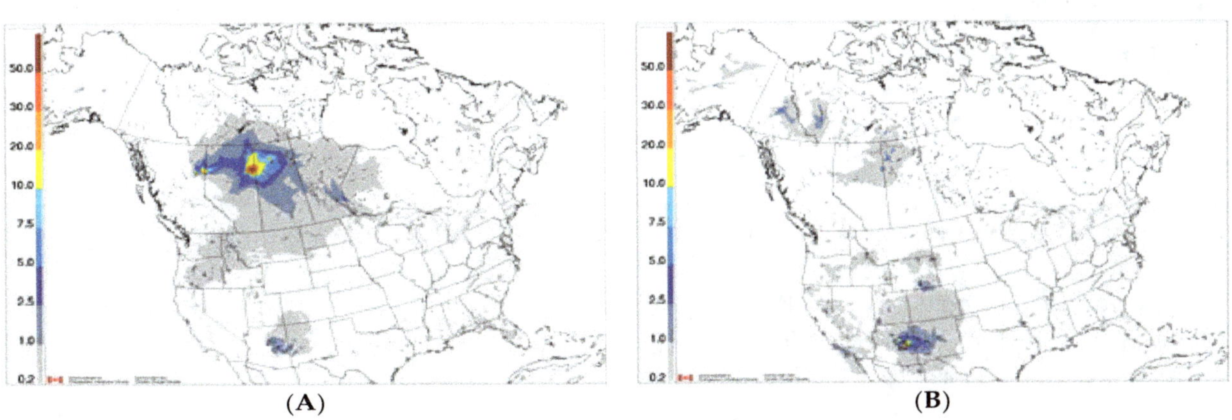

Figure 6. *Cont.*

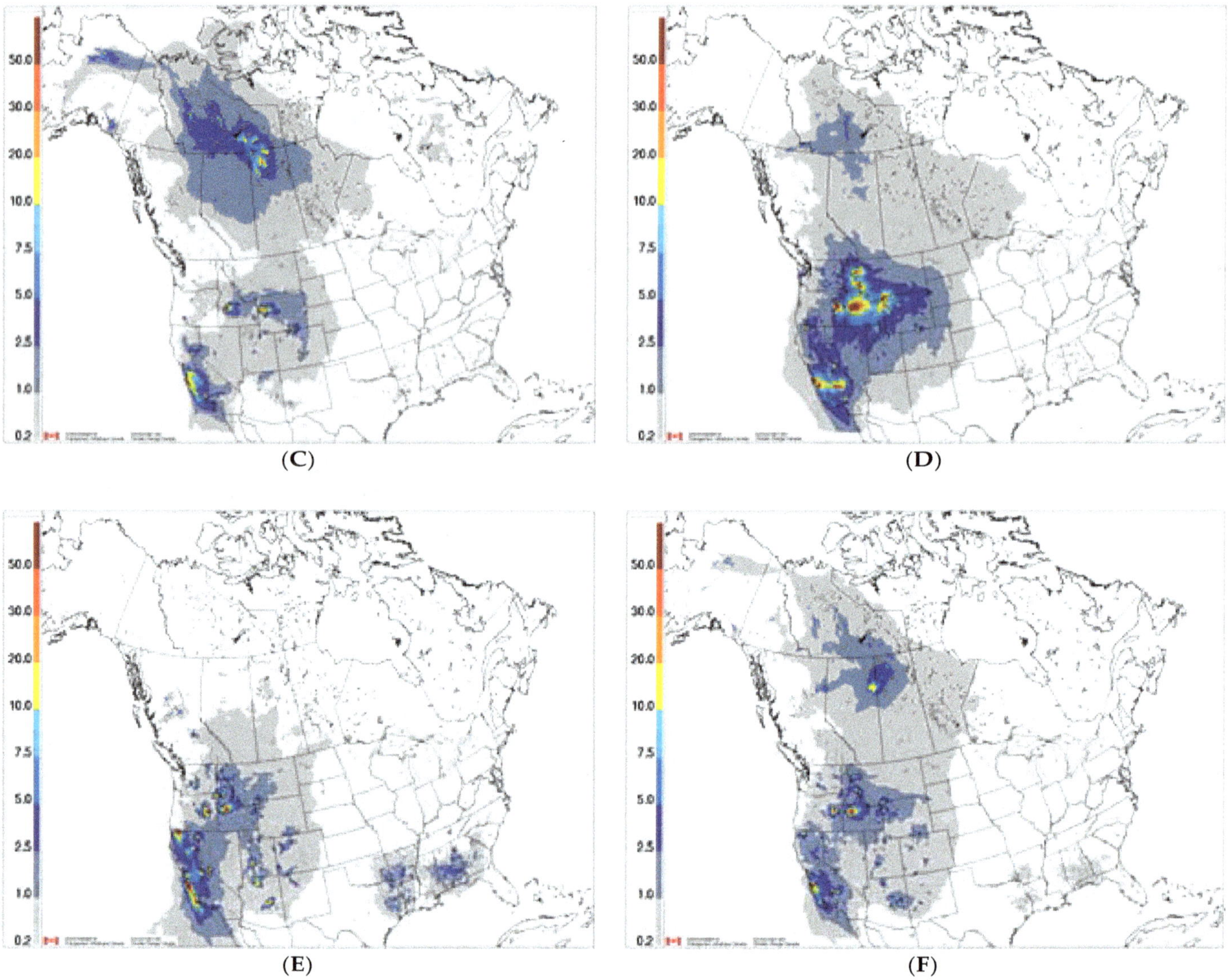

Figure 6. Same as Figure 3 but for 2016.

3.2. *Population Exposure to Wildfire Pollution*

In terms of population exposure to wildfire-related $PM_{2.5}$ pollution, for 17 of the 20 months considered, more than 1 million Canadians (3% of the population) were estimated to have been affected by average monthly fire-$PM_{2.5}$ > 0.2 μg/m^3 (Figure 7). During the same period, more than 14 million Canadians (39% of the population) were affected by average monthly fire-$PM_{2.5}$ > 0.2 μg/m^3 in 10 of the 20 months, and over 32 million Canadians (90% of the population) were affected in 7 months. The months affecting the most people were July and August (Figure 8 and Table S3), with August 2015 being the worst for average monthly fire-$PM_{2.5}$ > 10 μg/m^3. During August 2015 the proportion of the Canadian population affected by fire-$PM_{2.5}$ above thresholds of 0.2, 1, 5, 10, and 28 μg/m^3 were 97%, 21%, 11%, 8%, and 1.2%, respectively. The three periods in which the most Canadians were exposed to >28 μg/m^3 were: (a) August 2015 (417,171 people or 1.2% of the population) due to extreme wildfires in British Columbia and the northwestern USA; (b) May 2016 (69,909 people or 0.2% of the population) due to extreme wildfires in northern Alberta near Fort McMurray; and (c) August 2014 (27,160 people or 0.1% of the population) due to extreme wildfires in northwestern Canada. In all other months, less than 0.1% of the Canadian population was affected at this very high threshold (Table S3).

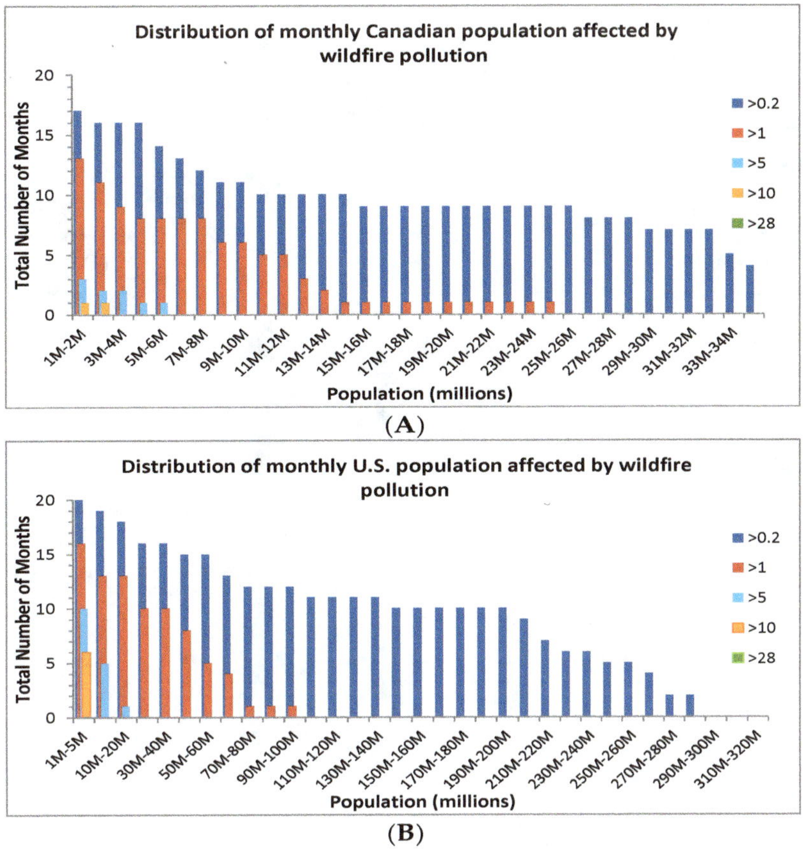

Figure 7. Cumulative distribution of number of months with the portions of the Canadian (**A**) and U.S. (**B**) population affected by monthly fire-$PM_{2.5}$ above five $PM_{2.5}$ concentration thresholds for four five-month wildfire seasons (2013–2016).

Based on the average seasonal statistics (Table S3), over 90% of the Canadian population was affected by seasonal fire-$PM_{2.5}$ > 0.2 μg/m^3 in 2013, 2014 and 2015. The 2016 wildfire season was much milder, with 19% of Canadian population affected by wildfire smoke above this threshold (Figure 8). The corresponding proportions for the 1 μg/m^3 threshold ranged from 0.4% (2016) to 26% (2014). The population affected by concentrations above the 10 μg/m^3 threshold reached its maximum in 2015, with over 100,000 people (Table S3).

In the USA, more than 200 million people (65% of the population) were exposed to average monthly fire-$PM_{2.5}$ > 0.2 μg/m^3 during nine of the 20 months considered (Figure 7). As well, a much greater proportion of the U.S. population was affected by wildfire pollution in 2013, 2014, and 2015 than in 2016, similar to Canada (Figure 8), and for both the USA and Canada, the proportion of the population exposed to wildfire pollution in September was larger than in June.

The total percentages of the U.S. population affected by seasonal fire-$PM_{2.5}$ > 0.2 μg/m^3 ranged from 21% (2016) to 90% (2015) (Figure 8 and Table S4), and the corresponding range above 1 μg/m^3 was 5% (2015) to 15% (2014). Based on the four-year average seasonal statistics for population exposure to fire-$PM_{2.5}$ > 0.2 μg/m^3, a smaller percentage of the U.S. population (69%) than the Canadian population (76%) was exposed to pollution above this threshold (Tables S3 and S4). For concentrations > 1 μg/m^3 the corresponding percentages were 10% in the USA and 12% in Canada. However, for average seasonal fire-$PM_{2.5}$ > 28 μg/m^3, a higher percentage of Americans than Canadian were exposed. The affected U.S. population ranged from 32,549 in 2014 to 56,442 in 2015 (Table S4), giving a four-year average exposure of 0.015% for the U.S. population compared with 0.004% for the Canadian population (Table S3).

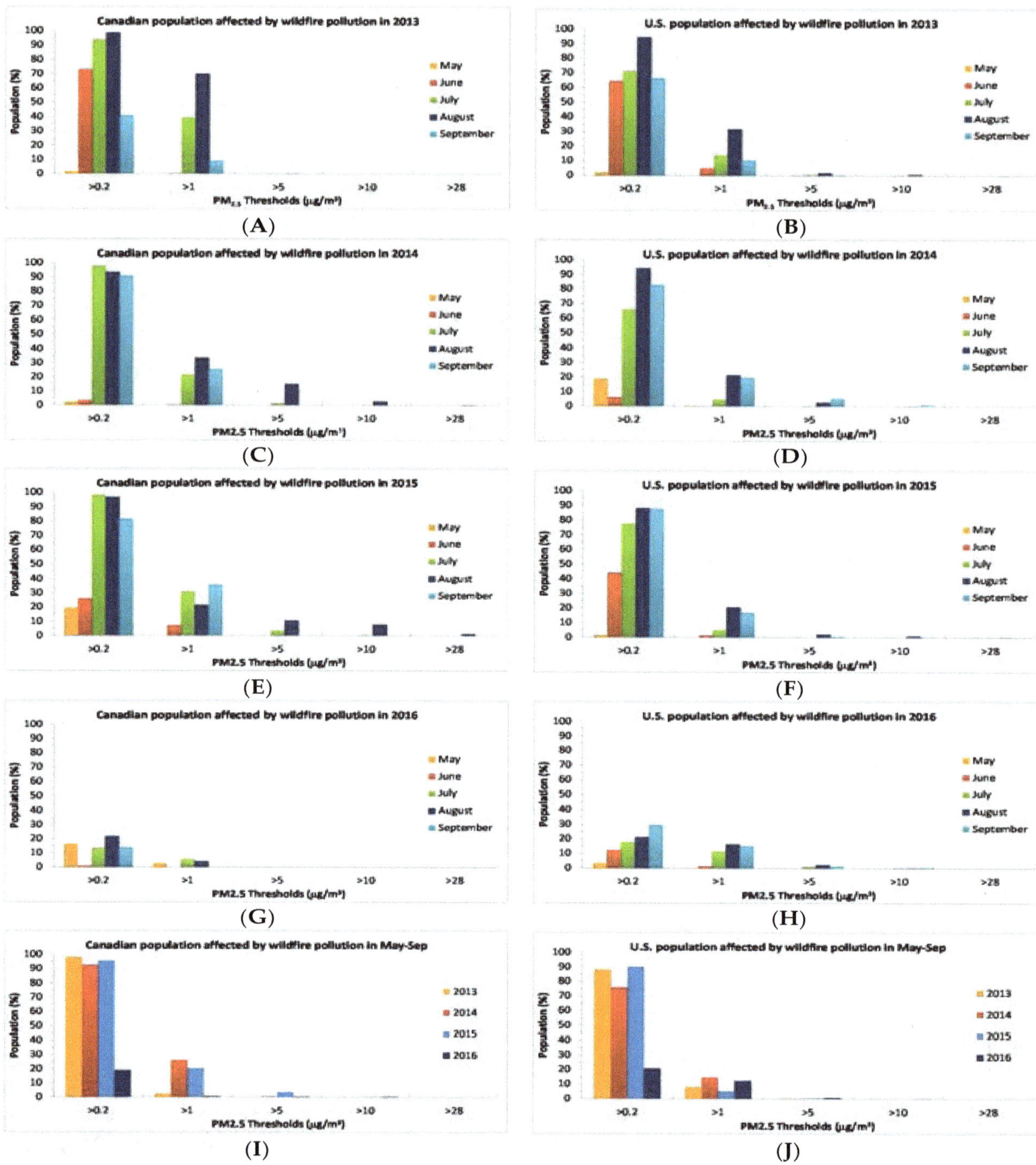

Figure 8. Percentage of population in Canada (left panels: **A, C, E, G,** and **I**) and the USA (right panels: **B, D, F, H,** and **J**), affected by wildfire pollution above five $PM_{2.5}$ concentration thresholds based on the average monthly and seasonal fire-$PM_{2.5}$ contribution to total average monthly and seasonal surface $PM_{2.5}$ concentrations for four wildfire seasons (2013–2016). The percentage of the affected population for Canada and the USA was calculated using the 2016 Canadian and the 2010 U.S. censuses, respectively. See also Tables S3 and S4.

3.3. Frequency of Wildfire-Related Pollution Events

In 2013–2016 most of North America (except Alaska) was affected by wildfire smoke on at least one day (Figure 9). The land-use map for Alaska used by National Resources Canada to determine total fuel consumption was only updated in 2015. Given the direct impact of wildfire emissions estimates from input land-use in FireWork, there may be an underestimation of wildfire emissions for regions of Alaska prior to 2015. The highest daily frequency of wildfire smoke occurred in 2014, where for most of western Canada and the northwestern USA more than 30% of the days from May to September had surface fire-$PM_{2.5}$ greater than 1 $\mu g/m^3$ for at least one hour. In parts of Alberta, Saskatchewan and the Northwest Territories, the daily frequency was above 40%, and in part of the southern Northwest Territories it was over 60%. For hourly frequency, in 2013 the majority of Canada and the USA had more than 10% of hourly forecasted fire-$PM_{2.5}$ above 1 $\mu g/m^3$. Hourly frequency was slightly lower in 2014 and 2015, but in 2016, only a portion of western North America had hourly frequencies over 10% (Figure 9).

The daily frequency of forecasted fire-$PM_{2.5}$ concentrations greater than 5 $\mu g/m^3$ was above 10% for the majority of western Canada in 2013, 2014 and 2015 (Figure 10). The same was true for northern Quebec in 2013. In the case of the USA, regions over 10% were found only in the west over the same period. In 2016, frequencies above 10% were less common when compared with other years and were limited to areas close to the wildfires in northwestern Canada and the northwestern USA and California. Frequencies above 30% were found in 2014 in northwestern Canada, while this percentage was limited to areas very close to the wildfires in the other years. The results of the hourly frequency analysis are similar to those of the daily frequency analysis, with values of 10–20% covering large areas of northwestern Canada and the western USA, especially for 2014. Hourly frequencies over 20% were not present over Canada in 2016 and were limited to areas close to wildfires in 2013, 2014, and 2015 (Figure 10).

The spatial patterns of daily and hourly frequencies for the 10 $\mu g/m^3$ concentration threshold were very similar to the patterns observed for the 5 $\mu g/m^3$ threshold (Figure 11). However, 2014 was the only year with daily and hourly frequencies above 30%, in northwestern Canada. For a threshold of 28 $\mu g/m^3$ (Figure 12), the frequency of days and hours with hourly forecasts above this threshold was generally below 10%.

We can also look at the number of days with elevated fire-$PM_{2.5}$ from a population exposure perspective. Tables S5 and S6 provide this information for Canada and the USA, respectively. In 2014 wildfire season, more than 14% of Canadians were exposed to a daily fire-$PM_{2.5}$ > 5 $\mu g/m^3$ on at least 30 days (i.e., 20% or more days) and 12% of Canadians were exposed to a daily fire-$PM_{2.5}$ > 10 $\mu g/m^3$ on at least 15 days (Table S5). For the USA, the corresponding values in 2014 were 2% and 3%, but interestingly they were higher (3% and 4%) in 2016 (Table S6), even though in other respects 2016 had a less active fire season (e.g., Table 2).

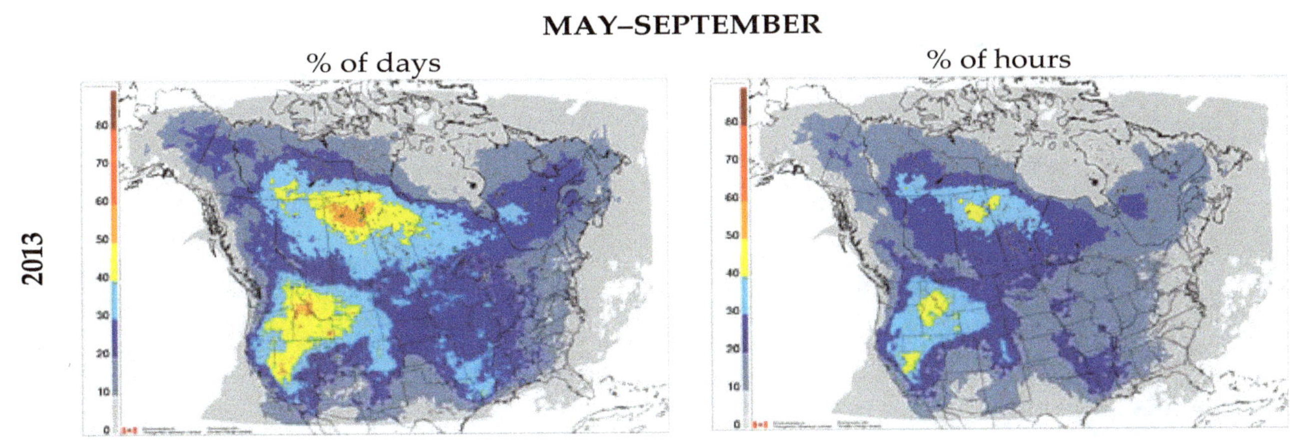

Figure 9. *Cont.*

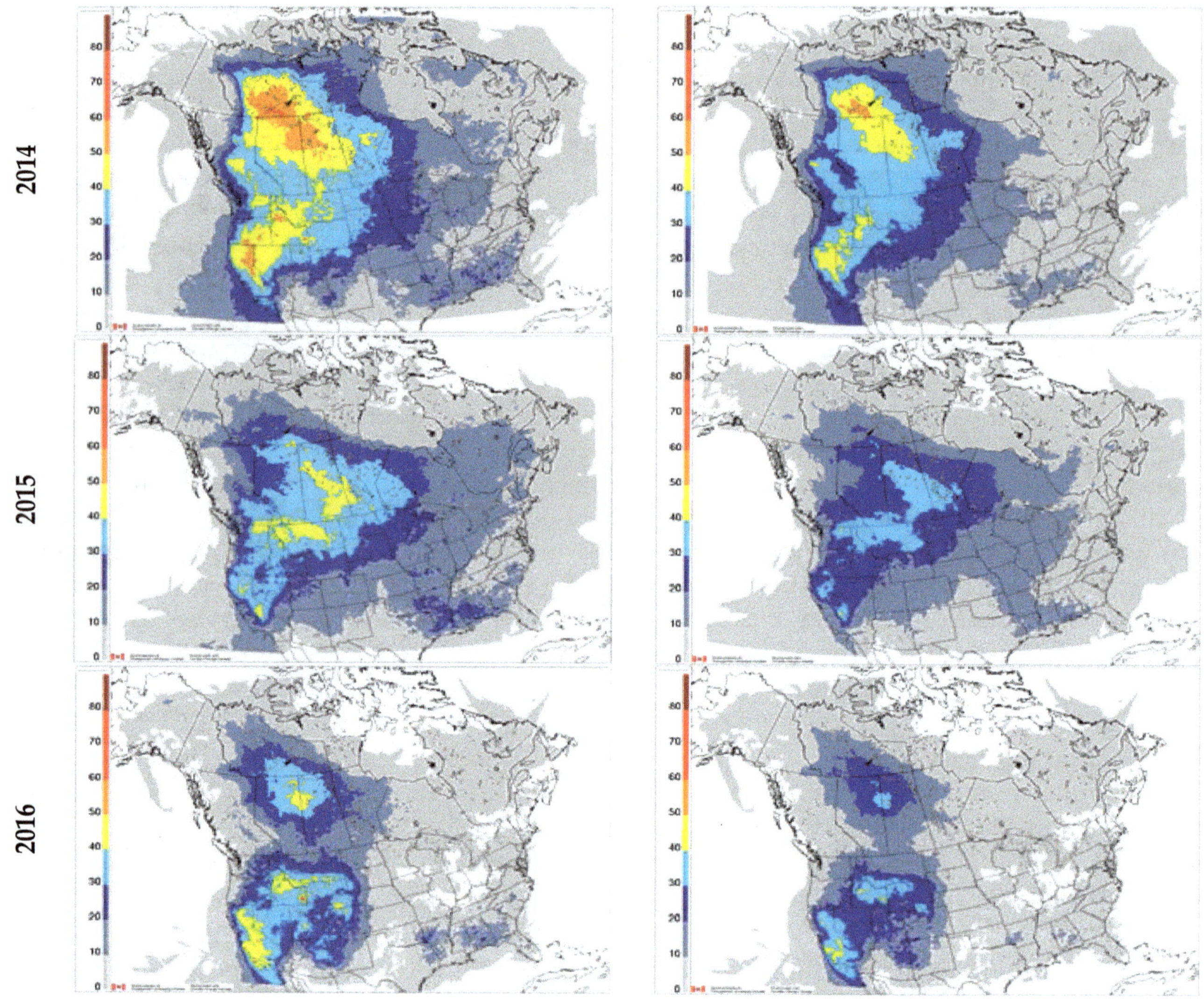

Figure 9. Percentage frequency of the number of days (**left**) and the number of hours (**right**) with forecasted 24-h moving average $PM_{2.5}$ concentration above 1 $\mu g/m^3$ from fire-$PM_{2.5}$ contribution for the period May–September for years 2013, 2014, 2015, and 2016. White areas indicate locations that experienced no days or hours above the threshold during the period.

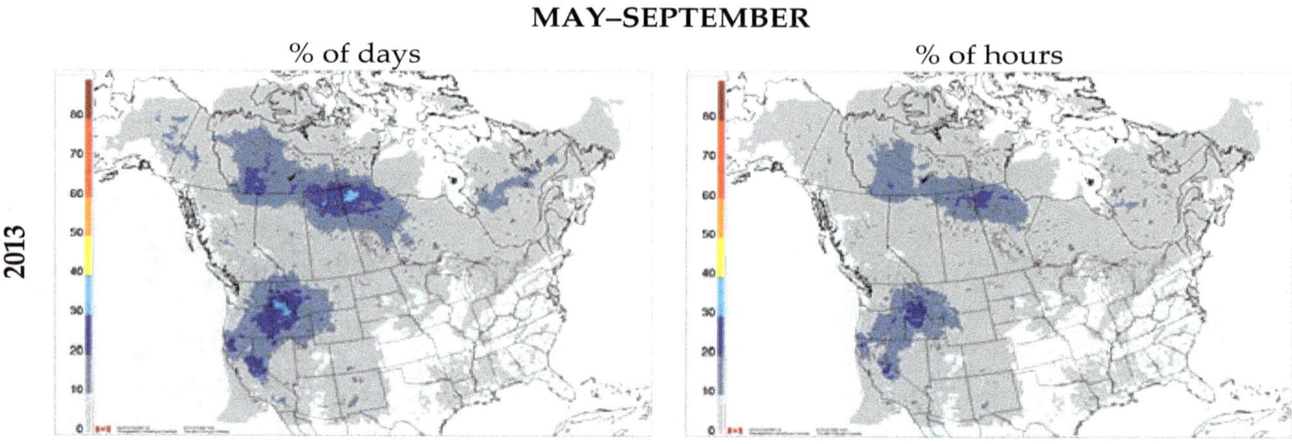

Figure 10. *Cont.*

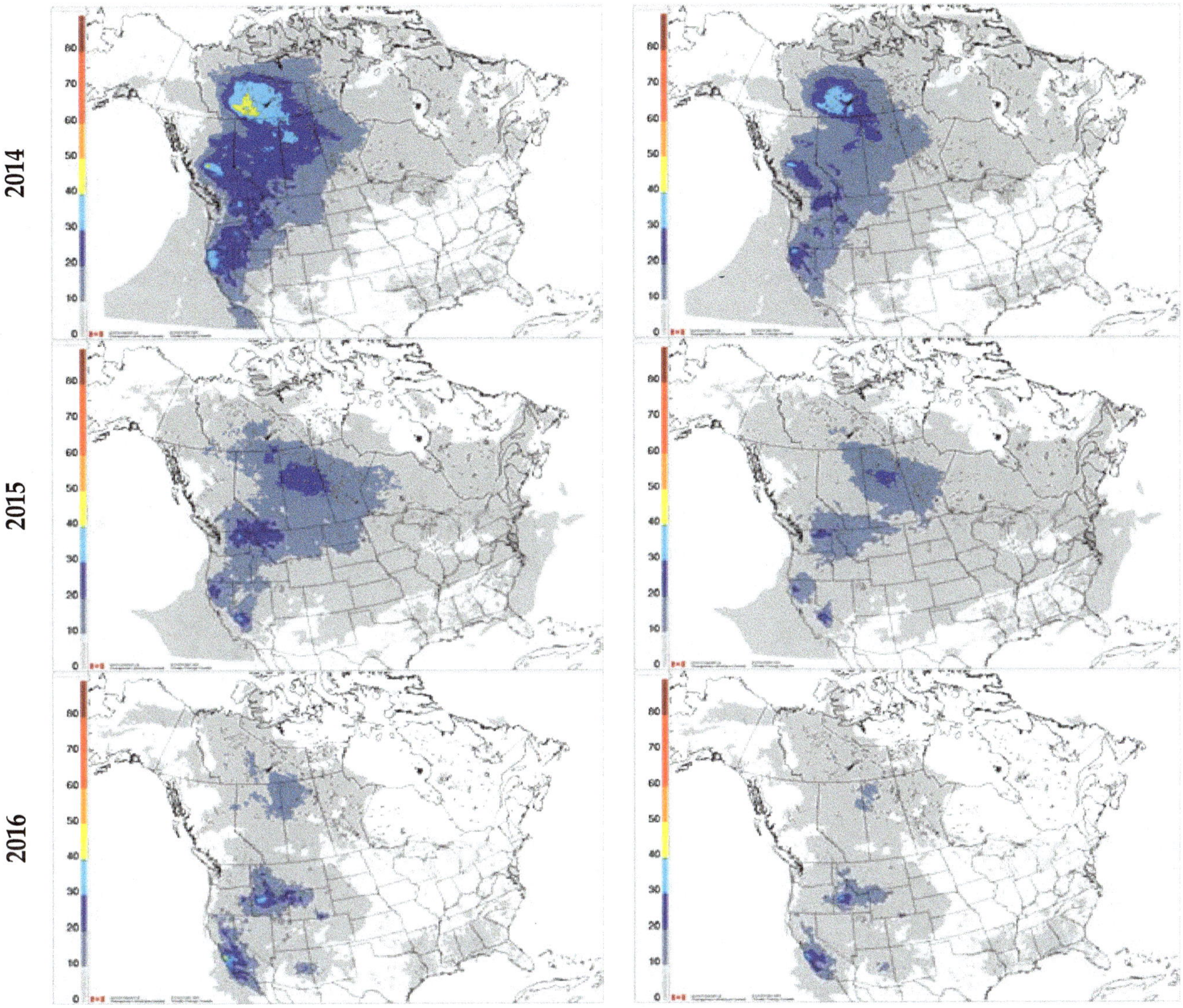

Figure 10. Same as Figure 9 but for 5 μg/m^3.

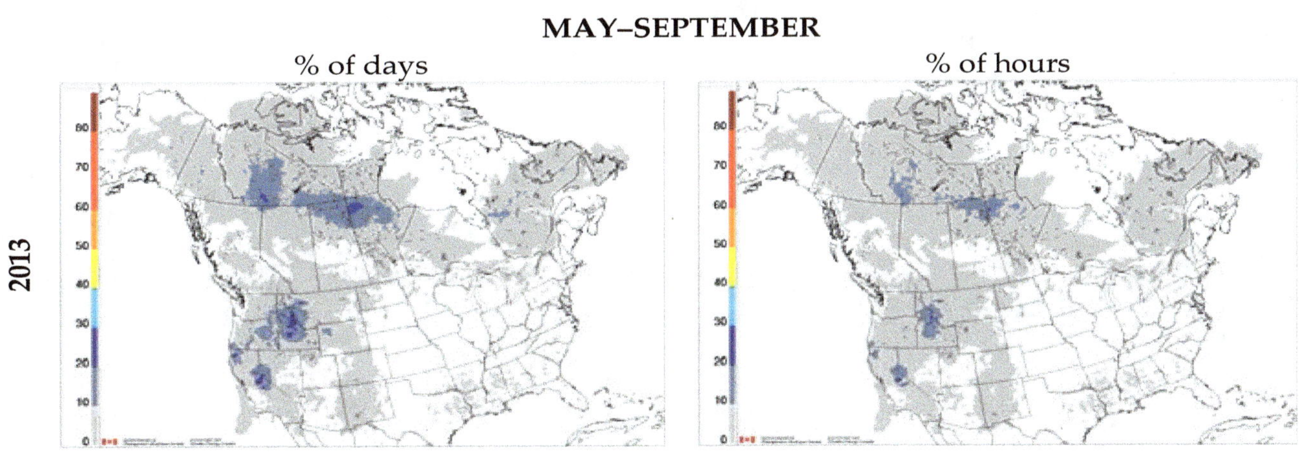

Figure 11. *Cont.*

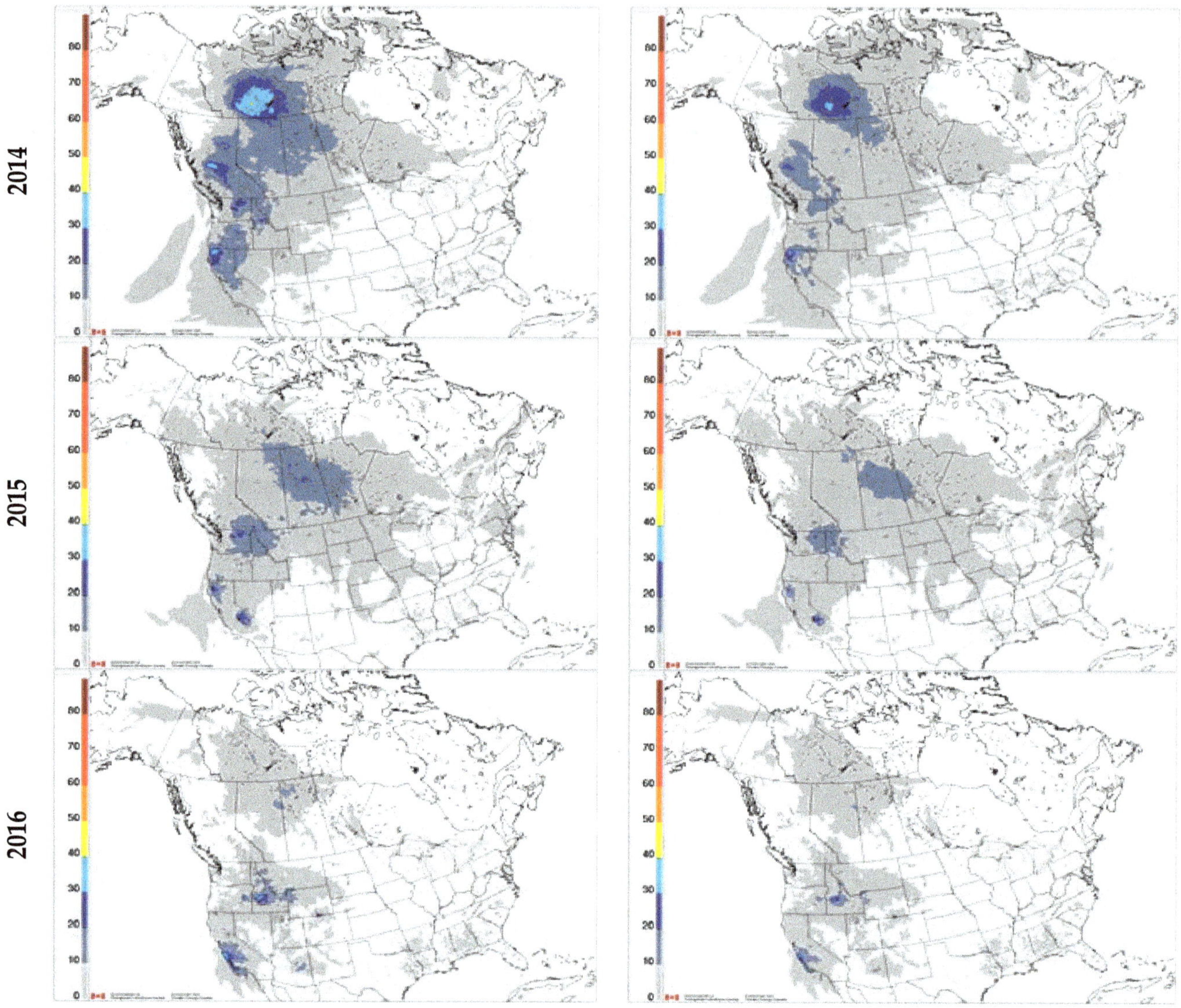

Figure 11. Same as Figure 9 but for 10 μg/m^3.

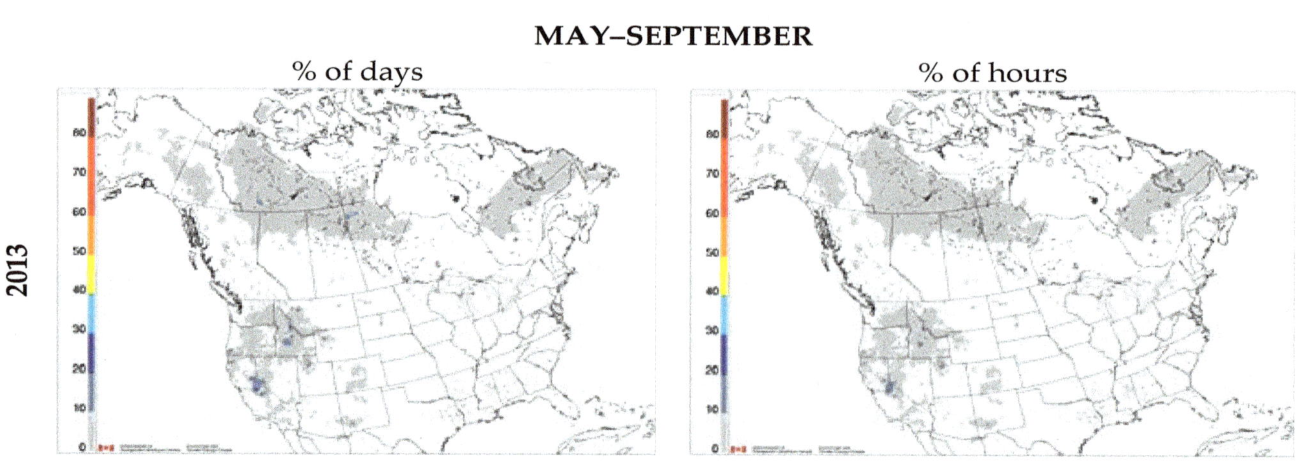

Figure 12. *Cont.*

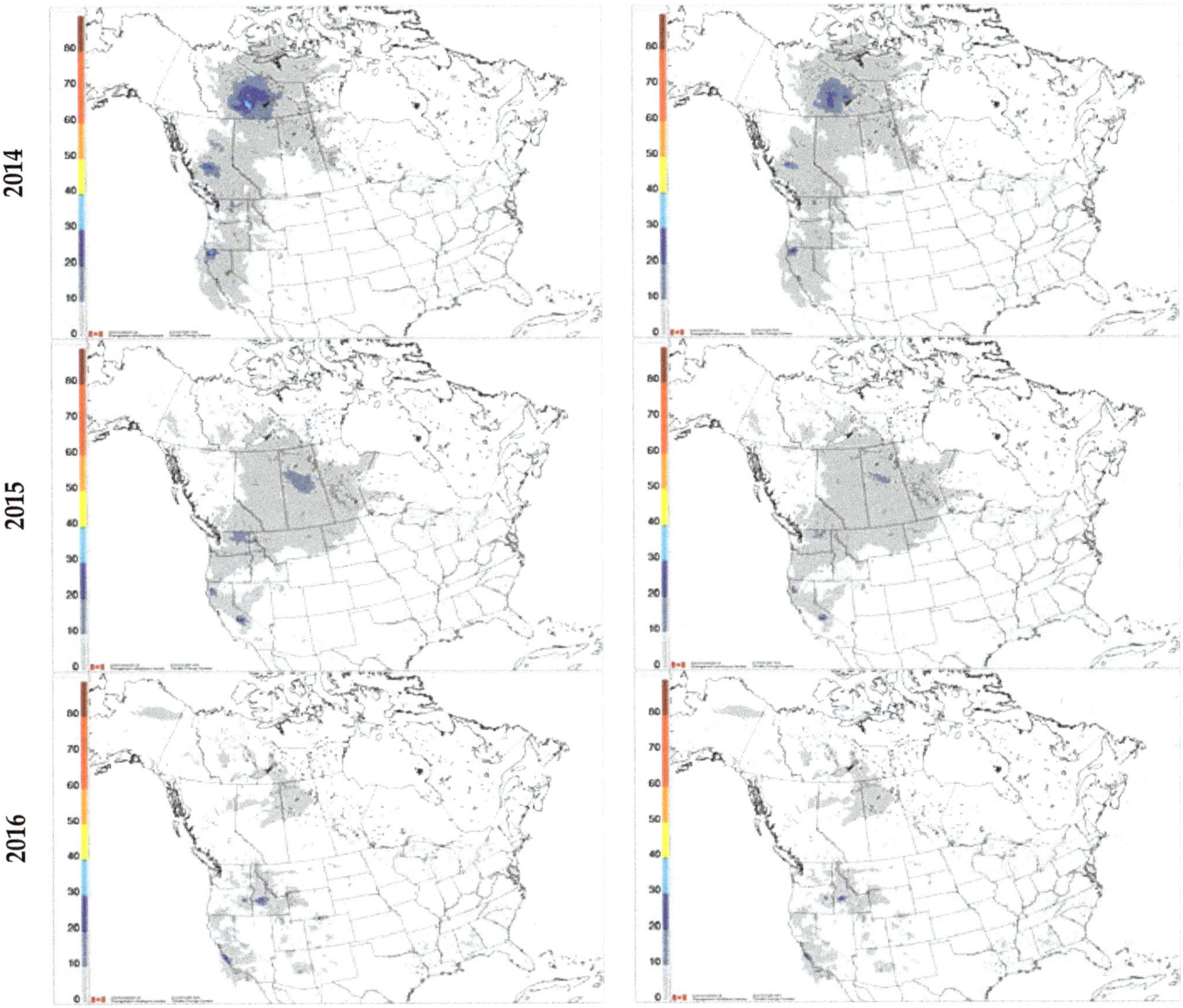

Figure 12. Same as Figure 9 but for 28 μg/m^3.

4. Discussion

The spatial distributions of monthly and seasonal wildfire plumes illustrate the large spatial and temporal variability of wildfire occurrence in North America (Figures 3–6). Our results also highlight how common wildfires are each summer and the large spatial extent of their influence due to long-range transport of wildfire emissions. Analyses of daily elemental and organic carbon measurements from the U.S. IMPROVE speciated $PM_{2.5}$ measurement network support the analysis of the FireWork wildfire smoke forecasts presented here, indicating that smoke from wildfires contributes substantially to $PM_{2.5}$ levels in the western USA [43–46].

A related finding is that wildfires can sometimes impact the same location many times during a single season (Figures 9–12). The frequent presence of fire-$PM_{2.5}$ shown here for 2013–2016, especially in western North America, has implications for regional attainment of $PM_{2.5}$ regulatory objectives. Both Canadian and U.S. standards for $PM_{2.5}$ allow the exclusion of days with concentrations above the national standard due to wildfire smoke. However, to invoke such an exclusion, it is necessary to demonstrate that an unmanaged emission source such as wildfires is the cause of an elevated concentration. FireWork forecasts could provide useful evidence for this purpose.

It is also relevant to compare the distribution of population in North America (Figure 2) with the distribution of wildfire smoke (Figures 3–6). At the continental scale the majority of the North American population lives in the eastern half of the continent whereas the majority of large wildfires occur in the western half. This anticorrelation reduces the degree of population exposure to wildfire smoke. Nevertheless, 32% of the Canadian population lives west of Ontario and 41% of the U.S. population lives west of the Mississippi River, closer to western wildfires. As an example of this difference between wildfire location and population location, consider that for the 2013–2016 period (a) August 2014 was identified as the month with the greatest areal extent of average monthly fire-$PM_{2.5}$ > 28 $\mu g/m^3$, whereas (b) August 2015 was identified as the month with the largest U.S. population exposure to average monthly fire-$PM_{2.5}$ > 28 $\mu g/m^3$.

Short-term exposure to $PM_{2.5}$ from wildfire smoke puts the population at an increased risk of experiencing a wide range of acute health outcomes, particulary those with chronic conditions such as asthma [47] or heart disease [48]. Long-term exposure to $PM_{2.5}$ from all sources, including the contribution from wildfire smoke, increases the risk of developing chronic conditions, such as asthma or heart disease [9]. Global estimates suggest that approximately 340,000 deaths per year can be attributed to smoke from landscape fires, of which only 18% are due to short-term effects while 82% are due to the long-term effects [49]. Our results confirm that fire-$PM_{2.5}$ puts the Canadian and U.S. populations at risk from both short- and long-term exposure. Although our long-term averages covered the five-month fire season rather than the entire year, we found that up to 26.4% of the Canadian population and 14.7% of the U.S. populations were affected by increases of > 1 $\mu g/m^3$ in the extreme fire seasons. This likely indicates that the annual fire-$PM_{2.5}$ averages were > 0.2 $\mu g/m^3$ for these populations, which is defined by the U.S. EPA as a non-negligible impact [38]. Given that wildfires are becoming more frequent and intense across North America [2,42], tools such as FireWork can help to characterize their contribution to the long-term exposure most responsible for the burden of disease attributable to air pollution.

A similar wildfire pollution exposure study was recently published for the USA [23], in which fire-$PM_{2.5}$ contributions were estimated for an earlier five-year period (2008–2012) using paired retrospective simulations performed with another AQ modeling system. As in this study, one simulation considered wildfire emissions and one did not, and then a post-simulation subtraction of predicted paired surface $PM_{2.5}$ fields yielded the fire-$PM_{2.5}$ contribution estimate. One difference between the two studies was the fire seasons sampled: the average annual U.S. area burned during their study period (2008–2012) was 11% higher than that during our study period (2013–2016) [42]. Other differences were the concentrations thresholds (in $\mu g/m^3$) that they considered (0.15, 0.75, 1.5) compared with those that we considered (0.2, 1, 5, 10, 28), and the annual concentration values aggregated from 12 km × 12 km grid cells to the county level that they considered vs. the monthly and five-month values for 10 km × 10 km grid cells that we considered (i.e., higher temporal and spatial resolution). Nevertheless, a limited comparison of the two studies is possible. Based on the 2010 U.S. census (also used in this study), Rappold et al. [23] estimated that 10% of the U.S. population lived in areas where the contribution of fire-$PM_{2.5}$ was >1.5 $\mu g/m^3$. We estimated that between 5.3% (in 2015) and 14.7% (in 2014) of the U.S. population lived in areas where the five-month fire-$PM_{2.5}$ contribution was >1 $\mu g/m^3$. Rappold et al. [23] also estimated that 10.3 million individuals in the U.S. lived in areas having 10 or more days (between 2008 and 2012) with fire-$PM_{2.5}$ contribution > 35 $\mu g/m^3$, and we estimated that 10.6 million individuals in the USA lived in areas having 10 or more days (for all five-month seasons between 2013 and 2016) with fire-$PM_{2.5}$ > 28 $\mu g/m^3$. Considering the differences between these two studies, this basic comparison suggests that the results are comparable and consistent.

This study was an "analysis of opportunity" based on the availability of four years of daily North American wildfire smoke forecasts, and further improvements are likely possible. For example, wildfires in Siberia are known to affect Alaska and western Canada, but FireWork does not currently consider wildfire emissions external to North America. In our operational on-line FireWork performance evaluation, a negative bias in $PM_{2.5}$ concentration forecasts is observed for long-range wildfire pollution advection. As a consequence, the results presented here are likely to be conservative. Work is ongoing to examine potential improvements to the $PM_{2.5}$ emission factors and the plume-rise parameterization used by FireWork, both of which influence $PM_{2.5}$ concentration predictions. Given the horizontal grid spacing of 10 km used by FireWork, it is expected too that forecasts of near-source concentration will be underestimates due to the limitation of assuming uniform emissions within a grid cell. ECCC also has an AQ objective analysis system that combines AQ model predictions with AQ observations [50]. This system is now used with FireWork, opening the possibility of analyzing optimally-combined fields of model predictions and $PM_{2.5}$ measurements instead of relying solely on model forecast fields. Finally, our results might have differed slightly had we run the same FireWork version for all four years instead of using archived FireWork versions from each year (which ensured that our analyses reflected the information available to ECCC forecasters and stakeholders at the time). However, none of these changes are likely to alter our overall conclusions about the impact of wildfires in North America.

5. Conclusions

The FireWork AQ forecast system with near-real-time wildfire emissions has been run daily for a North American domain by ECCC from 2013 to 2016 during the May–September wildfire season. A multi-year analysis for this period showed the importance of accounting for contributions from wildfire $PM_{2.5}$ emissions to total $PM_{2.5}$ surface concentrations (denoted here as fire-$PM_{2.5}$) during the wildfire season. For both Canada and the USA, the months of July and August usually showed the maximum fire-$PM_{2.5}$, although intense wildfires can also occur in September in the western USA, likely due to a longer summer season [51].

Monthly and seasonal analyses of the mean forecasted fire-$PM_{2.5}$ suggested that, on average, over 76% of Canadians and 69% of Americans were at least minimally affected by wildfire smoke during the four-year study period. Comparison of average monthly fire-$PM_{2.5}$ showed large year-to-year variations in both timing and spatial locations of wildfires between 2013 and 2016. Wildfire impacts are often driven by a few major wildfire events that can lead to poor air quality for several consecutive weeks near the emission sources and beyond. In August 2015 approximately 3 million Canadians and 3 million Americans were exposed to mean monthly fire-$PM_{2.5} > 10\ \mu g/m^3$.

Calculations of the number of days and hours with forecasted fire-$PM_{2.5}$ above various concentration thresholds ranging from 1 $\mu g/m^3$ to 28 $\mu g/m^3$ for 2013–2016 showed that most wildfire events over North America occurred in the western part of the USA and in western, northern, and central Canada. During months of extreme wildfire activity, some areas in northwestern Canada and the western USA had up to 20% of days where the fire-$PM_{2.5}$ was > 28 $\mu g/m^3$. The eastern USA and eastern Canada had fewer days with threshold exceedances, but most of North America was affected by fire-$PM_{2.5} > 1\ \mu g/m^3$ on at least one day per year.

FireWork is a valuable prognostic tool used as guidance by AQ meteorologists to issue forecasts on a daily basis, allowing advance warnings to populations at risk to reduce their exposure. In addition, this study has shown that FireWork is also useful for retrospective analysis of past wildfire events. The statistical analyses of these forecasts over multiple years can be used by public health researchers, AQ regulators and policymakers, and others interested in wildfire impacts to understand and characterize exposure to wildfire smoke and its interannual and geographic variability.

Acknowledgments: The ECCC FireWork team is very grateful to the Natural Resources Canada team behind the Canadian Wildland Information System (CWFIS), who provided crucial assistance in the development of the FireWork system. We are particularly grateful to Kerry Anderson and Peter Englefield, who provided useful analyses and guidance on the CWFIS system. We also thank the U.S. Forest Service BlueSky team, particularly Susan O'Neill, for analyses and advice on wildfire emission estimates for the USA. We appreciate the assistance of Alexandru Lupu and Cassandra Bolduc for preparing tables presented in this article. Finally, we would like to acknowledge three anonymous reviewers for their valuable and constructive comments, which considerably improved this paper.

Author Contributions: Rodrigo Munoz-Alpizar and Radenko Pavlovic conceived the study; Rodrigo Munoz-Alpizar, Radenko Pavlovic and Sylvain Ménard backfilled missing FireWork forecasts over 2013–2015 five-month periods; Hugo Landry, Samuel Gilbert and Paul-André Beaulieu developed analysis tools; Jacinthe Racine and Annie Duhamel contributed to the results presentation; Rodrigo Munoz-Alpizar, Radenko Pavlovic, Michael D. Moran, Sarah B. Henderson, Jack Chen, Sylvie Gravel and Sylvain Ménard wrote the paper; and Didier Davignon, Sophie Cousineau, and Véronique Bouchet contributed with organizational support and scientific suggestions.

References

1. Beverly, J.L.; Bothwell, P. Wildfire evacuations in Canada 1980–2007. *Nat. Hazards* **2011**, *59*, 571–596. [CrossRef]
2. Canadian Interagency Forest Fire Centre. Canada Report 2016. Available online: http://www.ciffc.ca/images/stories/pdf/2016_canada_report_2017_05_15_v02.pdf (accessed on 15 May 2017).
3. Natural Resources Canada. Cost of Fire Protection. Available online: http://www.nrcan.gc.ca/forests/climate-change/forest-change/17783 (accessed on 15 May 2017).
4. Flannigan, M.; Tymstra, C. The Fort McMurray Wildfire: By the Numbers. Canadian Wildland Fire & Smoke Newsletter. 2016; pp. 1–2. Available online: https://sites.ualberta.ca/~wcwfs/CWFSN/newsletters/CWFSN_Fall_2016.pdf (accessed on 15 September 2017).
5. Urbanski, P.U.; Hao, W.M.; Baker, S. Chemical composition of wildland fire emissions. *Dev. Environ. Sci.* **2009**, *8*, 79–107.
6. Reid, C.E.; Brauer, M.; Johnston, F.H.; Jerrett, M.; Balmes, J.R.; Elliott, C.T. Critical Review of Health Impacts of Wildfire Smoke Exposure. *Environ. Health Perspect.* **2016**, *124*, 1334–1342. [CrossRef] [PubMed]
7. Liu, J.C.; Pereira, G.; Uhl, S.A.; Bravo, M.A.; Bell, M.L. A systematic review of the physical health impacts from non-occupational exposure to wildfire smoke. *Environ. Res.* **2015**, *136*, 120–132. [CrossRef] [PubMed]
8. Liu, J.C.; Mickley, L.J.; Sulprizio, M.P.; Dominici, F.; Yue, X.; Ebisu, K.; Anderson, G.B.; Khan, R.F.; Bravo, M.A.; Bell, M.L. Particulate air pollution from wildfires in the Western US under climate change. *Clim. Chang.* **2016**, *138*, 655–666. [CrossRef] [PubMed]
9. Thurston, G.D.; Kipen, H.; Annesi-Maesano, I.; Balmes, J.; Brook, R.D.; Cromar, K.; de Matteis, S.; Forastiere, F.; Forsberg, B.; Frampton, M.W.; et al. A joint ERS/ATS policy statement: What constitutes an adverse health effect of air pollution? An analytical framework. *Eur. Respir. J.* **2017**, *49*, 1600419. [CrossRef] [PubMed]
10. Canadian Council of Ministers of the Environment (CCME). Guidance Document on Achievement Determination Canadian Ambient Air Quality Standards for Fine Particulate Matter and Ozone. 2012; p. 43. Available online: http://www.ccme.ca/files/Resources/air/aqms/pn_1483_gdad_eng.pdf (accessed on 15 May 2017).
11. Sapkota, A.; Symons, J.; Kleissl, J.; Wang, L.; Parlange, M.; Ondov, J.; Breysse, P.N.; Diette, G.B.; Eggleston, P.A.; Buckley, T.J. Impact of the 2002 Canadian forest fires on particulate matter air quality in Baltimore city. *Environ. Sci. Technol.* **2005**, *39*, 24–32. [CrossRef] [PubMed]
12. Forster, C.; Wandinger, U.; Wotawa, G.; James, P.; Mattis, I.; Althausen, D.; Simmonds, P.; O'Doherty, S.; Jennings, I.M.; Kleefeld, C.; et al. Transport of boreal forest fire emissions from Canada to Europe. *J. Geophys. Res.* **2001**, *106*, 22887–22906. [CrossRef]
13. Adam, M.; Pahlow, M.; Kovalev, V.A.; Ondov, J.M.; Parlange, M.B.; Nair, N. Aerosol optical characterization by nephelometer and lidar: The Baltimore Supersite experiment during the Canadian forest fire smoke intrusion. *J. Geophys. Res.* **2004**, *109*, D16. [CrossRef]
14. Colarco, P.R.; Schoeberl, M.R.; Doddridge, B.G.; Marufu, L.T.; Torres, O.; Welton, E.J. Transport of smoke from Canadian forest fires to the surface near Washington, DC: Injection height, entrainment, and optical properties. *J. Geophys. Res.* **2004**, *109*, 1–12. [CrossRef]
15. Spichtinger, N.; Wenig, M.; James, P.; Wagner, T.; Platt, U.; Stohl, A. Satellite detection of a continental-scale plume of nitrogen oxides from boreal forest fires. *Geophys. Res. Lett.* **2001**, *28*, 4579–4582. [CrossRef]
16. Morris, G.A.; Hershey, S.; Thompson, A.M.; Stohl, A.; Colarco, P.R.; McMillan, W.W.; Warner, J.; Johnson, B.J.; Witte, J.C.; Kucsera, T.L.; et al. Alaskan and Canadian forest fires exacerbate ozone pollution in Houston, Texas, on 19 and 20 July. *J. Geophys. Res.* **2004**, *111*, 1–10. [CrossRef]

17. DeBell, L.J.; Talbot, R.W.; Dibb, J.E.; Munger, J.W.; Fischer, E.V.; Frolking, S.E. A major regional air pollution event in the northeastern United States caused by extensive forest fires in Quebec, Canada. *J. Geophys. Res.* **2004**, *109*, 1–16. [CrossRef]
18. Fiore, A.M.; Pierce, R.B.; Dickerson, R.R.; Lin, M.; Bradley, R. Detecting and attributing episodic high background ozone events. *J. Air Waste Manag. Assoc.* **2014**, *64*, 22–28.
19. Dreessen, J.; Sullivan, J.; Delgado, R. Observations and impacts of transported Canadian wildfire smoke on ozone and aerosol air quality in the Maryland region on June 9–12. *J. Air Waste Manag. Assoc.* **2016**, *66*, 842–862. [CrossRef] [PubMed]
20. Cottle, P.; Strawbridge, K.; McKendry, I. Long-range transport of Siberian wildfire smoke to British Columbia: Lidar observations and air quality impacts. *Atmos. Environ.* **2014**, *90*, 71–77. [CrossRef]
21. Teakles, A.D.; So, R.; Ainslie, B.; Nissen, R.; Schiller, C.; Vingarzan, R.; McKendry, I.; Macdonald, A.M.; Jaffe, D.A.; Bertram, A.K.; et al. Impacts of the July 2012 Siberian fire plume on air quality in the Pacific Northwest. *Atmos. Chem. Phys.* **2017**, *17*, 2593–2611. [CrossRef]
22. Centre for Disease Control. Evidence Review: Exposure Measures for Wildfire Smoke Surveillance. 31 March 2014. Available online: http://www.bccdc.ca/resource-gallery/Documents/Guidelines%20and%20Forms/Guidelines%20and%20Manuals/Health-Environment/WFSG_EvidenceReview_Smokesurveillance_FINAL_v2_edstrs.pdf (accessed on 15 May 2017).
23. Rappold, A.G.; Reyes, J.; Pouliot, G.; Cascio, W.E.; Diaz-Sanchez, D. Community Vulnerability to Health Impacts of Wildland Fire Smoke Exposure. *Environ. Sci. Technol.* **2017**, *51*, 6674–6682. [CrossRef] [PubMed]
24. Zhang, Y.; Bocquet, M.; Mallet, V.; Seigneur, C.; Baklanov, A. Real-time air quality forecasting, part I: History, techniques, and current status. *Atmos. Environ.* **2012**, *60*, 632–655. [CrossRef]
25. Zhang, Y.; Bocquet, M.; Mallet, V.; Seigneur, C.; Baklanov, A. Real-time air quality forecasting, part II: State of the science, current research needs, and future prospects. *Atmos. Environ.* **2012**, *60*, 656–676. [CrossRef]
26. Larkin, N.K.; O'Neill, S.M.; Solomon, R.; Raffuse, S.; Strand, T.; Sullivan, D.C.; Krull, C.; Rorig, M.; Peterson, J.; Ferguson, S.A. The BlueSky smoke modeling framework. *Int. J. Wildland Fire* **2010**, *18*, 906–920. [CrossRef]
27. Pavlovic, R.; Chen, J.; Anderson, K.; Moran, M.D.; Beaulieu, P.-A.; Davignon, D.; Cousineau, S. The FireWork air quality forecast system with near-real-time biomass burning emissions: Recent developments and evaluation of performance for the 2015 North American wildfire season. *J. Air Waste Manage. Assoc.* **2016**, *66*, 819–841. [CrossRef] [PubMed]
28. Pavlovic, R.; Chen, J.; Davignon, D.; Moran, M.D.; Beaulieu, P.A.; Landry, H.; Sassi, M.; Gilbert, S.; Munoz-Alpizar, R.; Anderson, K.; et al. FireWork—A Canadian Operational Air Quality Forecast Model With Near-Real-Time Biomass Burning Emissions. Canadian Wildland Fire & Smoke Newsletter. 2016; pp. 18–29. Available online: https://sites.ualberta.ca/~wcwfs/CWFSN/newsletters/CWFSN_Fall_2016.pdf (accessed on 15 September 2017).
29. Yuchi, W.; Yao, J.; McLean, K.E.; Stull, R.; Pavlovic, R.; Davignon, D.; Moran, M.D.; Henderson, S.B. Blending forest fire smoke forecasts with observed data can improve their utility for public health applications. *Atmos. Environ.* **2016**, *145*, 308–317. [CrossRef]
30. Moran, M.D.; Ménard, S.; Pavlovic, R.; Anselmo, D.; Antonopoulos, S.; Makar, P.A.; Gong, W.; Stroud, C.; Zhang, J.; Zheng, Q.; et al. Recent advances in Canada's National Operational AQ Forecasting System. In Proceedings of the 32nd NATO/SPS ITM on Air Pollution Modelling and Its Application, Utrecht, The Netherlands, 7–11 May 2012; pp. 215–220.
31. Moran, M.; Zheng, Q.; Zhang, J.; Pavlovic, R. RAQDPS Version 013: Upgrades to the CMC Operational Regional Air Quality Deterministic Prediction System Released in June 2015. 2015; p. 57. Available online: http://collaboration.cmc.ec.gc.ca/cmc/cmoi/product_guide/docs/lib/op_systems/doc_opchanges/Technical_Note_GEM-MACH10_v1.5.3+SET2.1.1_Emissions_9Nov2015.pdf (accessed on 15 May 2017).
32. Im, U.; Bianconi, R.; Solazzo, E.; Kioutsioukis, I.; Badia, A.; Balzarini, A.; Baro, R.; Bellasio, R.; Brunner, D.; Chemel, C.; et al. Evaluation of operational online-coupled regional air quality models over Europe and North America in the context of AQMEII phase 2. Part I: Ozone. *Atmos. Environ.* **2015**, *115*, 404–420. [CrossRef]
33. Lee, B.S.; Alexander, M.E.; Hawkes, B.C.; Lynham, T.J.; Stocks, B.J.; Englefield, P. Information systems in support of wildland fire management decision making in Canada. *Comput. Electron. Agric.* **2002**, *37*, 185–198. [CrossRef]

34. Anderson, K.R.; Englefield, P.; Little, J.M.; Reuter, G. An approach to operational forest fire growth predictions for Canada. *Int. J. Wildland Fire* **2009**, *18*, 893–905. [CrossRef]
35. Moran, M.; Pavlovic, R.; Chen, J. FireWork 2016: Release of the Initial Operational Version of the CMC Regional Air Quality Deterministic Prediction System with Near-Real-Time Satellite-Derived Wildfire Emissions. 2016; p. 4. Available online: http://collaboration.cmc.ec.gc.ca/cmc/CMOI/product_guide/docs/tech_notes/technote_raqdps015fw_20160428_e.pdf (accessed on 15 May 2017).
36. Kaiser, J.W.; Heil, A.; Andreae, M.O.; Benedetti, A.; Chubarova, N.; Jones, L.; Morcrette, J.-J.; Razinger, M.; Schultz, M.G.; Suttie, M.; et al. Biomass burning emissions estimated with a global fire assimilation system based on observed fire radiative power. *Biogeosciences* **2012**, *9*, 527–554. [CrossRef]
37. Xing, J.; Pleim, J.; Mathur, R.; Pouliot, G.; Hogrefe Gan, C.-M.; Wei, C. Historical gaseous and primary aerosol emissions in the United States from 1990–2010. *Atmos. Chem. Phys.* **2013**, *13*, 7531–7549. [CrossRef]
38. EPA. Draft Guidance for Comment: Significant Impact Levels for Ozone and Fine Particle in the Prevention of Significant Deterioration Permitting Program. 2016; p. 13. Available online: https://www.epa.gov/sites/production/files/2016-08/documents/pm2_5_sils_and_ozone_draft_guidance.pdf (accessed on 15 May 2017).
39. Statistics Canada. Dissemination Areas, Census 2016. Available online: http://www12.statcan.gc.ca/census-recensement/2016/dp-pd/hlt-fst/pd-pl/Comprehensive.cfm (accessed on 15 May 2017).
40. Statistics Canada. Boundary Files, Census 2016. Available online: http://www12.statcan.gc.ca/census-recensement/2011/geo/bound-limit/bound-limit-eng.cfm (accessed on 15 May 2017).
41. United States Census Bureau. Census 2010. Available online: https://www.census.gov/geo/maps-data/data/tiger-data.html (accessed on 15 May 2017).
42. National Intergancy Fire Center. Available online: https://www.nifc.gov/fireInfo/fireInfo_stats_totalFires.html (accessed on 15 May 2017).
43. Jaffe, D.; Hafner, W.; Chand, D.; Westerling, A.; Spracklen, D. Interannual variations in $PM_{2.5}$ due to wildfires in the Western United States. *Environ. Sci. Technol.* **2008**, *42*, 2812–2818. [CrossRef] [PubMed]
44. Mao, Y.H.; Li, Q.B.; Zhang, L.; Chen, Y.; Randerson, J.T.; Chen, D.; Liou, K.N. Biomass burning contribution to black carbon in the Western United States Mountain Ranges. *Atmos. Chem. Phys.* **2011**, *11*, 11253–11266. [CrossRef]
45. Spracklen, D.V.; Logan, J.A.; Mickley, L.J.; Park, R.J.; Yevich, R.; Westerling, A.L.; Jaffe, D.A. Wildfires drive interannual variability of organic carbon aerosol in the western U.S. in summer. *Geophys. Res. Lett.* **2007**, *34*, L16816. [CrossRef]
46. Zeng, T.; Wang, Y. Nationwide summer peaks of OC/EC ratios in the contiguous United States. *Atmos. Environ.* **2011**, *45*, 578–586. [CrossRef]
47. Henderson, S.B.; Johnston, F.H. Measures of forest fire smoke exposure and their associations with respiratory health outcomes. *Curr. Opin. Allergy Clin. Immunol.* **2012**, *12*, 221–227. [CrossRef] [PubMed]
48. Dennekamp, M.; Straney, L.D.; Erbas, B.; Abramson, M.J.; Keywood, M.; Smith, K.; Sim, M.R.; Glass, D.C.; Monaco, A.D.; Haikerwal, A.; et al. Forest Fire Smoke Exposures and Out-of-Hospital Cardiac Arrests in Melbourne, Australia: A Case-Crossover Study. *Environ. Health Perspect.* **2015**, *123*, 959–964. [CrossRef] [PubMed]
49. Johnston, F.H.; Henderson, S.B.; Chen, Y.; Randerson, J.T.; Marlier, M.; DeFries, R.S.; Kinney, P.; Bowman, D.M.; Brauer, M. Estimated Global Mortality Attributable to Smoke from Landscape Fires. *Environ. Health Perspect.* **2012**, *120*, 695–701. [CrossRef] [PubMed]
50. Robichaud, A.; Ménard, R.; Zaïtseva, Y.; Anselmo, D. Multi-pollutant surface objective analyses and mapping of air quality health index over North America. *Air Qual. Atmos. Health* **2016**, *9*, 743–759. [CrossRef] [PubMed]
51. Jolly, W.M.; Cochrane, M.A.; Freeborn, P.H.; Holden, Z.A.; Brown, T.J.; Williamson, G.J.; Bowman, D.M. Climate-induced variations in global wildfire danger from 1979 to 2013. *Nat. Commun.* **2015**, *6*, 7537. [CrossRef] [PubMed]

12

Impact of the '13th Five-Year Plan' Policy on Air Quality in Pearl River Delta, China: A Case Study of Haizhu District in Guangzhou City using WRF-Chem

Juanming Zhan [1,2,3], Minyi Wang [1,2,3], Yonghong Liu [1,2,3], Chunming Feng [4], Ting Gan [1,2,3], Li Li [1,2,3], Ruiwen Ou [4] and Hui Ding [1,2,3,*]

1 School of Intelligent Systems Engineering, Sun Yat-sen University, Guangzhou 510275, China; zjmxian@126.com (J.Z.); wangmy68@mail2.sysu.edu.cn (M.W.); liu_its@163.com (Y.L.); gt09032206happy@163.com (T.G.); lylee1990@126.com (L.L.)

2 Guangdong Provincial Key Laboratory of Intelligent Transport System, Guangzhou 510275, China

3 Guangdong Provincial Engineering Research Center for Traffic Environmental Monitoring and Control, Guangzhou 510275, China

4 Guangzhou Haizhu District Environmental Monitoring Station, Guangzhou 510220, China; fengcming@126.com (C.F.); sherman_au@126.com (R.O.)

* Correspondence: hui_gnid@126.com

Abstract: Due to increasingly stringent control policy, air quality has generally improved in major cities in China during the past decade. However, the standards of national regulation and the World Health Organization are yet to be fulfilled in certain areas (in some urban districts among the cities) and/or certain periods (during pollution episode event). A further control policy, hence, has been issued in the 13th Five-Year Plan (2016–2020, hereafter 13th FYP). It will be of interest to evaluate the air quality before the 13th FYP (2015) and to estimate the potential air quality by the end of the 13th FYP (2020) with a focus on the area of an urban district and the periods of severe pollution episodes. Based on observation data of major air pollutants, including SO_2 (sulphur dioxide), NO_2 (nitrogen dioxide), CO (carbon monoxide), PM_{10} (particulate matter with aerodynamic diameter equal to or less than 10 μm), $PM_{2.5}$ (particulate matter with aerodynamic diameter equal to or less than 2.5 μm) and O_3 (Ozone), the air quality of Haizhu district [an urban district in the Pearl River Delta (PRD), China] in 2015 suggested that typical heavy pollution occurred in winter and the hot season, with NO_2 or $PM_{2.5}$ as the key pollutants in winter and O_3 as the key pollutant in the hot season. We also adopted a state-of-the-art chemical transport model, the Weather Research and Forecasting model coupled with Chemistry (WRF-Chem), to predict the air quality in Haizhu District 2020 under different scenarios. The simulation results suggested that among the emission control scenarios, comprehensive measures taken in the whole of Guangzhou city would improve air quality more significantly than measures taken just in Haizhu, under all conditions. In the urban district, vehicle emission control would account more than half of the influence of all source emission control on air quality. Based on our simulation, by the end of the 13th FYP, it is noticeable that O_3 pollution would increase, which indicates that the control ratio of volatile organic compounds (VOCs) and nitrogen oxides (NO_x) may be unsuitable and therefore should be adjusted. Our study highlights the significance of evaluating the efficacy of current policy in reducing the air pollutants and recommends possible directions for further air pollution control for urban areas during the 13th FYP.

Keywords: 13th Five-Year Plan; pollution episode; WRF-Chem; emission control scenario; PRD

1. Introduction

The quality of ambient air is vital to human health. Air quality management is important for many authorities around the world [1]. The Chinese government has put great effort into mitigating the elevated level of air pollutants in the past decade, especially since the Air Pollution Control Action Plan (APCAP) was issued by the State Council in 2013 [2]. The annual levels of $PM_{2.5}$, PM_{10}, SO_2, and NO_2 decreased by 12%, 11%, 20%, and 5%, respectively, in major cities during 2013–2015 [3,4].

However, the effect was highly heterogeneous both spatially and temporally since the national standards of air quality were violated in some districts/towns in cities and during severe pollution episode events. Observation stations in the urban area of Beijing [5], Shanghai [6], and Guangzhou [7,8] recorded heavy pollution episodes in winter (PM_{10}, $PM_{2.5}$, and NO_2 as key pollutants) and/or hot seasons (summer and autumn, O_3 as key pollutant) [9,10].

To further tackle the air pollution issue, authorities of national and local level have formulated a series of regulations for the 13th Five-Year Plan (13th FYP) [11,12]. Targets have been set such that, by the end of 13th FYP, the emission of SO_2 and NO_2 should decrease 15% compared to that in 2015, and the ratio of heavy pollution days in 2020 should reduce 25% compared to that in 2015 [11]. The three national typical air pollution city clusters (Beijing-Tianjin-Hebei, the Yangtze River Delta and the Pearl River Delta (PRD)) have their own targets. For example, for the cities of the PRD region, concentrations of air pollutants ($PM_{2.5}$, PM_{10}, SO_2, NO_2, CO, and O_3) should meet national secondary air quality standards and the number of heavy pollution day should be zero [11]. Although the emission of the air pollutant from large industrial point sources has been decreasing in the past decade, the emission from traffic has been significantly increasing, which leads to critical challenges in air pollution control. It will be of significance to assess the air quality level in the year before the 13th FYP (2015) and to predict the impact of 13th FYP on air quality by 2020.

Some modelling studies have been conducted to evaluate the potential influence of the 13th FYP policy or the impact of policies on air quality during some major public events. Wang et al. (2016) [13] evaluated the impact of emission control measures on the air quality in the PRD region with WRF-CMAQ (The Weather Research and Forecasting Model-Community Multiscale Air Quality Model) model simulation on emission scenarios (a base case in 2010, two cases in 2020). Liu et al. (2017) [14] assessed the cobenefits (air quality and climate change) of vehicle emission control measures for 2015–2020 in the PRD region. Maji et al. (2018) [15] reported the $PM_{2.5}$-related mortality under air pollution control policies for China 2020. Yang et al. (2016) [16] analysed the effect of the coal control strategy in China on carbon mitigation and pollutants control for 2020 and 2030. Wang et al. (2015) [17] assessed the air quality situation under the pollution control policy of thermal power plants in China for 2020 with MM5-CMAQ (Fifth-Generation Mesoscale Regional Weather Model-Community Multiscale Air Quality Model). Qui et al. (2017) [18] studied the effect of emission control strategies on air quality of Baotou, China. Cai et al. (2017) [19] researched the impact of the 'Air Pollution Prevention and Control Action Plan' on $PM_{2.5}$ in the Jing-Jin-Ji region from 2012–2020 with WRF-CMAQ. Li et al. (2017) [20] estimated the effect of policies in the '13th Five-Year Plan' period on air pollutants emission of China's electric power sector. Wei et al. (2017) [21] analysed the impact of policies in Shanxi province, China. Guo et al. (2016) [22] researched the impact of emission control measures on air quality during APEC (Asia–Pacific Economic Cooperation summit) China 2014 with the Weather Research and Forecasting coupled with Chemistry (WRF-Chem). Xu et al. (2013) [23] evaluated the effect of air pollution control policies on air quality during the 16th Asian Games with CMAQ (Community Multiscale Air Quality Model). Shen et al. (2016) [24] analysed the influence of emission control policies on air quality during China's V-Day parade in 2015. Air quality numerical models (WRF-CMAQ, MM5-CMAQ, WRF-Chem, etc.) and scenario analyses were used widely to evaluate the effectiveness of policies on air quality. These studies only focused on the impact of policies on annual air quality in 2020 or on air quality during some major public events and did not investigate air quality on heavy pollution events.

For heavy pollution, previous works focused on the characteristics and formation of pollution episodes. Tan et al. (2009) [25] investigated the chemical characteristics of haze in Guangzhou (2002–2003, summer and winter) with PM_{10} samples and gas chromatography-mass spectrometry (GC-MS). Wang et al. (2015) [26] researched the formation process of a severe haze episode in the Yangtze River Delta (2013 winter) based on visibility and meteorological parameters, and backward trajectories of the air mass. Zhang et al. (2015) [27] applied WRF-Chem to simulate a severe haze in Beijing (2013 winter) and discussed the meteorological impacts on haze. Zhan et al. (2017) [7] analysed the spatial and temporal association of $PM_{2.5}$ pollution events between typical cities of PRD (2014 winter). Ding et al. (2004) [28] discussed the effects of sea–land breezes on the transport of air pollution during an ozone episode in PRD (2001 autumn) with the MM5 (Fifth-Generation Mesoscale Regional Weather) model. Shen et al. (2015) [29] researched the source of an ozone episode in PRD (2008 autumn) with the CAMx (Comprehensive Air-quality Model with extensions) model. Zhao et al. (2015) [30] investigated the chemical characteristics of ozone episodes in Shanghai (2010–2013, O_3 peaked in summer) with the differential optical absorption spectroscopy (DOAS) and the hybrid single particle Lagrangian integrated trajectory (HYSPLIT) model. Xu et al. (2008) [31] simulated typical summertime ozone episodes in Beijing (2000) with the WRF-CAMx (Weather Research and Forecasting Model Comprehensive Air-quality Model with extensions) model to analyse the process, and Qu et al. (2014) [32] used the CMAQ-MADRID (Community Multiscale Air Quality Model-Model of Aerosol Dynamics, Reaction, Ionization, and Dissolution) model to evaluate the effects of NO_x and VOCs emissions on ozone pollution in Beijing (2007 summer). According to previous studies, meteorological conditions and emission significantly affected pollution, and numerical models were widely used to analyse the characteristics of pollution. However, the model evaluation of the impact of the 13th FYP on the air quality in 2020 during the pollution episode has not attracted much attention.

The air pollution control strategy in China is now at the new stage that the manufacturing industry in megacities in China (e.g., Guangzhou) is no longer the dominant emission source; control on vehicle emissions is becoming the primary subject [33]; the control strategy is transforming from urban/regional control to district/town grid control [34]. The PRD region is one of the three national typical air pollution city clusters [2]. PRD has experienced the major problem of transferring from haze to complex (haze and photochemical) pollution. Moreover, it is the first city cluster to achieve the goal in the APCAP, i.e., compared with the levels in 2012, the concentration of fine particles ($PM_{2.5}$) in Beijing-Tianjin-Hebei, the Yangtze River Delta, and the Pearl River Delta (PRD) reduced by 25%, 20%, and 15%, respectively, by 2017 [4]. As the centre of PRD and the capital city in Guangdong province, Guangzhou is one of the cities with the poorest air quality in PRD, more comprehensive and more stringent air pollution control measures were implemented in Guangzhou than in other cities of PRD, so that Guangzhou is a role model in air quality management [35]. Haizhu district is an island in the urban area of Guangzhou, with a typical urban landscape (residential and commercial areas) comprising many high rises, shops, residential apartments, highways, and major roads that pass through the urban area, and its air quality was relatively poor [36–38]. It is an ideal testbed to determine how the emission control measures affect air quality in urban areas. Besides, manufacturing industries have been moved out from Haizhu due to the 'limitation of high-pollution production' policy [39,40], which has been a typical trend in cities of China recently [3]. Therefore, our study focused on the Guangzhou central district (Haizhu district) (Figure 1).

In this study, we first compiled an evaluation on the observational data in Haizhu district in 2015 with a focus on the periods of pollution episodes. Then we designed four scenarios based on the 13th FYP regulations, and utilized the WRF-Chem model to evaluate the effect of the emission control measures and the influence of Haizhu policy and Guangzhou (except Haizhu) policy on air quality in Haizhu district 2020 under those scenarios. The research framework is shown in Figure S1 in the Supplementary Materials.

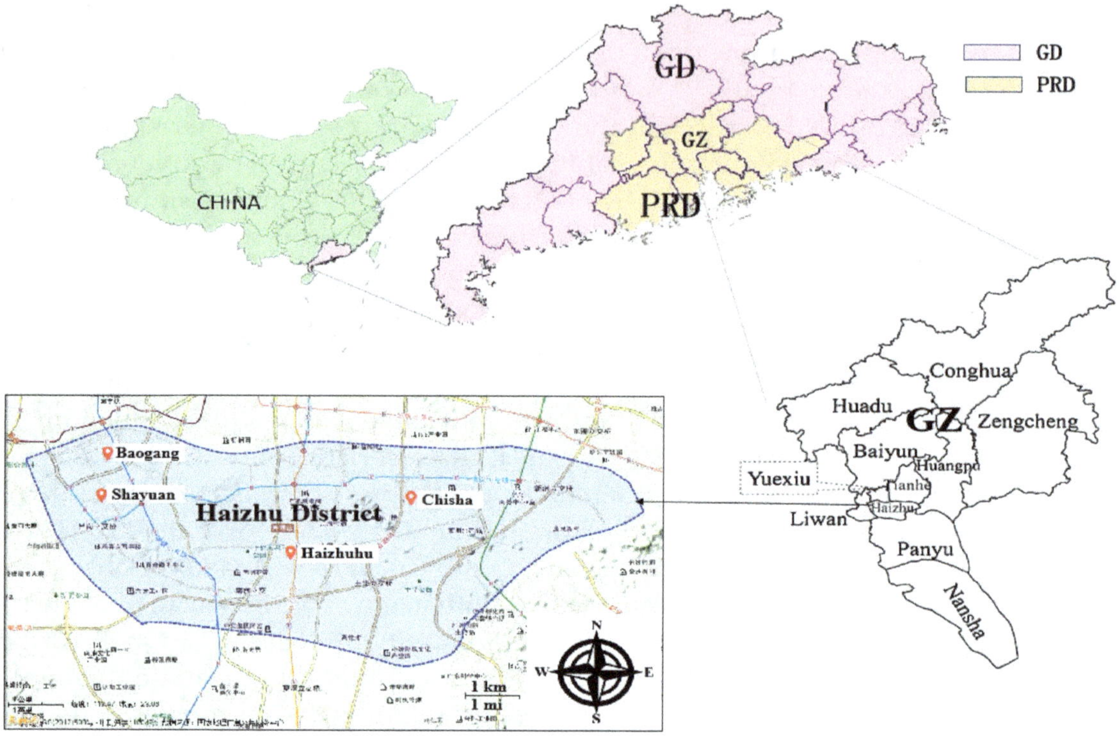

Figure 1. The Haizhu region and its location (GD: Guangdong province; PRD: Pearl River Delta; GZ: Guangzhou) [14,41,42].

2. Data and Model Experiment

2.1. Observation Data

Air quality monitoring data, including the hourly concentrations of pollutants (CO, SO_2, $PM_{2.5}$, PM_{10}, NO_2, O_3) and daily AQI (Air Quality Index), were collected from Guangzhou Environmental Monitoring Centre data network (http://210.72.1.216:8080/gzaqi_new/RealTimeDate.html). Hourly and daily meteorological data [wind speed, wind direction, relative humidity (RH), temperature, satellite cloud images, etc.] were downloaded from the Guangzhou meteorological data network (http://data.tqyb.com.cn/weather/index.jsp). These websites are regularly maintained by the government, and the data are released to the public. The data mentioned above were used to analyse the air quality status and pollution episodes in 2015, and to evaluate the WRF-Chem model's performance.

2.2. Emission Inventory and Scenarios

2.2.1. Emission Inventory

The anthropogenic emissions from the 2012-based Multiresolution Emission Inventory for China (MEIC) [43] (http://www.meicmodel.org/), biogenic emission parameterisation [44], dust emission parameterisation [45], and sea salt emission parameterisation [46], and the marine emission from EDGAR (Emission Database for Global Atmospheric Research, http://edgar.jrc.ec.europa.eu/) were applied in this study. The 2012-based MEIC was a Chinese national emission inventory with a resolution of 0.25° × 0.25° based on 2012 emission status, developed by Tsinghua University. It included the emissions in major sectors, such as transport, industry and human residential, etc. [13]. Previous studies reported that MEIC was well developed by a technology-based emission model [47] and similar emission datasets have been widely used in the numerical simulation for cities in PRD [27,48].

The emissions in the key research area, Guangzhou, were obtained based on the Guangzhou ambient air quality plan (2016–2025) [49]. The emission of air pollutants from different sources in Haizhu district in 2015 was retrieved from the Guangzhou environmental protection bureau's data network (http://www.gzepb.gov.cn/infoindex.htm). Based on the statistics yearbook (http://210.72.4.52/gzStat1/chaxun/njsj.jsp) and the Guangzhou environmental protection bureau's data network, the emission inventory was generated according to Chinese national technical guidelines [50,51] and relevant studies (Zheng et al., 2009 [52]; Zhao et al., 2015 [53]). Spatial allocation of emission inventory depended on the source characteristics. Haizhu district is an urban area, and the major emission sectors are industry, residential, and (road) transport. Emission sectors, corresponding inventory technical guidelines, and special allocation rules in this research are shown in Table S1.

According to the 2015-based Haizhu district emission inventory, the 2020 Haizhu district emission inventory was predicted with the extrapolation function [13] [EI_2020 = f_x (EI_2015 × activity factor), where EI refers to the emission inventory, and the activity factor refers to the trend of emission sections, gained from emission control policies for the 13th Five-Year period [12,49,54] and the trend of city development in the past five to ten years (e.g., society, economy, vehicle and population, etc.) according to the statistical yearbook network (http://210.72.4.52/gzStat1/chaxun/njsj.jsp)]. We note that the emission inventory in Haizhu district is compiled on an annual resolution. The resolution of these inventories was 0.01° × 0.01°. All of the emission inventories have the same major sectors (transport, industry, and residential, etc.).

2.2.2. Emission Scenarios

Because the major and easily controlled sections had been regulated during the 12th Five-Year Plan period (12th FYP), the 13th FYP emission control policy was set more aggressively: industries generating air pollution will be moved out from the central city area, ultraclean/effective technology will be widely used, the ratio of public transport will rise to 70% of motorised travel [12], electric buses will be applied widely, accounting for 63% of public buses [54], vehicle emission standards and fuel standards will be strengthened (specifically, eliminating high pollution vehicles, implementing the new national emission standard of vehicle and the new national vehicle fuel standard, etc.), and VOC emission will be controlled entirely in particular industries (e.g., chemical industry, paint industry, and printing industry, etc.) [49].

Four scenarios were designed to evaluate the impact of the 13th FYP on the air quality in Haizhu district in 2020. The meteorological conditions were assumed to be unchanged, which means that the meteorological conditions in 2015 were used for all scenarios in 2020. The four 2020 scenarios are as follows (also in Table S2):

Scenario 2020A was proposed such that both Haizhu emission control policy and Guangzhou (except Haizhu) emission control policy would follow the 2015 emission control policy tendency. This scenario is a baseline scenario.

Scenario 2020B was designed such that Haizhu emission control would be implemented based on the 13th FYP emission control plan, while Guangzhou (except Haizhu) would still adhere to the 2015 emission control policy.

Scenario 2020C was a 13th FYP policy scenario in which both Haizhu emission control policy and Guangzhou (except Haizhu) emission control policy would be implemented based on the 13th FYP emission control plan.

Scenario 2020D was a scenario for vehicle control in Haizhu due to the significance of vehicle emission in Haizhu. In this scenario, vehicle control in Haizhu would be implemented based on the 13th FYP emission control plan, while other emission source controls in the Haizhu and Guangzhou (except Haizhu) emission control policy would remain as the 2015 emission control policy tendency. We note that the contribution of vehicle emission in Haizhu district is <10% of that in Guangzhou city.

Table 1 shows the changes in emissions of SO_2, NO_x, CO, PM_{10}, $PM_{2.5}$, and VOCs for the different scenarios in the whole city of Guangzhou. The 2020A, 2020B, and 2020D scenario emissions would be higher than the 2015 emissions (SO_2, NO_x, CO, PM_{10}, $PM_{2.5}$, and VOCs would increase 57.2–58.9%, 47.4–48.6%, 60.1–60.2%, 14.6–15.1%,25.8–26.4%, and 24.9–25.0%, respectively), whereas the 2020C scenario emission would be lower than or roughly equal to the 2015 emission. The difference of emissions among each scenarios indicates that the emission control of the whole Guangzhou city would significantly affect the 2020 emission. Additionally, the emission ratio (VOCs/NO_x) would be around 0.72 in scenarios 2020A, 2020B, and 2020D and 0.60 in scenario 2020C, lower than 0.86 in 2015.

Table 1. Changes in pollutant emissions of the different scenarios in the whole city (Guangzhou) (A, B, C, and D are scenarios 2020A, 2020B, 2020C, and 2020D, respectively).

	SO_2	NO_x	CO	PM_{10}	$PM_{2.5}$	VOCs
(A-2015)/2015	58.9%	48.6%	60.2%	15.1%	26.4%	25.0%
(B-2015)/2015	57.2%	47.4%	60.1%	14.6%	25.8%	24.9%
(C-2015)/2015	−72.9%	−14.2%	1.0%	−36.7%	−41.8%	−33.2%
(D-2015)/2015	58.9%	48.1%	60.1%	15.0%	26.2%	25.0%

The changes in emissions for each sector are shown in Figure 2. SO_2 emission from the sectors residential, vehicle, and others, and NO_x emission from the residential sector in the four 2020 scenarios would be lower than those in the 2015 emission. $PM_{2.5}$ emission and PM_{10} emission from the residential sector in scenarios 2020B and 2020C would also be lower than those in the 2015 emission. In scenarios 2020C, NO_x emission from the vehicle and others sectors, CO emission from the industry, residential, and vehicle sectors, and VOCs emission from the residential sector would be higher than those in the 2015 emission. Overall pollutants emissions in scenarios 2020A, 2020B and 2020D would be higher than those in 2015, pollutants emissions in scenarios 2020C would be lower than in 2015, except that CO emission in scenarios 2020C would be slightly higher than that in 2015.

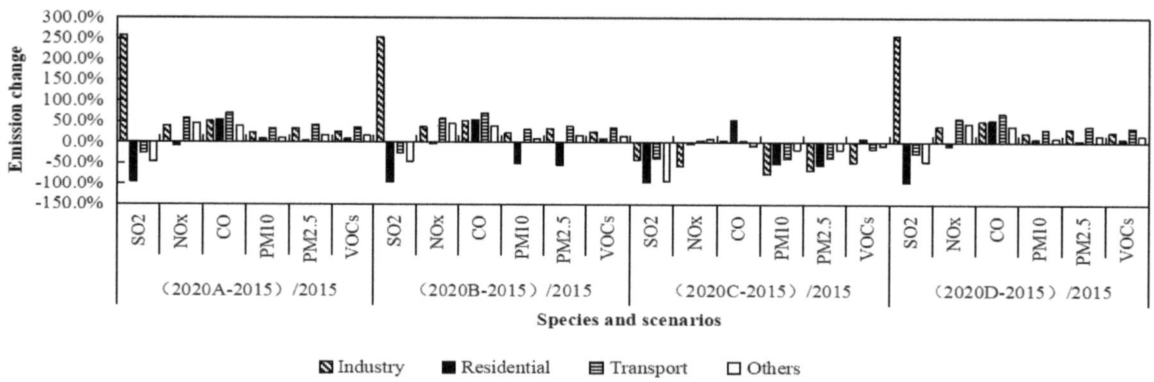

Figure 2. Emission changes in sectors in different scenarios in the whole city (Guangzhou).

2.3. WRF-Chem Model Setup

2.3.1. Description of WRF-Chem Model

WRF-Chem is a chemical transport model developed by the community led by NOAA/ESRL (Earth System Research Laboratory, The United States National Oceanic and Atmospheric Administration) [55,56], and it is widely used for analysing heavy pollution processes and the effectiveness of emission control measures [22,27,57].

2.3.2. Configuration of WRF-Chem

In this study, a three-nested domain was applied for model set-up (Figure S2), with grid cell areas of 9 × 9 km, 3 × 3 km, and 1 × 1 km, respectively. The biggest domain covered the southern China area, the middle domain covered PRD, and the smallest one covered the Guangzhou central area. There were 27 sigma levels for all domains, and in this study, the data of the ground level was mainly used. The NCEP (the United States' National Centers for Environmental Prediction) 6-h FNL (Final Operational Global Analysis data) meteorological data and the emission inventory mentioned above were input for model setup. The simulation data from MOZART (Model for Ozone and Related

Chemical Tracers, https://www.acom.ucar.edu/wrf-chem/mozart.shtml) was used as the initial and boundary chemistry data [58]. The related parameterisation schemes in the simulation were the Regional Acid Deposition Model, version 2 (RADM2) gas-phase chemical mechanism [59] and the MADE/SORGAM (Modal Aerosol Dynamics Model for Europe/Secondary Organic Aerosol Model) aerosol chemical mechanism [60,61]. We have not activated the feedback from chemistry on meteorology. The first 120 h in the simulation were used as the model spin-up.

2.3.3. Cases Setup

(1) Evaluation of overall air quality in 2020

The month of October was used as a proxy of the entire year of 2015 since the pollutant concentrations in October 2015 were very similar to those in the whole year (details are shown in Section 3.1). WRF-Chem simulations were conducted with the emissions in four 2020 scenarios and the emissions in 2015. The simulated concentrations of $PM_{2.5}$, PM_{10}, SO_2, NO_2, CO, and O_3 were compared between results of the four 2020 scenarios and the result of 2015, in order to evaluate the effectiveness of the 13th FYP policy on air quality. The meteorological conditions for 2020 scenarios were assumed to be the same as the meteorological conditions in 2015. The results are shown in Section 3.3.

(2) Evaluation of air quality during pollution episodes

To identify the air quality status, we analysed the annual air quality situation and air pollution events based on the data observed in four urban stations (Baogang, Chisha, Shayuan, and Haizhuhu) in Guangzhou Haizhu district (Figure 1). A pollution episode was defined as a short period consisting of subsequent days (at least one day) with the AQI $\geq$ 101 [62]. A heavy pollution episode was a pollution episode with AQI $\geq$ 201 [62].

The pollution episodes in 2015 are shown in Table 2 (more information is shown in Table S3). The table shows that the number and the pollution level of pollution episodes had seasonal characters. The pollution episodes occurred more often in winter, and heavy pollution was observed in the summer and winter. The Shayuan station and Haizhuhu station suffered pollution episodes more than other stations in 2015, and Shayuan was the only station where the heavy pollution episodes were observed in both winter and summer (14–28 January 2015, and 3–8 August 2015, respectively). The two heavy pollution episodes were typical pollution ($PM_{2.5}$ and NO_2 pollution in winter, O_3 pollution in summer [63]). Therefore, the two heavy pollution episodes in Shayuan were analysed with pollution and meteorology progress (results shown in Section 3.1.2). Also, the meteorology conditions of these two episodes were applied in the 2020 scenarios numerical simulations with WRF-Chem to evaluate the effectiveness of the 13th FYP policy on air quality of pollution episodes (specifically, $PM_{2.5}$ and NO_2 pollution for the winter episode, O_3 pollution for the summer episode). The simulation results are shown in Section 3.4.

Table 2. Pollution episodes observed at Haizhu monitoring stations in 2015.

Station	Season	Event Frequency (Units)	key Pollutant	Max Pollution Level
Baogang	Spring	5	$PM_{2.5}$, NO_2, O_3	Moderate pollution
	Summer	4	O_3	**Heavy pollution**
	Autumn	3	$PM_{2.5}$, NO_2, O_3	Moderate pollution
	Winter	9	$PM_{2.5}$, NO_2	Moderate pollution
Chisha	Spring	4	$PM_{2.5}$, NO_2, O_3	Moderate pollution
	Summer	4	O_3	**Heavy pollution**
	Autumn	3	NO_2, O_3	Moderate pollution
	Winter	7	$PM_{2.5}$, NO_2	Moderate pollution
Shayuan	Spring	7	$PM_{2.5}$, O_3	Moderate pollution
	Summer	5	O_3	**Heavy pollution**
	Autumn	12	$PM_{2.5}$, NO_2, O_3	Moderate pollution
	Winter	7	$PM_{2.5}$, NO_2	**Heavy pollution**
Haizhuhu	Spring	6	$PM_{2.5}$, NO_2, O_3	Moderate pollution
	Summer	3	O_3	Light pollution
	Autumn	10	NO_2, O_3	Moderate pollution
	Winter	12	$PM_{2.5}$, NO_2, O_3	Moderate pollution

3. Results and Discussion

3.1. Air Quality in Haizhu in 2015

3.1.1. Overview

The annually-averaged concentrations of $PM_{2.5}$, PM_{10}, SO_2, and NO_2 were 40 μg/m^3, 61 μg/m^3, 14 μg/m^3, and 49 μg/m^3, respectively, in Haizhu district in 2015, which were higher than those in the whole of Guangzhou city (39 μg/m^3, 59 μg/m^3, 13 μg/m^3, and 47 μg/m^3, respectively [64]). The 95th percentile of the CO daily concentration was 0.9 mg/m^3, in Haizhu district in 2015, which was lower than that in the whole of Guangzhou city (1.5 mg/m^3, [64]), The routine pollutants (except for O_3) peaked in January (winter), among which $PM_{2.5}$ and NO_2 were the major pollutants. The 90th percentile of the daily 8 h maximum O_3 concentration (O_3-8h) in 2015 was 139 μg/m^3, slightly lower than that in Guangzhou (145 μg/m^3; [64]). O_3-8h peaked in September (summer to autumn, or hot season). O_3 was also the key pollutant in summer and autumn in Haizhu district [64]. These observational results suggested that air quality in Haizhu was still harmful to human health (the thresholds: $PM_{2.5} \leq 10$ μg/m^3 annual mean, $PM_{10} \leq 20$ μg/m^3 annual mean, $NO_2 \leq 40$ μg/m^3 annual mean, O_3-8h ≤ 100 μg/m^3 8 h mean) [65] and it was worse than the Guangzhou average level in 2015, which highlights the necessity of district/town regulation in air quality control. Please note that this study does not intend to estimate the influence of air quality on human health which is sensitive to the exposure levels (the levels of air pollutants). The pollutant concentrations in October 2015 were very similar to those in the year of 2015 (a-SO_2, a-NO_2, a-PM_{10}, a-$PM_{2.5}$, a-O_3-8h, a-CO, in Figure 3a).

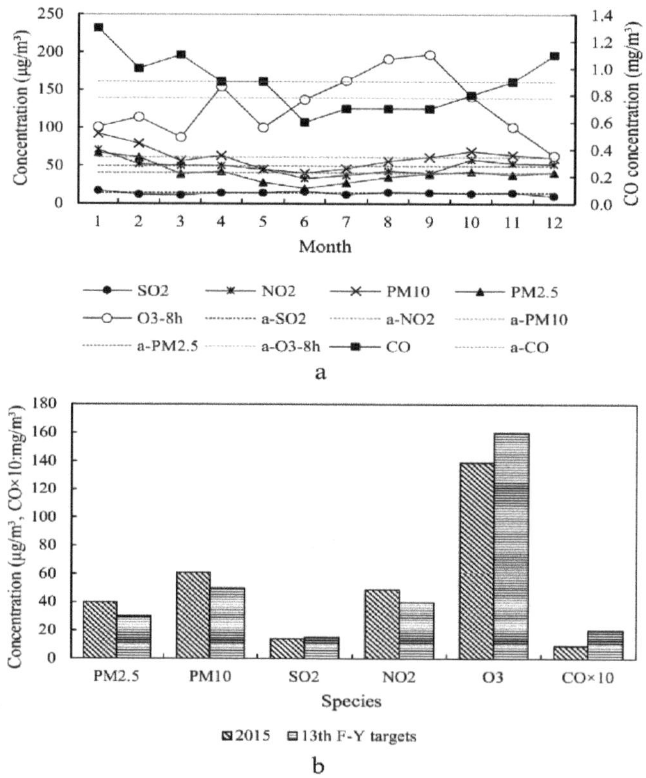

Figure 3. (**a**) Air pollutant monthly concentrations in 2015 averaged in four stations (Baogang, Chisha, Shayuan, and Haizhuhu) in Haizhu district, (**b**) comparison of air quality in 2015 in Haizhu district and the 13th Five-Year targets. (In (a), a-SO_2, a-NO_2, a-PM_{10}, a-$PM_{2.5}$ means 2015 annual value for SO_2, NO_2, PM_{10}, $PM_{2.5}$, respectively; a-O_3-8h means the 90th percentile of O_3-8h in 2015; a-CO means the 95th percentile of the CO daily concentration in 2015; other pollutants: monthly-averaged concentrations. In (b), O_3: the 90th percentile of O_3-8h in a year; CO: the 95th percentile of the CO daily concentration in a year; other pollutants: annually-averaged concentrations. These parameters were adopted according to the national standards in China [66].).

Compared with the 13th FYP targets in 2020 ($PM_{2.5} \leq 30$ μg/m^3, $PM_{10} \leq 50$ μg/m^3, $SO_2 \leq 15$ μg/m^3, $NO_2 \leq 40$ μg/m^3, O_3-8h ≤ 160 μg/m^3, and $CO \leq 2$ mg/m^3) in Guangzhou [49] (Figure 3b), $PM_{2.5}$, PM_{10}, and NO_2 concentrations in 2015 exceeded the targets, and the air pollutant emission could even rise in the coming years (Table 1). Haizhu district could face worsening air quality, and more emphasis should be placed on the control measures and effectiveness assessment for $PM_{2.5}$ and NO_2 during the 13th FYP so that the measures can be duly adjusted to achieve the 13th FYP targets.

3.1.2. Air Quality during Pollution Episodes

Table 2 summarizes the pollution episodes in Haizhu district in 2015. Shayuan station was the only station where heavy pollution episodes were observed in both winter and summer, and the two pollution episodes were also the typical pollution ($PM_{2.5}$ and NO_2 pollution in winter, O_3 pollution in summer). Therefore, these two typical pollution episodes in Shayuan will be analysed in detail below.

(1) Heavy pollution episode in winter

A winter heavy pollution episode in Haizhu district occurred during 14–28 January 2015 (Table S4 shows the statistical data). In this pollution episode, the air quality was light pollution in the beginning (15–19 January), then the pollution level peaked during 20–21 January. It finally decreased to light pollution again in the third period (22–27 January). On 28 January, the air quality was good again. This episode was a process in which the air quality decreased gradually and then increased gradually. During this episode, the wind speed was in the range of 1.1–2.9 m/s, the RH was in the range of 52–85%, and there was no precipitation. The temperature was in the range of 10.3–18.0 °C, and the surface pressure was in the range of 1009.1–1016.1 hPa. The low wind speed during this winter episode (~1.5 m/s) indicated that the atmosphere was stable, and as reported in previous studies, low or calm wind speed generally leads to high pollution [67,68], because stable atmospheric condition favours the accumulation of pollutants. There is another possibility that the existence of temperature inversion layer could also lead to the accumulation of the air pollutants (e.g., Malek et al., 2006 [69]).

(2) Heavy pollution episode in summer

A summer heavy pollution episode in Haizhu district happened during 3–8 August 2015 (Table S5). The whole episode covered only six days, including the two days of light pollution (4 and 7 August), one day of moderate pollution (5 August), and one day of heavy pollution (6 August). The pollution occurred and disappeared within a short time, but it exhibited high intensity. In this episode, the wind speed was in the range of 1.1–4.4 m/s, the RH was between 65 and 76%, and there was no precipitation. In the first period (3–6 August 2015), the wind speed was in the range of 1.1–1.4 m/s. There was no precipitation nor cloud. The temperature was in the range of 27.6–30.3 °C, and the surface pressure was in the range of 991.7–1002.4 hPa. The hourly temperatures were between 23.4 and 35.6 °C. Sunny weather, low wind, and high temperatures would create favourable conditions for O_3 production and accumulation [10]. Then, the wind speed increased to 3.9–4.4 m/s during 7–8 August 2015. Strong wind transported the air pollutants out of Haizhu, improving the air quality in Haizhu to light pollution. The temperature inversion layer might also enhance the accumulation of the air pollutants (e.g., Malek et al., 2006 [69]).

3.2. Evaluation of WRF-Chem Model Performance

October was the month when the concentrations of $PM_{2.5}$, NO_2 and O_3 (the major pollutants in Haizhu) were all at elevated levels. Therefore, we used the observed data in October to evaluate the performance of the WRF-Chem simulation.

(1) Meteorological simulation evaluation

The simulated surface meteorological parameters (wind speed, wind direction, temperature, and RH) in Guangzhou were validated with the observed data (Table S6 and Figure 4). The temperature, RH, and wind speed in the simulation were generally in line with the observations (Table S6), and the simulated wind directions covered the observed wind directions (Figure 4, wind direction). Similar model behaviour was reported in some previous studies [9,70,71]. The over-prediction of the wind speed in the present study is probably due to the underestimation of the roughness in the cities. For example, the constructions in the urban area are not adequately represented in our model simulation although a very fine resolution of 1 km is already used. Overall, the simulation generally reproduced the meteorological condition in 2015.

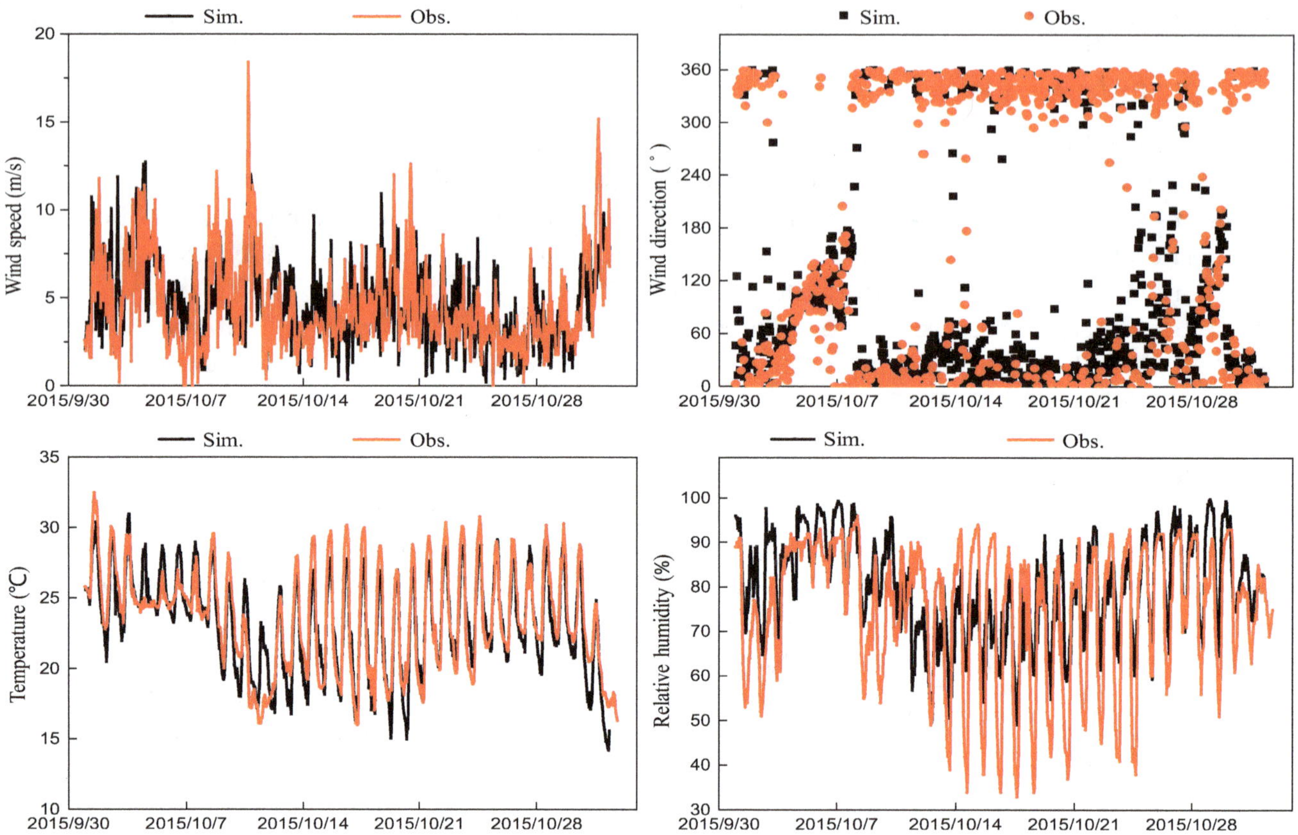

Figure 4. Comparison of simulated and observed meteorological parameters in Guangzhou, 2015. (Hourly statistical data in October 2015).

(2) Chemical simulation evaluation.

Simulated pollutant concentrations were also compared to observed data, including O_3 daily 8 h maximum concentration and other pollutants' daily-averaged concentrations (CO, $PM_{2.5}$, PM_{10}, SO_2, and NO_2). Metrics of evaluation on model performance are shown in Table S7. The ranges of the mean-bias (MB) and root-mean-square-error (RMSE) on pollutant concentrations were -8.07 μg/m^3 to 24.34 μg/m^3, 9.81 μg/m^3 to 48.35 μg/m^3, respectively, except CO. The MB and RMSE of CO were -0.32 mg/m^3 and 0.43 mg/m^3, respectively. Compared to previous studies [13,70–72], the model in this paper showed similar behaviour. Figure 5 shows that the general characteristics of the routine pollutants were captured by the simulation. Monthly averaged simulated concentrations of NO_2, PM_{10}, SO_2, CO, O_3-8h, and $PM_{2.5}$ at the locations of the monitoring sites were 147.6%, 101.1%, 112.0%, 65.1%, 91.4%, and 80.9% of the observations, respectively. Concentrations of NO_2, PM_{10}, and SO_2 were overpredicted in the simulation, and concentrations of CO, O_3-8h, and $PM_{2.5}$ were underestimated. It is noticeable that there was an observed peak in pollution on 14 October which was underestimated in the simulation. Concentrations of NO_2, PM_{10}, SO_2, CO, O_3-8h, and $PM_{2.5}$ in the simulation were 39%, 49%, 42%, 52%, 49%, and 65% less than that of the observation on 14 October, respectively. The most probable reason for the discrepancy is the meteorological condition was not well captured. Other possible causes include the uncertainty of the emission inventory, the uncertainty of the chemical scheme in simulating the formation of secondary air pollutants, etc. The uncertainty of emission inventories results from the uncertainty in the activity level, the emission factor, and the different grid size [73–75]. The simulation of secondary air pollutants (O_3, secondary inorganic and organic aerosol) is a hot research topic and still contains some uncertainty as reported in Ahmadov et al. (2012) [76], Li et al., (2018) [77], etc. Besides, the planetary boundary (PBL) scheme also plays an important role in predicting the level of air pollutants at the surface and the reader is referred to relevant studies (e.g., Banks and Baldasano, 2016 [78]; Shin and Hong, 2011 [79]; Hu et al., 2010 [80]) for the evaluation of various PBL schemes including the one used in the present study, i.e., the YSU (Yonsei University) scheme [81]. There is also the possibility that the representative area of the measurement site differs from the grid cell that contains the site and such difference might lead to the difficulty of directly comparing the simulation with observation (e.g., Schutgens et al. 2016 [82]).

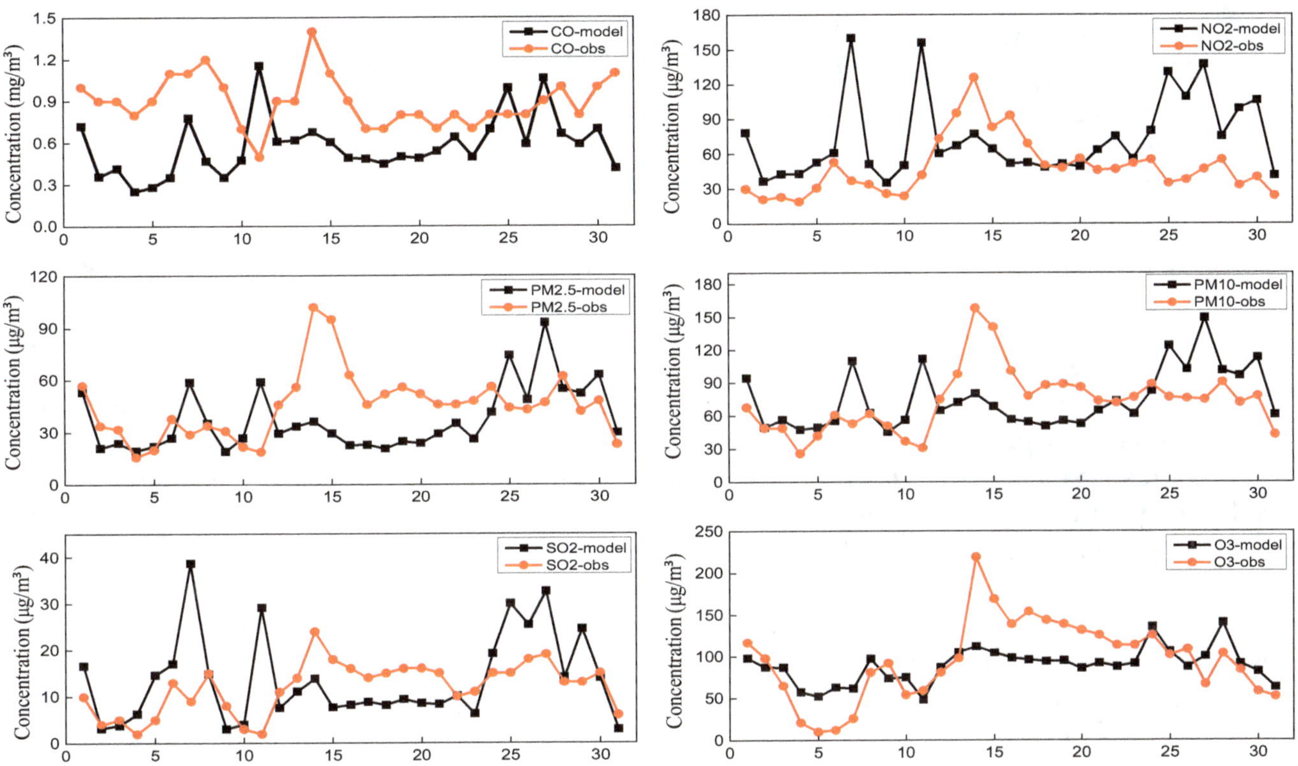

Figure 5. Comparisons of simulated and observed CO, NO_2, $PM_{2.5}$, PM_{10}, SO_2, and O_3. (Daily statistical data in October 2015 in Haizhu district; O_3: daily 8-h maximum concentration; other pollutants: daily-averaged concentration).

In the present study, we follow the relevant air quality modelling studies, e.g., those studies included in Table S7, to use the statistic tools to evaluate the WRF-Chem model performance. While the statistic method can provide some insights into the model abilities, it would help if some advanced diagnostic tools, e.g., the simulation error apportioning techniques proposed by Solazzo et al. (2017a) [83] and Solazzo et al. (2017b) [84], are adopted in future study to identify the critical processes that demand most urgent attention.

3.3. *Effect of 13th FYP on Overall Air Quality in Haizhu in 2020*

Since the pollutant concentrations in October 2015 were very similar to those in the whole year, the simulated results in the four 2020 scenarios in October are used for evaluating the effect of the 13th FYP on the air quality in Haizhu by comparing to the simulation for October 2015 (Figure 6).

Figure 6 shows that simulated concentrations of five pollutant species (CO, NO_2, $PM_{2.5}$, PM_{10}, and SO_2) in scenario 2020C are lower than pollutants concentrations simulated in 2015. For O_3, 2020 scenarios' simulation results are higher than 2015 result. CO, $PM_{2.5}$, PM_{10}, and SO_2 concentrations in 2020 scenarios A, B, D are higher than the pollutants concentration in 2015 simulation. NO_2 concentrations in the four 2020 scenarios are slightly lower than the concentration in 2015 simulation.

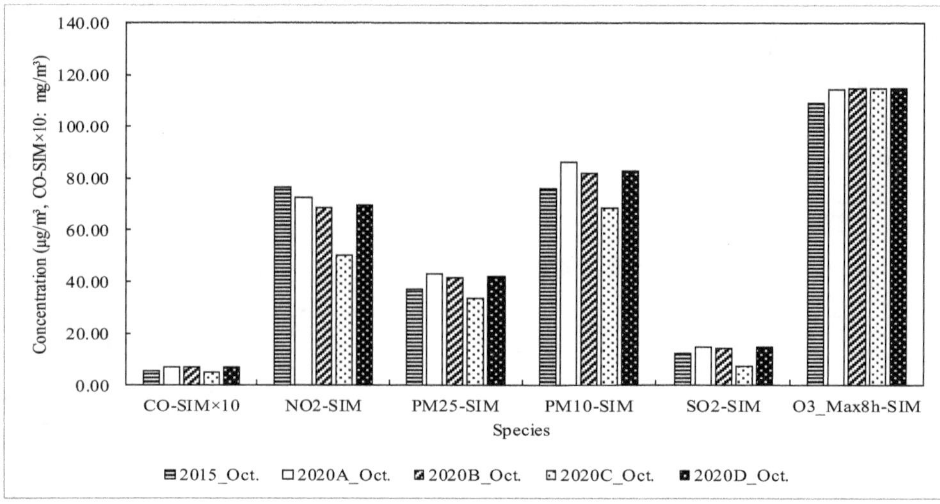

Figure 6. Comparison between simulation on October 2015 and 2020 scenarios simulation on October in Haizhu district (monthly-averaged surface concentrations).

It is noticeable that O_3 concentrations in all of the four 2020 scenarios are higher than those in 2015 simulation, which means that ozone pollution will probably get worse by the end of 13th FYP. Although scenario 2020A had the slightly less pollution than other scenarios had, the level of O_x (O_3 + NO_2) in scenario 2020A was higher than that of other scenarios (Figure 6), which means that atmospheric oxidation in scenario 2020A was more elevated [13,85]. It would cause the more intensive formation of secondary aerosols, which would lead to regional pollution [86]. Meanwhile, the level of O_x in scenario 2020C was the lowest.

In Section 3.1, the observed concentrations of $PM_{2.5}$, PM_{10}, and NO_2 in 2015 exceeded the targets. Meanwhile, the WRF-Chem simulations suggested that from 2015 to 2020, the concentrations of $PM_{2.5}$ and PM_{10} would increase in scenarios 2020A, 2020B, and 2020D. Therefore, the real concentrations of aerosols in 2020 could be higher than those in 2015 and further exceed the targets. The observed O_3-8h concentration in 2015 did not exceed the target (Section 3.1), but with the increasing trend from 2015 to 2020 as simulated in all scenarios, the real O_3-8h concentration in 2020 should be considered as a cause for concern.

According to the simulations, by the end of 13th FYP, if the emission control policy just follows 12th FYP policy (scenario 2020A), the air pollution would increase and the control on all emission sectors in Haizhu (scenario 2020B) and control just on vehicle emissions in Haizhu (scenario 2020D) would have a similar impact on air quality, and could not improve air quality very much. The scenario 2020C (the comprehensive emission control measures taken in the whole Guangzhou city) would have the better effect on the emission control than other scenarios would do. Among the four 2020 scenarios, the scenario 2020C is the better scenario.

According to Figure 6, pollutant concentrations between scenario 2020B and 2020D are similar. The effect of controlling every emission sectors in Haizhu district would be almost equal with the effect of only controlling the traffic emission in Haizhu district, suggesting that traffic source would be a key source in Haizhu district.

Wang et al. (2016) [13] used WRF-CMAQ model (a base case in 2010, two cases in 2020) and evaluated the impact of emission control measures on the air quality in the PRD region, and their results showed that reducing NO_x emissions would cause rising $PM_{2.5}$ levels in certain areas, although it would benefit with reduction of regional $PM_{2.5}$. They also noted that O_3 formation in PRD was generally VOCs-limited and cutting VOCs emission could benefit the reduction of overall O_3. Liu et al. (2017) [14] assessed the influence of vehicle emission control measures for 2015–2020 in the PRD region on air quality and climate and their results suggested that, if vehicle emission wasn't controlled, most air pollutants and GHG would increase by 20–64% by 2020.

3.4. Effect of 13th FYP on Air Quality during Heavy Pollution Episodes in Haizhu in 2020

To evaluate the effect of 13th FYP on NO_2 and $PM_{2.5}$ in the winter heavy pollution episode and O_3 in the summer heavy pollution episode, we conducted the WRF-Chem simulations with the four 2020 pollution control scenarios (2020A, 2020B, 2020C, and 2020D). Table 3 shows the changes in pollutant concentrations of the four scenarios. Since Table 3 has summarized the overall impacts (in numbers) of each emission reduction scenarios on the concentration of $PM_{2.5}$, NO_2, and O_3, we will present the spatial variation of the impact in Figures 7–9 in the following text. According to the results of Section 3.2, the simulation during the observational peak was underestimated compared to the measurement, implying that the actual concentrations of pollutants in 2020 could be higher than our simulations.

Table 3. Changes in pollutant concentrations in Haizhu district of scenarios (A, B, C, and D are scenarios 2020A, 2020B, 2020C, and 2020D, respectively; winter episode: 14–28 January; summer episode: 3–8 August).

	Species	Changes in Pollutant Concentrations		
		(B-A)/A	(C-A)/A	(D-A)/A
Winter episode	$PM_{2.5}$	−7.15%	−23.42%	−3.18%
	NO_2	−6.46%	−28.29%	−4.59%
Summer episode	O_3	0.45%	0.35%	0.14%

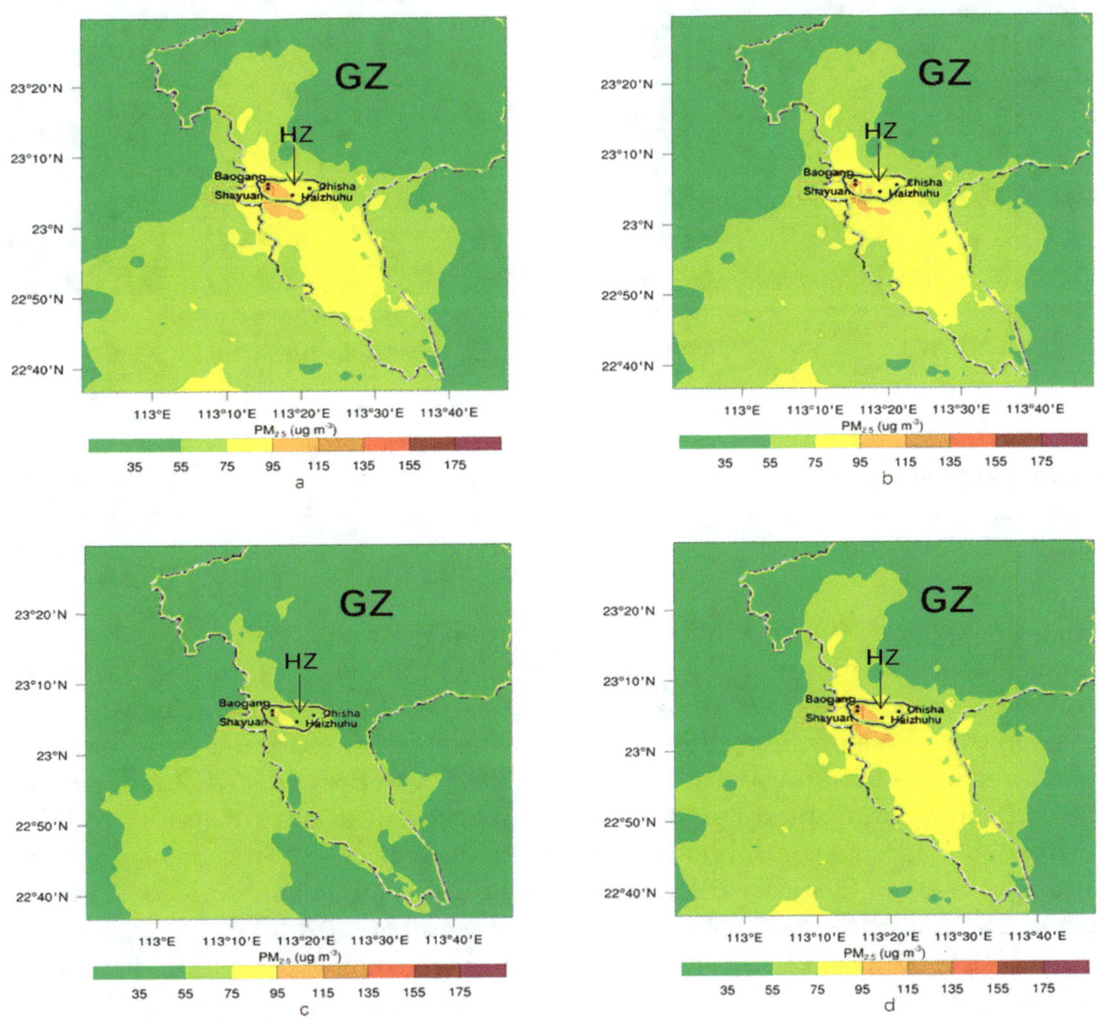

Figure 7. Maps of the simulated mean $PM_{2.5}$ concentration distribution during the period 0000 BJT, 20 January, to 2300 BJT, 21 January, over Haizhu for 2020 scenarios (a: 2020A; b: 2020B; c: 2020C; d: 2020D; HZ: Haizhu; GZ: Guangzhou; BJT: Beijing Time).

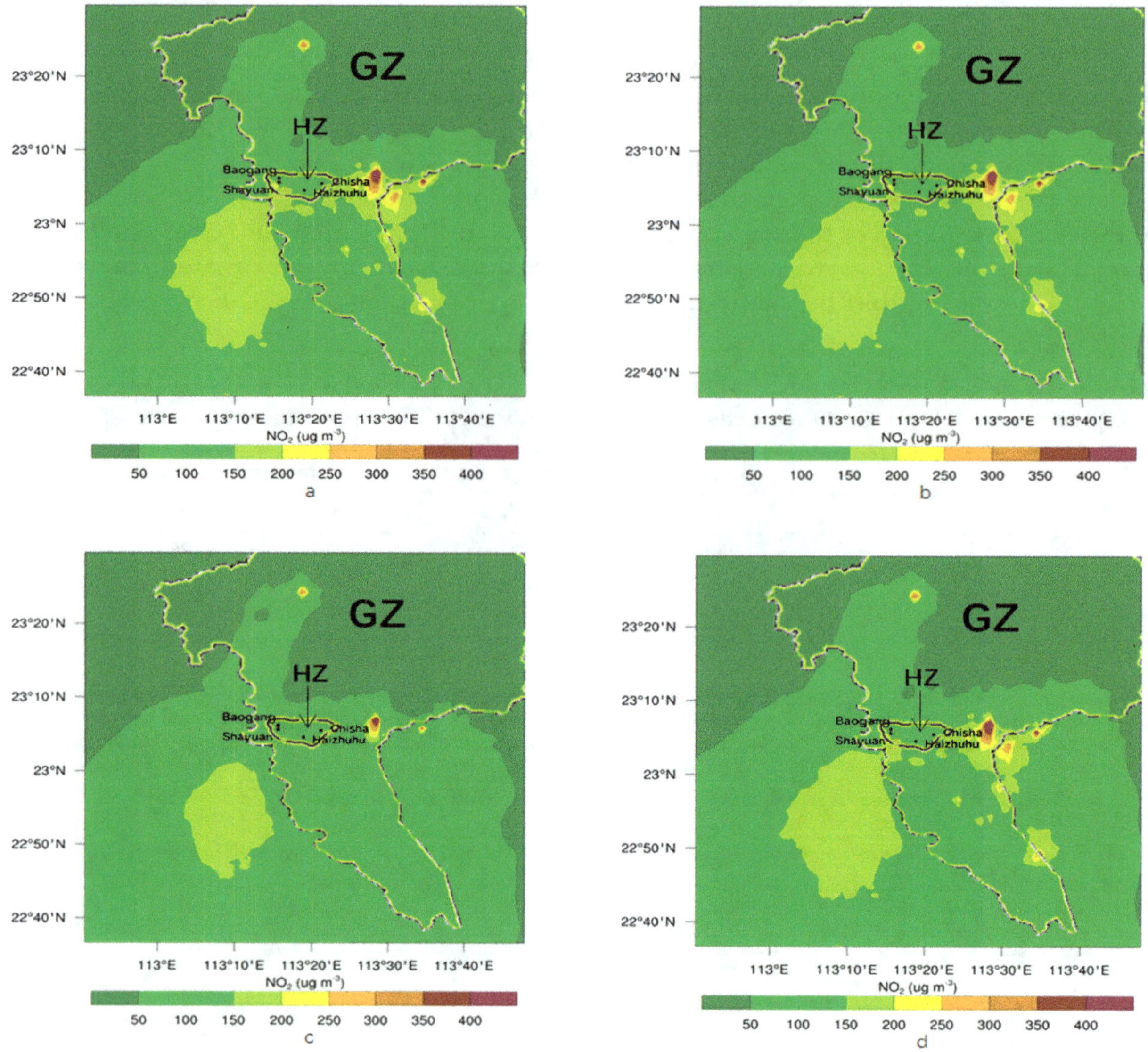

Figure 8. Maps of the simulated mean NO_2 concentration distribution during the period 0000 BJT, 20 January, to 2300 BJT, 21 January, over Haizhu for 2020 scenarios (a: 2020A; b: 2020B; c: 2020C; d: 2020D; HZ: Haizhu; GZ: Guangzhou).

(1) Effect on $PM_{2.5}$ and NO_2 in winter heavy pollution episode

For the simulated concentrations of $PM_{2.5}$ and NO_2 in Shayuan station, the concentrations of $PM_{2.5}$ and NO_2 were in the order of 2020A > 2020D > 2020B > 2020C. Comprehensive emission control measures taken in the whole of Guangzhou (2020C) would be able to reduce concentrations of $PM_{2.5}$ and NO_2 two times more than those just in Haizhu (2020B). In previous research, emission control from outside of Beijing contributed 8–14% to the improvement of the air quality in Beijing during APEC [57]. Therefore, regional control policies are needed. Regional control measures in a whole city could affect the air quality of the urban district, even if the district is just an isolated island. Referring to the measures in Haizhu, the effect of vehicle emission control (2020D) on $PM_{2.5}$ (NO_2) reduction would be ~44% (~71%) of the effect of comprehensive measures (2020B), meaning that vehicles are the major $PM_{2.5}$ (NO_2) emission source in Haizhu. Vehicle emission is a typical problem in urban areas of China [87]. Previous studies showed that vehicle emission accounted for 21%, 10%, and 19.3% of the total $PM_{2.5}$ in Dongguan (2014) [68], Guangzhou (2014) [88], and PRD (2012) [89], respectively.

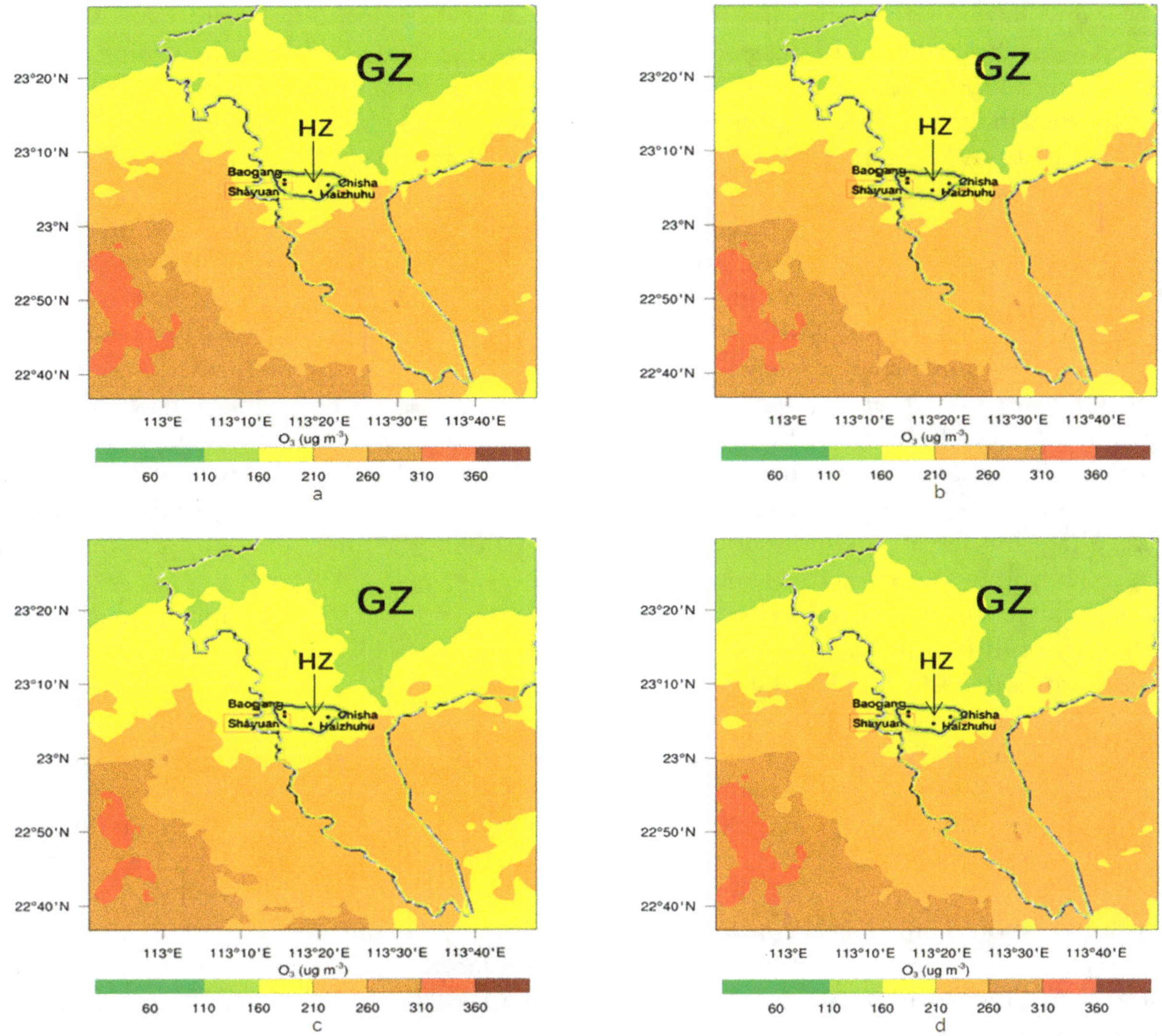

Figure 9. Maps of the simulated O_3 mean daily 8 h maximum concentration distribution over Haizhu for 2020 scenarios on 6 August (a: 2020A; b: 2020B; c: 2020C; d: 2020D; HZ: Haizhu; GZ: Guangzhou).

In the worst meteorological condition, $PM_{2.5}$ pollution on Haizhu district would be improved, as shown in Figures 7c and 7b, which means that the effects on regional air quality of scenarios 2020C and 2020B were relatively significant. In particular, compared with 2020D (Figure 7d), the improvement on regional air quality in 2020B was more obvious, although the $PM_{2.5}$ pollution in the particular station (Shayuan) seemed to have no significant difference. In Figure 8, NO_2 pollution shows a similar situation. It indicates that regional and comprehensive measures could improve regional air quality during heavy pollution days.

(2) Effect on O_3 in summer heavy pollution episode

For the Shayuan station, the O_3 pollution in scenario 2020B would be higher than in other scenarios, and the O_3 pollution in the other scenarios were in the order 2020C > 2020D > 2020A (Table 3). The simulation results suggested that the O_3 level increases from 2015 to 2020 in all emission scenarios even in 2020C in which the rest of air pollutants decrease. Such changes of O_3 concentration is probably because the Haizhu district is a VOC-limited region in which the O_3 level increases (decreases) with the reduction of NO_x (VOCs) emission (e.g., Sillman, 1999 [90]). There are generally two pathways that can be adopted to relieve the O_3 pollution in VOC-limited region like Haizhu district. (1) To significantly reduce the NO_x level and push the Haizhu district into NO_x-limited region in which O_3 level decreases with the reduction of NO_x and VOCs. However, in this pathway, there will be a period in which O_3 level significantly increases before it starts to decrease. (2) To adopt a proper reduction ratio of

VOCs/NO_x and flatten the O_3 increase to the largest extent during the transient from VOC-limited region to NO_x-limited region. But this pathway requires a great deal of effort to determine a favourable reduction ratio of VOCs/NO_x and cut the emission of VOCs from various sectors.

The findings in the present study highlight the complexity and nonlinear chemistry of O_3 formation and call for further investigations. For example, the reduction ratio of VOCs/NO_x for anthropogenic sources was generally suggested to be 1:2 in PRD [10]. However, considering the change of precursor emissions and meteorological conditions, the ratio of VOCs/NO_x might need to be further studied. Besides, the NO_2 is overpredicted in the present study implying that the VOC-limited nature in Haizhu district might be overpredicted, although there is very low chance that Haizhu district is a NOx-limited region in 2015 because the NO_x is still at an elevated level and this region (PRD) has been repeatedly diagnosed as a VOC-limited region (e.g., Xue et al., 2014 [91]). Moreover, the formation of O_3 depends on the local mixture of NO_x and VOCs. Therefore, the location/sector of the emission of NO_x and VOC, apart from the overall reduction of NO_x and VOC in an area, might have an impact on the changes of O_3. Although we have made some effort on such topic by evaluating the effect of only controlling traffic emission on air quality (scenario 2020D), it would be of interest to adopt more sophisticated tools, e.g., a tagging technique (Grewe, 2013 [92]), to evaluate the effect of various combinations of emission reduction in sectors in future study.

The simulated O_3-8h distribution on 6 August is shown in Figure 9 to demonstrate the difference of regional O_3 pollution in Haizhu district. Ozone pollution would be relieved in scenario 2020C (Figure 9c) compared with that in other scenarios (Figure 9a,b,d). It is noticeable that ozone pollution in the southwest area (red area, suburban and rural area) in 2020C (Figure 9c) could be smaller than those in other scenarios (Figure 9a,b,d), which may be because of the suitable ratio of precursors in that area of scenario 2020C, although the pollution in Haizhu (urban area) seemed to have no significant difference. The difference of regional O_3 pollution shows that in the southwest area (suburban and rural area), the policy in scenario 2020C could decrease the O_3 significantly (less red area), and in other regions, e.g., in Haizhu (urban area), the effect is less noticeable, which indicated that the control of ozone pollution should be taken according to local conditions, i.e., measures in urban areas should be different from those in rural areas. We note that transboundary transport of O_3 and its precursors (CO, VOCs, and NO_x) due to mesoscale dynamics complicates O_3 pollution control and requires strong and efficient cooperation among the adjacent regions.

4. Conclusions

In this study, the air quality in the year of 2015 and during pollution episodes in Haizhu district were analysed, and the impacts of emission control scenarios by the year of 2020 on air quality were evaluated using the WRF-Chem numerical simulation.

For the air quality in Haizhu 2015, the annually-averaged concentrations of $PM_{2.5}$, PM_{10}, SO_2, and NO_2 were higher than those in the entire Guangzhou city. O_3-8h was slightly lower than that in Guangzhou. Pollution episodes in Haizhu in 2015 primarily occurred in summer and winter. The typical winter pollution episode (14 to 28 January) was a process of gradual accumulation of pollution and dissipation, with NO_2 or $PM_{2.5}$ as the key pollutants, associated with the wind. The heavy pollution episode in the hot season (3 to 8 August) was a process of pollution that occurred and disappeared quickly, with O_3 as the key pollutant, due to suitable local pollution and strong sunshine.

The WRF-Chem simulation generally captured the observed chemical characteristics, suggesting that the WRF-Chem model could be used to simulate air quality in the research case.

Emission control scenario 2020C (comprehensive measures taken in the whole of Guangzhou city) would improve air quality more significantly than other scenarios (measures taken in Haizhu) under all conditions (heavy pollution conditions and annual level). For urban areas, scenario 2020D (vehicle emission control) would account for more than half of the influence of 2020B (all source emission controls) on air quality. By the end of the 13th FYP, it is noticeable that O_3 pollution would increase, which indicates that the control ratio of VOCs and NO_x may be unfavourable and requires further assessment. It would be of interest to perform simulations with the meteorology/chemistry interactions to investigate the influence of policies on O_3 and other air pollutants in future research.

Our study suggested that control measures should be strengthened for NO_2, $PM_{2.5}$, and PM_{10}, and control ratio of VOCs and NO_x should be adjusted for controlling O_3. The urban area should focus on vehicle emission control, strengthen regional cooperation on pollution control, and establish short-term measures for heavy pollution conditions.

Supplementary Materials: Figure S1: Research framework, Figure S2: WRF-Chem simulation domain, Table S1: Emission sectors, corresponding inventory technical guidelines and special allocation rules, Table S2: Description of emission control scenarios, Table S3: Detailed information for pollution episodes observed in Haizhu monitoring stations, 2015, Table S4: Statistical data for heavy pollution episode in Haizhu district during 14–28 January 2015, Table S5: Statistical data for heavy pollution episode in Haizhu district during 3–8 August 2015, Table S6: Performance statistics for meteorological simulation (hourly statistical data in October 2015), Table S7: Statistical comparison of model evaluation in this paper and previous study.

Author Contributions: J.Z. and H.D. designed the study. Y.L. and C.F. collected and analysed data. T.G. and L.L. did emission data preparation and generated the figures. R.O. and M.W. did the policy investigation. J.Z. contributed to literature search, numerical simulation and writing the manuscript. All authors have read and agreed to the published version of the manuscript.

Acknowledgments: The author greatly appreciated Guangzhou Haizhu District Environmental Monitoring Station for providing the data in this work and Tsinghua University for providing MEIC. Thanks also go to Qinyi Li in Spanish National Research Council and Jinpu Zhang in Guangzhou Environmental Monitoring Center for valuable suggestions and fruitful discussion.

References

1. Gulia, S.; Shiva, N.S.M.; Khare, M.; Khanna, I. Urban air quality management—A review. *Atmos. Pollut. Res.* **2015**, *6*, 286–304. [CrossRef]
2. SC (State Council of People's Republic of China. Air Pollution Control Action Plan. 2013. Available online: http://www.gov.cn/zwgk/2013-09/12/content_2486773.htm (accessed on 23 November 2018).
3. CAA (Clean Air Asia). China Air 2015—Air Pollution Prevention and Control Progress in Chinese Cities. 2015. Available online: http://cleanairasia.org/wp-content/uploads/2016/03/ChinaAir2015-report.pdf (accessed on 23 November 2018).
4. CAA (Clean Air Asia). China Air 2016—Air Pollution Prevention and Control Progress in Chinese Cities. 2016. Available online: http://www.allaboutair.cn/a/cbw/bg/2016/0822/472.html (accessed on 23 November 2018).
5. BJMBEE (Beijing Municipal Bureau of Ecoloy and Environment). Beijing Environmental Statement 2015. 2015. Available online: http://www.bjepb.gov.cn/bjhrb/xxgk/ywdt/hjzlzk/hjzkgb65/index.html (accessed on 23 November 2018).
6. SHEP (Shanghai Environmental Protection). Shanghai Air Quality Report. 2015. Available online: http://www.sepb.gov.cn/hb/fa/cms/shhj/list_login.jsp?channelId=5157 (accessed on 23 November 2018).
7. Zhan, J.; Liu, Y.; Lin, L.; Ding, H.; Xu, W. Investigation of spatial and temporal association of $PM_{2.5}$ pollution during the winter of 2014 in typical cities of Pearl River Delta. *Res. Environ. Sci.* **2017**, *30*, 110–120. [CrossRef]
8. Ma, X.; Jia, H. Particulate matter and gaseous pollutions in three megacities over China: Situation and implication. *Atmos. Environ.* **2016**, *140*, 476–494. [CrossRef]
9. Li, M.; Song, Y.; Mao, Z.; Liu, M.; Huang, X. Impacts of thermal circulations induced by urbanization on ozone formation in the Pearl River Delta region, China. *Atmos. Environ.* **2016**, *127*, 382–392. [CrossRef]
10. Wang, T.; Xue, L.; Brimblecombe, P.; Lam, Y.F.; Li, L.; Zhang, L. Ozone pollution in China: A review of concentrations, meteorological influences, chemical precursors, and effects. *Sci. Total Environ.* **2017**, *575*, 1582–1596. [CrossRef]

11. SC (State Council of People's Republic of China). Environmental Protection 13th Five-Year Plan. 2016. Available online: http://www.gov.cn/zhengce/content/2016-12/05/content_5143290.htm (accessed on 23 November 2018).
12. PGGZ (People's Government of Guangzhou). Guangzhou Environmental Protection 13th Five-Year Plan. 2016. Available online: http://www.gz.gov.cn/gzgov/s2812/201612/467e946dfcb048c0a7bff56655808c60.shtml (accessed on 23 November 2018).
13. Wang, N.; Lyu, X.P.; Deng, X.J.; Guo, H.; Deng, T.; Li, Y.; Yin, C.Q.; Li, F.; Wang, S.Q. Assessment of regional air quality resulting from emission control in the Pearl River Delta region, southern China. *Sci. Total Environ.* **2016**, *573*, 1554–1565. [CrossRef]
14. Liu, Y.H.; Liao, W.Y.; Lin, X.F.; Li, L.; Zeng, X.L. Assessment of Co-benefits of vehicle emission reduction measures for 2015–2020 in the Pearl River Delta region, China. *Environ. Pollut.* **2017**, *223*, 62–72. [CrossRef]
15. Maji, K.J.; Dikshit, A.K.; Arora, M.; Deshpande, A. Estimating premature mortality attributable to $PM_{2.5}$ exposure and benefit of air pollution control policies in China for 2020. *Sci. Total Environ.* **2018**, *612*, 683–693. [CrossRef]
16. Yang, X.; Teng, F. The air quality co-benefit of coal control strategy in China. *Resour. Conserv. Recycl.* **2016**, *129*, 373–382. [CrossRef]
17. Wang, Z.; Pan, L.; Li, Y.; Zhang, D.; Ma, J.; Sun, F.; Xu, W.; Wang, X. Assessment of air quality benefits from the national pollution control policy of thermal power plants in China: A numerical simulation. *Atmos. Environ.* **2015**, *106*, 288–304. [CrossRef]
18. Qiu, X.; Duan, L.; Cai, S.; Yu, Q.; Wang, S.; Chai, F.; Gao, J.; Li, Y.; Xu, Z. Effect of current emission abatement strategies on air quality improvement in China: A case study of Baotou, a typical industrial city in Inner Mongolia. *J. Environ. Sci.* **2017**, *57*, 383–390. [CrossRef] [PubMed]
19. Cai, S.; Wang, Y.; Zhao, B.; Wang, S.; Chang, X.; Hao, J. The impact of the "air pollution prevention and control action plan" on PM2. 5 concentrations in jing-jin-ji region during 2012–2020. *Sci. Total Environ.* **2017**, *580*, 197–209. [CrossRef] [PubMed]
20. Li, M.; Patiño-Echeverri, D. Estimating benefits and costs of policies proposed in the 13th FYP to improve energy efficiency and reduce air emissions of China's electric power sector. *Energy Policy* **2017**, *111*, 222–234. [CrossRef]
21. Wei, W.; Li, P.; Wang, H.; Song, M. Quantifying the effects of air pollution control policies: A case of Shanxi province in China. *Atmos. Pollut. Res.* **2017**, *9*, 429–438. [CrossRef]
22. Guo, J.; He, J.; Liu, H.; Miao, Y.; Liu, H.; Zhai, P. Impact of various emission control schemes on air quality using WRF-Chem during APEC China 2014. *Atmos. Environ.* **2016**, *140*, 311–319. [CrossRef]
23. Xu, H.M.; Tao, J.; Ho, S.S.H.; Ho, K.F.; Cao, J.J.; Li, N.; Chow, J.C.; Wang, G.H.; Han, Y.M.; Zhang, R.J.; et al. Characteristics of fine particulate non-polar organic compounds in Guangzhou during the 16th Asian Games: Effectiveness of air pollution controls. *Atmos. Environ.* **2013**, *76*, 94–101. [CrossRef]
24. Shen, X.J.; Sun, J.Y.; Zhang, X.Y.; Zhang, Y.M.; Zhang, L.; Fan, R.X.; Zhang, Z.X.; Zhang, X.L.; Zhou, H.G.; Zhou, L.Y.; et al. The influence of emission control on particle number size distribution and new particle formation during China's V-Day parade in 2015. *Sci. Total Environ.* **2016**, *573*, 409–419. [CrossRef]
25. Tan, J.H.; Duan, J.C.; Chen, D.H.; Wang, X.H.; Guo, S.J.; Bi, X.H.; Sheng, G.Y.; He, K.B.; Fu, J.M. Chemical characteristics of haze during summer and winter in Guangzhou. *Atmos. Res.* **2009**, *94*, 238–245. [CrossRef]
26. Wang, M.; Cao, C.; Li, G.; Singh, R.P. Analysis of a severe prolonged regional haze episode in the Yangtze River Delta, China. *Atmos. Environ.* **2015**, *102*, 112–121. [CrossRef]
27. Zhang, L.; Wang, T.; Lv, M.; Zhang, Q. On the severe haze in Beijing during January 2013: Unraveling the effects of meteorological anomalies with WRF-Chem. *Atmos. Environ.* **2015**, *104*, 11–21. [CrossRef]
28. Ding, A.; Wang, T.; Zhao, M.; Wang, T.; Li, Z. Simulation of sea-land breezes and a discussion of their implications on the transport of air pollution during a multi-day ozone episode in the Pearl River Delta of China. *Atmos. Environ.* **2004**, *38*, 6737–6750. [CrossRef]
29. Shen, J.; Zhang, Y.; Wang, X.; Li, J.; Chen, H.; Liu, R.; Zhong, L.; Jiang, M.; Yue, D.; Chen, D.; et al. An ozone episode over the Pearl River Delta in October 2008. *Atmos. Environ.* **2015**, *122*, 852–863. [CrossRef]
30. Zhao, H.; Wang, S.; Wang, W.; Liu, R.; Zhou, B. Investigation of ground-level ozone and high-pollution episodes in a megacity of Eastern China. *PLoS ONE.* **2015**, *10*, e0131878. [CrossRef] [PubMed]

31. Xu, J.; Zhang, Y.; Fu, J.S.; Zheng, S.; Wang, W. Process analysis of typical summertime ozone episodes over the Beijing area. *Sci. Total Environ.* **2008**, *399*, 147–157. [CrossRef]
32. Qu, Y.; An, J.; Li, J.; Chen, Y.; Li, Y.; Liu, X.; Hu, M. Effects of NOx and VOCs from five emission sources on summer surface O_3 over the Beijing-Tianjin-Hebei region. *Adv. Atmos. Sci.* **2014**, *31*, 787–800. [CrossRef]
33. MEE (Ministry of Ecology and Environment of People's Republic of China). China Vehicle Environmental Management Annual Report 2017. 2017. Available online: http://dqhj.mee.gov.cn/jdchjgl/zhgldt/201706/P020170605550637870889.pdf (accessed on 23 November 2018).
34. EPGD (Environmental Protection of Guangdong Province). Guangzhou Environmental Protection's Implementation of Grid Control. 2016. Available online: http://www.gdep.gov.cn/zwxx_1/hbxx/201607/t20160725_213142.html (accessed on 23 November 2018).
35. EPGD (Environmental Protection of Guangdong Province). 2017 Guangdong Cities Air Quality Ranking Released, Guangdong Achieved Targets for Three Years. 2018. Available online: http://mp.weixin.qq.com/s/YKKBj9pncWaq67lFq0l6mQ (accessed on 23 November 2018).
36. Wang, J.; Kwan, M.P.; Ma, L. Delimiting service area using adaptive crystal-growth Voronoi diagrams based on weighted planes: A case study in Haizhu District of Guangzhou in China. *Appl. Geogr.* **2014**, *50*, 108–119. [CrossRef]
37. Zhang, Z.; Li, H.; Liu, H.; Ni, R.; Li, J.; Deng, L.; Lu, D.; Cheng, X.; Duan, P.; Li, W. A preliminary analysis of the surface chemistry of atmospheric aerosol particles in a typical urban area of Beijing. *J. Environ. Sci.* **2016**, *47*, 71–81. [CrossRef]
38. Ling, Z.H.; Zhao, J.; Fan, S.J.; Wang, X.M. Sources of formaldehyde and their contributions to photochemical O_3 formation at an urban site in the Pearl River Delta, southern China. *Chemosphere* **2017**, *168*, 1293–1301. [CrossRef]
39. HZEPB (Haizhu Environmental Protection Bureau). Haizhu Environmental Protection Bureau 2015 Work Report and 2016 Key Work. 2016. Available online: http://zwgk.haizhu.gov.cn/HZ12/201605/t20160526_346735.html (accessed on 23 November 2018).
40. HZEPB (Haizhu Environmental Protection Bureau). Haizhu Environmental Protection Bureau 2016 Work Report and 2017 Key Work. 2017. Available online: http://zwgk.haizhu.gov.cn/HZ12/201703/t20170324_392921.html (accessed on 23 November 2018).
41. CNGIPSP (China National Geographic Information Public Service Platform). GS(2017)508-1100471 Map Data. 2017. Available online: http://map.tianditu.com/ (accessed on 23 November 2018).
42. Gong, J.; Hu, Z.; Chen, W.; Liu, Y.; Wang, J. Urban expansion dynamics and modes in metropolitan Guangzhou, China. *Land Use Policy.* **2018**, *72*, 100–109. [CrossRef]
43. He, K. Multi-resolution Emission Inventory for China (MEIC): Model framework and 1990–2010 anthropogenic emissions. In *AGU Fall Meeting Abstracts*; Fall Meeting 2012 of American Geophysical Union; American Geophysical Union: San Francisco, CA, USA, 3–7 December 2012; abstract id. A32B-05; Available online: http://adsabs.harvard.edu/abs/2012AGUFM.A32B..05H (accessed on 23 November 2018).
44. Guenther, A.; Zimmerman, P.; Wildermuth, M. Natural volatile organic compound emission rate estimates for US woodland landscapes. *Atmos. Environ.* **1994**, *28*, 1197–1210. [CrossRef]
45. Shaw, W.J.; Allwine, K.J.; Fritz, B.G.; Rutz, F.C.; Rishel, J.P.; Chapman, E.G. An evaluation of the wind erosion module in DUSTRAN. *Atmos. Environ.* **2008**, *42*, 1907–1921. [CrossRef]
46. Gong, S.L.; Barrie, L.A.; Lazare, M. Canadian Aerosol Module (CAM): A size-segregated simulation of atmospheric aerosol processes for climate and air quality models: 2. Global sea-salt aerosol and its budgets. *J. Geophys. Res. Atmos.* **2002**, *107*, 4779. [CrossRef]
47. Lei, Y.; Zhang, Q.; He, K.B.; Streets, D.G. Primary anthropogenic aerosol emission trends for China, 1990–2005. *Atmos. Chem. Phys.* **2011**, *11*, 931–954. [CrossRef]
48. Li, Q.; Zhang, L.; Wang, T.; Tham, Y.J.; Ahmadov, R.; Xue, L.; Zhang, Q.; Zheng, J. Impacts of heterogeneous uptake of dinitrogen pentoxide and chlorine activation on ozone and reactive nitrogen partitioning: Improvement and application of the WRF-Chem model in southern China. *Atmos. Chem. Phys.* **2016**, *16*, 14875. [CrossRef]
49. PGGZ (People's Government of Guangzhou). Guangzhou Air Quality Targets Plan (2016–2025). 2017. Available online: http://www.gz.gov.cn/gzgov/s2811/201712/57727a1d77354f5dbc22bb5831aa7d93.shtml (accessed on 23 November 2018).

50. MEE (Ministry of Ecology and Environment of People's Republic of China). Announcement on the publication of four technical guidelines (Technical Guidelines for the Primary Source Emission Inventory of Atmospheric Fine Particles (Trial), Technical Guidelines for the Emission Inventory of Atmospheric Volatile Organic Compounds (Trial), Technical Guidelines for the Emission Inventory of Atmospheric Ammonia Source (Trial), and Technical Guidelines for Priority Control of Atmospheric Pollution Sources (Trial)). 2014. Available online: http://www.mee.gov.cn/gkml/hbb/bgg/201408/t20140828_288364.htm (accessed on 23 November 2018).
51. MEE (Ministry of Ecology and Environment of People's Republic of China). Announcement on the Publication of Five Technical Guidelines (Technical Guidelines for the Primary Source Emission Inventory of Inhalable Particulate Matter (Trial), Technical Guidelines for the Air Pollutant Emission Inventory for Road Vehicles (Trial), Technical Guidelines for the Air Pollutant Emission Inventory of Non-Road Mobile Source (Trial), Technical Guidelines for the Air Pollutant Emission Inventory for Biomass Combustion Sources (Trial), Technical Guidelines for the Emission Inventory of Dust Particles Discharge (Trial)). 2014. Available online: http://www.mee.gov.cn/gkml/hbb/bgg/201501/t20150107_293955.htm (accessed on 23 November 2018).
52. Zheng, J.Y.; Zhang, L.J.; Che, W.W.; Zheng, Z.Y.; Yin, S.S. A highly resolved temporal and spatial air pollutant emission inventory for the Pearl River Delta region, China and its uncertainty assessment. *Atmos. Environ.* **2009**, *43*, 5112–5122. [CrossRef]
53. Zhao, X.Y.; Hu, Q.H.; Wang, X.M.; Ding, X.; He, Q.F.; Zhang, Z.; Shen, R.Q.; Lü, S.J.; Liu, T.Y.; Fu, X.X.; et al. Composition profiles of organic aerosols from Chinese residential cooking: Case study in urban Guangzhou, south China. *J. Atmos. Chem.* **2015**, *72*, 1–18. [CrossRef]
54. PGGZ (People's Government of Guangzhou). Guangzhou Transportation Development 13th Five-Year Plan. 2016. Available online: http://www.gz.gov.cn/gzgov/s2812/201611/a8ee48d726b649caaba3e4f12572feae.shtml (accessed on 23 November 2018).
55. NOAA (National Oceanic and Atmospheric Administration). Weather Research and Forecasting Model Coupled to Chemistry (WRF-Chem). 2017. Available online: https://ruc.noaa.gov/wrf/wrf-chem/ (accessed on 23 November 2018).
56. NCAR (National Center for Atmospheric Research). Atmospheric Chemistry Observation & Modeling: WRF-CHEM. 2018. Available online: https://www2.acom.ucar.edu/wrf-chem (accessed on 23 November 2018).
57. He, H.; Tie, X.; Zhang, Q.; Liu, X.; Gao, Q.; Li, X.; Gao, Y. Analysis of the causes of heavy aerosol pollution in Beijing, China: A case study with the WRF-Chem model. *Particuology* **2015**, *20*, 32–40. [CrossRef]
58. Emmons, L.K.; Walters, S.; Hess, P.G.; Lamarque, J.F.; Pfister, G.G.; Fillmore, D.; Granier, C.; Guenther, A.; Kinnison, D.; Laepple, T.; et al. Description and evaluation of the Model for Ozone and Related chemical Tracers, version 4 (MOZART-4). *Geosci. Model Dev.* **2010**, *3*, 43–67. [CrossRef]
59. Stockwell, W.R.; Middleton, P.; Chang, J.S.; Tang, X. The second generation regional acid deposition model chemical mechanism for regional air quality modeling. *J. Geophys. Res. Atmos.* **1990**, *95*, 16343–16367. [CrossRef]
60. Ackermann, I.J.; Hass, H.; Memmesheimer, M.; Ebel, A.; Binkowski, F.S.; Shankar, U.M.A. Modal aerosol dynamics model for Europe: Development and first applications. *Atmos. Environ.* **1998**, *32*, 2981–2999. [CrossRef]
61. Schell, B.; Ackermann, I.J.; Hass, H.; Binkowski, F.S.; Ebel, A. Modeling the formation of secondary organic aerosol within a comprehensive air quality model system. *J. Geophys. Res. Atmos.* **2001**, *106*, 28275–28293. [CrossRef]
62. MEE (Ministry of Ecology and Environment of People's Republic of China). Technical Regulation on Ambient Air Quality Index (on trial) (HJ 633-2012). 2016. Available online: http://www.gov.cn/zwgk/2012-03/02/content_2081374.htm (accessed on 23 November 2018).
63. EPGD (Environmental Protection of Guangdong Province). Guangdong-Hong Kong-Macao Pearl River Delta Regional Air Quality Monitoring Network: Report of Monitoring Results 2015. 2016. Available online: http://www.gdep.gov.cn/hjjce/kqjc/index_1.html (accessed on 23 November 2018).
64. GZEP (Guangzhou Environmental Protection Bureau). Guangzhou Air Quality Report 2015. 2016. Available online: http://www.gzepb.gov.cn/zwgk/hjgb/ (accessed on 23 November 2018).
65. WHO (World Health Organization). WHO Air Quality Guidelines (Global Update 2005). 2005. Available online: http://apps.who.int/iris/bitstream/handle/10665/69477/WHO_SDE_PHE_OEH_06.02_eng.pdf;jsessionid=0757FBBE71232DE20E0BB5E7C64E45A1?sequence=1 (accessed on 23 November 2018).

66. MEE (Ministry of Ecology and Environment of People's Republic of China) Technical Regulations for Urban Environmental Air Quality Ranking. 2018. Available online: http://www.mee.gov.cn/gkml/sthjbgw/bgtwj/201808/W020180815576019704248.pdf (accessed on 23 November 2018).
67. Tong, C.H.M.; Yim, S.H.L.; Rothenberg, D.; Wang, C.; Lin, C.Y.; Chen, Y.D.; Lau, N.C. Assessing the impacts of seasonal and vertical atmospheric conditions on air quality over the Pearl River Delta region. *Atmos. Environ.* **2018**, *180*, 69–78. [CrossRef]
68. Zou, B.B.; Huang, X.F.; Zhang, B.; Dai, J.; Zeng, L.W.; Feng, N.; He, L.Y. Source apportionment of $PM_{2.5}$ pollution in an industrial city in southern China. *Atmos. Pollut. Res.* **2017**, *8*, 1193–1202. [CrossRef]
69. Malek, E.; Davis, T.; Martin, R.S.; Silva, P.J. Meteorological and environmental aspects of one of the worst national air pollution episodes (January, 2004) in Logan, Cache Valley, Utah, USA. *Atmos. Res.* **2006**, *79*, 108–122. [CrossRef]
70. Cui, H.; Chen, W.; Dai, W.; Liu, H.; Wang, X.; He, K. Source apportionment of $PM_{2.5}$ in Guangzhou combining observation data analysis and chemical transport model simulation. *Atmos. Environ.* **2015**, *116*, 262–271. [CrossRef]
71. Song, H.; Wang, K.; Zhang, Y.; Hong, C.; Zhou, S. Simulation and evaluation of dust emissions with WRF-Chem (v3. 7.1) and its relationship to the changing climate over East Asia from 1980 to 2015. *Atmos. Environ.* **2017**, *167*, 511–522. [CrossRef]
72. Gao, J.; Zhu, B.; Xiao, H.; Kang, H.; Hou, X.; Shao, P. A case study of surface ozone source apportionment during a high concentration episode, under frequent shifting wind conditions over the Yangtze River Delta, China. *Sci. Total Environ.* **2016**, *544*, 853–863. [CrossRef]
73. Hong, C.; Zhang, Q.; He, K.; Guan, D.; Li, M.; Liu, F.; Zheng, B. Variations of China's emission estimates: Response to uncertainties in energy statistics. *Atmos. Chem. Phys.* **2017**, *17*, 1227–1239. [CrossRef]
74. Huo, H.; Yao, Z.; Zhang, Y.; Shen, X.; Zhang, Q.; He, K. On-board measurements of emissions from diesel trucks in five cities in China. *Atmos. Environ.* **2012**, *54*, 159–167. [CrossRef]
75. Zheng, B.; Zhang, Q.; Tong, D.; Chen, C.; Hong, C.; Li, M.; Geng, G.; Lei, Y.; Huo, H.; He, K. Resolution dependence of uncertainties in gridded emission inventories: A case study in Hebei, China. *Atmos. Chem. Phys.* **2017**, *17*, 921–933. [CrossRef]
76. Ahmadov, R.; McKeen, S.A.; Robinson, A.L.; Bahreini, R.; Middlebrook, A.M.; De Gouw, J.A.; Meagher, J.; Hsie, E.Y.; Edgerton, E.; Shaw, S.; et al. A volatility basis set model for summertime secondary organic aerosols over the eastern United States in 2006. *J. Geophys. Res. Atmos.* **2012**, *117*, D6. [CrossRef]
77. Li, Q.; Zhang, L.; Wang, T.; Wang, Z.; Fu, X.; Zhang, Q. "New" Reactive Nitrogen Chemistry Reshapes the Relationship of Ozone to Its Precursors. *Environ. Sci. Technol.* **2018**, *52*, 2810–2818. [CrossRef] [PubMed]
78. Banks, R.F.; Baldasano, J.M. Impact of WRF model PBL schemes on air quality simulations over Catalonia, Spain. *Sci. Total Environ.* **2016**, *572*, 98–113. [CrossRef] [PubMed]
79. Shin, H.H.; Hong, S.Y. Intercomparison of planetary boundary-layer parametrizations in the WRF model for a single day from CASES-99. *Bound. Layer Meteorol.* **2011**, *139*, 261–281. [CrossRef]
80. Hu, X.M.; Nielsen-Gammon, J.W.; Zhang, F. Evaluation of three planetary boundary layer schemes in the WRF model. *J. Appl. Meteorol. Climatol.* **2010**, *49*, 1831–1844. [CrossRef]
81. Hong, S.Y.; Noh, Y.; Dudhia, J. A new vertical diffusion package with an explicit treatment of entrainment processes. *Mon. Weather Rev.* **2006**, *134*, 2318–2341. [CrossRef]
82. Schutgens, N.A.; Gryspeerdt, E.; Weigum, N.; Tsyro, S.; Goto, D.; Schulz, M.; Stier, P. Will a perfect model agree with perfect observations? The impact of spatial sampling. *Atmos. Chem. Phys.* **2016**, *16*, 6335–6353. [CrossRef]
83. Solazzo, E.; Bianconi, R.; Hogrefe, C.; Curci, G.; Tuccella, P.; Alyuz, U.; Balzarini, A.; Baró, R.; Bellasio, R.; Bieser, J.; et al. Evaluation and error apportionment of an ensemble of atmospheric chemistry transport modeling systems: Multivariable temporal and spatial breakdown. *Atmos. Chem. Phys.* **2017**, *17*, 3001–3054. [CrossRef]
84. Solazzo, E.; Hogrefe, C.; Colette, A.; Garcia-Vivanco, M.; Galmarini, S. Advanced error diagnostics of the CMAQ and Chimere modelling systems within the AQMEII3 model evaluation framework. *Atmos. Chem. Phys.* **2017**, *17*, 10435–10465. [CrossRef]
85. Koch, R.; Knispel, R.; Elend, M.; Siese, M.; Zetzsch, C. Consecutive reactions of aromatic-OH adducts with NO, NO_2 and O_2: Benzene, naphthalene, toluene, m-and p-xylene, hexamethylbenzene, phenol, m-cresol and aniline. *Atmos. Chem. Phys.* **2007**, *7*, 2057–2071. [CrossRef]

86. Barthelmie, R.J.; Pryor, S.C. Secondary organic aerosols: Formation potential and ambient data. *Sci. Total Environ.* **1997**, *205*, 167–178. [CrossRef]
87. Li, Y.J.; Sun, Y.; Zhang, Q.; Li, X.; Li, M.; Zhou, Z.; Chan, C.K. Real-time chemical characterization of atmospheric particulate matter in China: A review. *Atmos. Environ.* **2017**, *158*, 270–304. [CrossRef]
88. Tao, J.; Zhang, L.; Cao, J.; Zhong, L.; Chen, D.; Yang, Y.; Chen, D.; Chen, L.; Zhang, Z.; Wu, Y.; et al. Source apportionment of PM2.5 at urban and suburban areas of the Pearl River Delta region, south China-With emphasis on ship emissions. *Sci. Total Environ.* **2017**, *574*, 1559–1570. [CrossRef] [PubMed]
89. Yin, X.; Huang, Z.; Zheng, J.; Yuan, Z.; Zhu, W.; Huang, X.; Chen, D. Source contributions to $PM_{2.5}$ in Guangdong province, China by numerical modeling: Results and implications. *Atmos. Res.* **2017**, *186*, 63–71. [CrossRef]
90. Sillman, S. The relation between ozone, NO_x and hydrocarbons in urban and polluted rural environments. *Atmos. Environ.* **1999**, *33*, 1821–1845. [CrossRef]
91. Xue, L.K.; Wang, T.; Gao, J.; Ding, A.J.; Zhou, X.H.; Blake, D.R.; Wang, X.F.; Saunders, S.M.; Fan, S.J.; Zuo, H.C.; et al. Ground-level ozone in four Chinese cities: Precursors, regional transport and heterogeneous processes. *Atmos. Chem. Phys.* **2014**, *14*, 13175–13188. [CrossRef]
92. Grewe, V. A generalized tagging method. *Geosci. Model Dev.* **2013**, *6*, 247–253. [CrossRef]

13

Air Quality and Control Measures Evaluation during the 2014 Youth Olympic Games in Nanjing and its Surrounding Cities

Hui Zhao [1], Youfei Zheng [1,2,*] and Ting Li [2]

1 Key Laboratory for Aerosol-Cloud-Precipitation of China Meteorological Administration, Nanjing University of Information Science and Technology, Nanjing 210044, China; zhaohui_nuist@163.com

2 Key Laboratory of Atmospheric Environment Monitoring and Pollution Control, Collaborative Innovation Center of Atmospheric Environment and Equipment Technology, Nanjing University of Information Science and Technology, Nanjing 210044, China; c2015liting@163.com

* Correspondence: zhengyf@nuist.edu.cn

Academic Editors: Pius Lee, Rick Saylor and Jeff McQueen

Abstract: Air pollution had become a vital concern for the 2014 Youth Olympic Games in Nanjing. In order to control air pollutant emissions and ensure better air quality during the Games, the Nanjing municipal government took a series of aggressive control measures to reduce pollutant emissions in Nanjing and its surrounding cities during the Youth Olympic Games. The Air Quality Index (AQI) is an index of air quality which is used to inform the public about levels of air pollution and associated health risks. In this study, we use the AQI and air pollutant concentrations data to evaluate the effectiveness of the implementation of control measures. The results suggest that the emission reduction measures significantly improved air quality in Nanjing. In August 2014, the mean concentrations of $PM_{2.5}$, PM_{10}, SO_2, NO_2, CO and O_3 were 42.44 $\mu g \cdot m^{-3}$, 59.01 $\mu g \cdot m^{-3}$, 11.12 $\mu g \cdot m^{-3}$, 31.09 $\mu g \cdot m^{-3}$, 0.76 $mg \cdot m^{-3}$ and 38.39 $\mu g \cdot m^{-3}$, respectively, and fell by 35.92%, 36.75%, 20.40%, 15.05%, 8.54% and 47.15%, respectively, compared to the prophase mean before the emission reduction. After the emission reduction, the mean concentrations of $PM_{2.5}$, PM_{10}, SO_2, NO_2, and O_3 increased by 20.81%, 41.84%, 22.84%, 21.16% and 60.93%, respectively, which is due to the cancellation of temporary atmospheric pollution control measures. The air pollutants diurnal variation curve during the emission reduction was lower than the other two periods, except for CO. In addition, the AQI of Nanjing and its surrounding cities showed a downward trend, compared with July 2014. The most of effective method to control air pollution is to implement the measures of regional cooperation and joint defense and control, and reduce local emissions during the polluted period, such as airborne dust, coal-burning, vehicle emissions, mobile sources and industrial production.

Keywords: air pollutants; $PM_{2.5}$; Youth Olympic Games; meteorological conditions; emission reduction; air quality

---

1. Introduction

As the largest developing country in the world, China has suffered from serious air pollution due to the rapid economic development and industrial reconstruction in the past three decades. It has become one of the major environmental concerns in some Chinese cities, including Beijing, Shanghai, Guangzhou and Nanjing [1]. Some major air pollutants in the atmosphere, including particulate matter (PM), sulfur dioxide (SO_2), carbon monoxide (CO), nitrogen dioxide (NO_2) and ozone (O_3), have attracted increasing attention due to their impacts on air quality, visibility reduction, human health and global climate [2–6]. From June 2000, the China National Environmental Monitoring Center started reporting the status of ambient air quality by using an air pollution index (API), which is calculated based on the highest index of 24-h average concentrations of PM_{10}, SO_2 and NO_2. From March 2012, the Ministry of Environmental Protection (MEP) started using the air quality index (AQI), which is calculated based on the concentrations of the six major pollutants including $PM_{2.5}$, PM_{10}, SO_2, NO_2, CO and O_3. The AQI ranges from 0 to 500. The greater the value of the AQI, the higher the level of air pollution. Currently, the AQI is widely used to describe the quality of the air.

Nanjing, the capital city of Jiangsu Province, is located in the western Yangtze River Delta and has a population of over 8.2 million. As a highly industrialized and urbanized city, Nanjing suffers from severe air pollution. It hosted the 2nd Summer Youth Olympic Games (YOG) from 17 to 28 August 2014. In view of the significant air pollution in Nanjing, to ensure good air quality during the YOG, the Nanjing municipal government actively took temporary control measures to reduce pollutant emissions. These measures mainly included dust control, coal-burning control, vehicle emission control, industrial production halts, construction site shutdowns and regional joint prevention and control. In addition, other cities in the surrounding area cooperated with Nanjing to guarantee good air quality during the same period.

In recent years, China has hosted many major international events. The local government has implemented numerous stringent local emission standards, which significantly reduced the emissions and concentrations of air pollutants in the city. During the 2008 Beijing Olympics Games, the mean concentration of SO_2, $PM_{2.5}$ and NO_2 were reduced by 51.0%, 43.7% and 13% compared to the period before Olympic Games in Beijing and its surrounding area [7]. Wang et al. [8] demonstrated that the concentration of O_3, SO_2, CO and NOx decreased 23%, 61%, 25% and 21%, respectively, compared to previous years. Wang et al. [9] found that the average concentrations of $PM_{2.5}$, PM_{10}, SO_2 and NO_2 during the Asia-Pacific Economic Cooperation period decreased by 47%, 36%, 62% and 41% respectively, whereas concentrations of O_3 increased by 102%. Meanwhile, emission control measures which were implemented in Shanghai and Guangzhou also successfully improved air quality for the World Expo and the Asian Games [10,11].

In this study, we used air quality and air pollutants measurement data at nine sites in Nanjing City from July to September 2014 to quantify the efficiency of the emission control measures in YOG. The objectives of this study were (i) to show the temporal and spatial characteristics of air quality and air pollutants during the YOG, and (ii) to give some suggestions for controlling air pollution in large cities.

2. Data and Methods

In this paper, the AQI and hourly concentrations data for six air pollutants ($PM_{2.5}$, PM_{10}, SO_2, NO_2, CO and O_3) were provided by the China National Environmental Monitoring Center (CNEMC). The AQI approach is based on the National Ambient Air Quality Standards of China (NAAQS-2012) [12]. There are nine sites located in Nanjing, including Maigao Bridge (MG), Caochangmen (CC), Shanxi Road (SX), Zhonghuamen (ZH), Ruijin Road (RJ), Xuanwu Lake (XW), Pukou (PK), Olympic Stadium (OS) and Xianlin University Town (XL). To show the effectiveness of the implementation of control measures, we acquired hourly monitoring data from July to September 2014. We divided the data into three time periods, before the period of the emission reduction (1 to 31 July 2014), during the period of the emission reduction (1 to 31 August 2014) and after the period of the emission reduction (1 to 30 September 2014), to assess the impacts of emission reduction measures on air quality. In addition, we also recorded the data of the AQI in the surrounding cities, including Shanghai (SH), Hangzhou (HZ1), Huzhou (HZ2), Jiaxing (JX), Hefei (HF), Maanshan (MAS), Wuhu (WH), Xuancheng (XC), Chuzhou (CZ1), Bengbu (BB), Suzhou (SZ), Wuxi (WX), Changzhou (CZ2), Zhenjiang (ZJ), Yangzhou (YZ), Huaian (HA), Taizhou (TZ), Nantong (NT), Xuzhou (XZ), Yancheng (YC), Suqian (SQ) and Lianyungang (LYG), which are shown in Figure 1.

Meteorological data, including air temperature (T), relative humidity (RH), wind speed (WS) and atmospheric pressure (AP), were obtained from the China Meteorological Data Sharing Service

System Administration (http://data.cma.cn/site/index.html), and solar radiation (SR) was recorded by Watchdog weather station (Watchdog 2900ET, Spectrum Technologies, Inc., USA) on the campus of Nanjing University of Information Science & Technology (NUIST) (32°14′ N, 118°42′ E).

This study analyzed temporal variation and characteristics of air quality and air pollutants in different periods using the measured time-series. All graphs were created using origin 9.0 software (Origin Lab, Northampton, MA, USA).

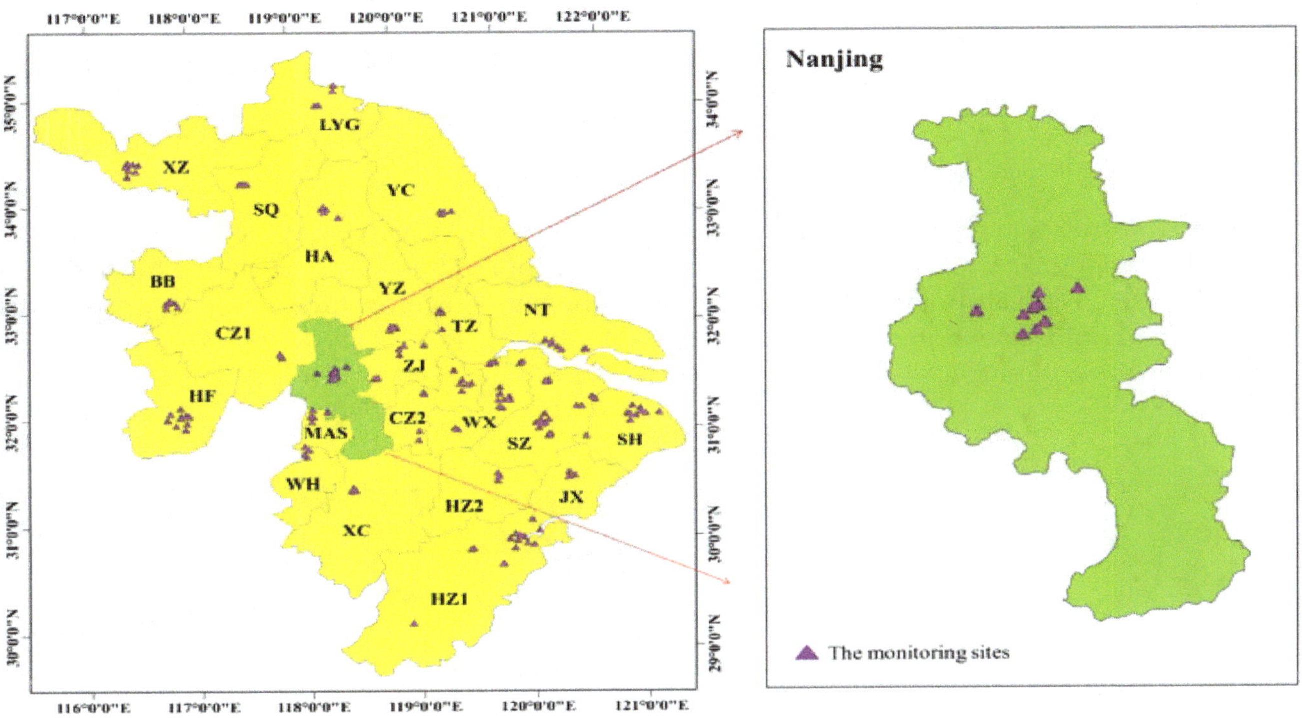

Figure 1. Distribution of air quality monitoring stations in Nanjing and surrounding cities.

3. Results and Discussion

3.1. Variation in Meteorological Conditions

Air temperature, relative humidity, wind speed, solar radiation and atmospheric pressure were important external factors affecting the concentration of pollutants. In order to evaluate the effectiveness of the implementation of control measures, we need to eliminate the interference of meteorological conditions, so we compare meteorological parameters at different periods. Table 1 shows the mean of meteorological parameters before the emission reduction (July, 2014), during the emission reduction (August, 2014) and during the same period in 2013 (August, 2013). The mean air temperature during the period of the emission reduction was 25.6 °C, which was slightly lower than the average level in July 2014. Relative humidity in August 2013 was far lower than the average level in August 2014, while relative humidity in July 2014 was close to the average level in August 2014. The average wind speed during the emission reduction period was 2.19 m/s, which was also very close to the period before the emission reduction. Solar radiation during the emission reduction was very much lower than that in 2013 for the same time, while solar radiation in July 2014 was higher than the average level in August 2014. Atmospheric pressure was not much different across the three periods. Overall, compared with the period during the emission reduction, solar radiation in July 2014 was higher, and other meteorological parameters were close to the average level during the emission reduction. By comparison, we found that the meteorological parameters in July 2014 were close to the average level during the emission reduction, but were very different from that in August 2013.

Thus, the meteorological conditions in July 2014 were more appropriate than those in August 2013 to evaluate the effect of air quality improvement during the period of the emission reduction.

Table 1. Comparison of meteorological conditions during the different periods.

	Air Temperature (°C)	Relative Humidity (%)	Wind Speed (m/s)	Solar Radiation (w/m^2)	Atmospheric Pressure (hPa)
August 2013	30.8	65.9	3.09	205.6	1000.9
July 2014	27.1	83.4	2.26	161.6	1001.5
August 2014	25.6	87.1	2.19	127.7	1003.5

3.2. Comparison of Air Quality and Air Pollutants in Different Periods

During the emission reduction, a number of control measures were adopted to guarantee good air quality in Nanjing. To understand the air quality and air pollutants variation characteristics during this period, the AQI and air pollutant concentrations were plotted against time using the average daily concentration of pollutants data in different periods (Figure 2). Table 2 shows the mean value of the AQI and the mean concentration of air pollutants during the different time periods in Nanjing.

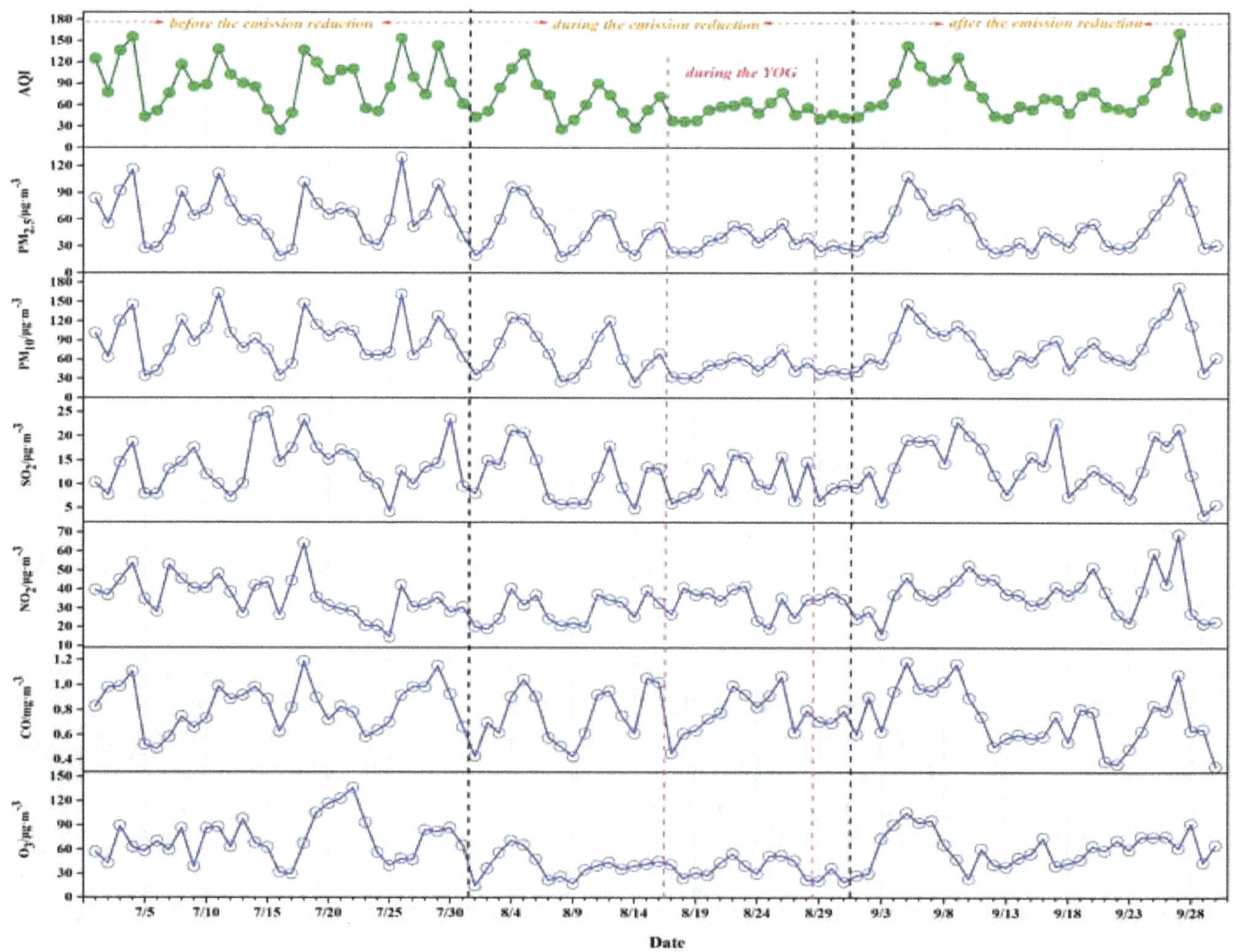

Figure 2. Daily variation of the AQI and air pollutant concentrations in different periods.

Table 2. Mean value of the AQI and mean air pollutant concentrations during the different periods.

Period		Air Quality Index	$PM_{2.5}$ ($\mu g \cdot m^{-3}$)	PM_{10} ($\mu g \cdot m^{-3}$)	SO_2 ($\mu g \cdot m^{-3}$)	NO_2 ($\mu g \cdot m^{-3}$)	CO ($mg \cdot m^{-3}$)	O_3 ($\mu g \cdot m^{-3}$)
Before the emission reduction	1–31 July 2014	93.35	66.23	93.29	13.97	36.60	0.831	72.64
During the emission reduction	1–31 August 2014	59.65	42.44	59.01	11.12	31.09	0.760	38.39
During the Youth Olympic Games	17–28 August 2014	53.40	37.84	49.60	10.86	32.89	0.778	39.12
After the emission reduction	1–30 September 2014	76.20	51.27	83.70	13.66	37.67	0.732	61.78

The AQI is an indicator of air quality, which is used to describe the air quality level of a place and ranges from 0 to 500, where higher values represent worse air quality. The air quality was the best in August 2014 with the AQI of 59.65, an average decrease of 36.10% compared with July 2014, reflecting a significant improvement in the air quality during the emission reduction.

During the emission reduction period, the average $PM_{2.5}$ mass concentration was 42.44 $\mu g \cdot m^{-3}$. Compared with the period before the emission reduction, it exhibited a decrease of 35.92%. The $PM_{2.5}$ levels were much lower than the mean $PM_{2.5}$ concentration of 106 $\mu g \cdot m^{-3}$ in August 2012 in Nanjing [13].

From Figure 2, we can see that the trend and variability of PM_{10} showed a similar trend to $PM_{2.5}$. PM_{10} concentration declined from 93.29 $\mu g \cdot m^{-3}$ in July 2014 to 59.01 $\mu g \cdot m^{-3}$ during the emission reduction period, a 36.75% reduction. It is suggested that the pollution control measures of halting work at construction sites and shutting down heavy-industry factories during the emission reduction yielded a significant effect on the concentration of PM_{10}. In addition, we calculated the ratios between $PM_{2.5}$ and PM_{10} and found that the higher ratios occurred in August 2014 (0.73) and the lower ratios occurred in August 2013 (0.50), indicating that $PM_{2.5}$ was a main source for PM_{10} during the emission reduction period.

SO_2 concentration was reduced from 13.97 $\mu g \cdot m^{-3}$ in July 2014 to 11.12 $\mu g \cdot m^{-3}$ during the emission reduction period, a decrease of 20.40%. When the control measures were terminated, some factories and enterprises returned to normal production in Nanjing and the surrounding cities after the Youth Olympic Games, thus the SO_2 level began to rise to 13.66 $\mu g \cdot m^{-3}$. The variation of SO_2 concentration was affected mainly by the emission sources and weather conditions. Due to low emission sources, high temperature and strong convection in the summer, the concentration of SO_2 was much lower than in other seasons due to the influence of gas-to-particle conversion and wet scavenging.

NO_2 and CO concentrations were reduced by 15.1% and 8.54%, respectively, in comparison to the levels recorded for July 2014. However, they showed a tendency to increase during the YOG, thus the concentrations of NO_2 and CO were 32.89 $\mu g \cdot m^{-3}$ and 0.778 $mg \cdot m^{-3}$, respectively, compared with August 2014, which was closely related to meteorological conditions. NO_2 and CO are important precursors for O_3 formation, and they were affected by the photochemical reaction which enhanced with the increase of solar radiation. According to the comparison of meteorological parameters, we found that solar radiation during the emission reduction (127.7 $w \cdot m^{-2}$) was higher than that that during the YOG (120.31 $w \cdot m^{-2}$). This indicated that NO_2 and CO levels in August 2014 were more likely to convert to O_3 than those of 17–28 August 2014.

Ground-level O_3 is a greenhouse gas, a strong oxidant and a secondary pollutant mainly produced via photochemical reactions of nitrogen oxides, carbon monoxide and volatile organic compounds [14]. Its concentration is affected by solar radiation, air temperature and mixing/transport [15]. O_3 concentration in July 2014 was 72.64 $\mu g \cdot m^{-3}$, and was 38.39 $\mu g \cdot m^{-3}$ during the emission reduction period, showing a 47.15% decrease, which was significantly greater than the those other air pollutants. As a secondary pollutant, the decrease of O_3 was due to the reduction of precursors and the weather conditions during the period of the emission reduction.

3.3. Diurnal Evolution of Air Pollutant Concentrations

Figure 3 shows the diurnal variation of air pollutant concentrations in different periods. Although the air pollutant concentrations of each period are different, the basic pattern shows that the air pollutants diurnal variation curve in August 2014 was lower than those of the other two periods, except for CO.

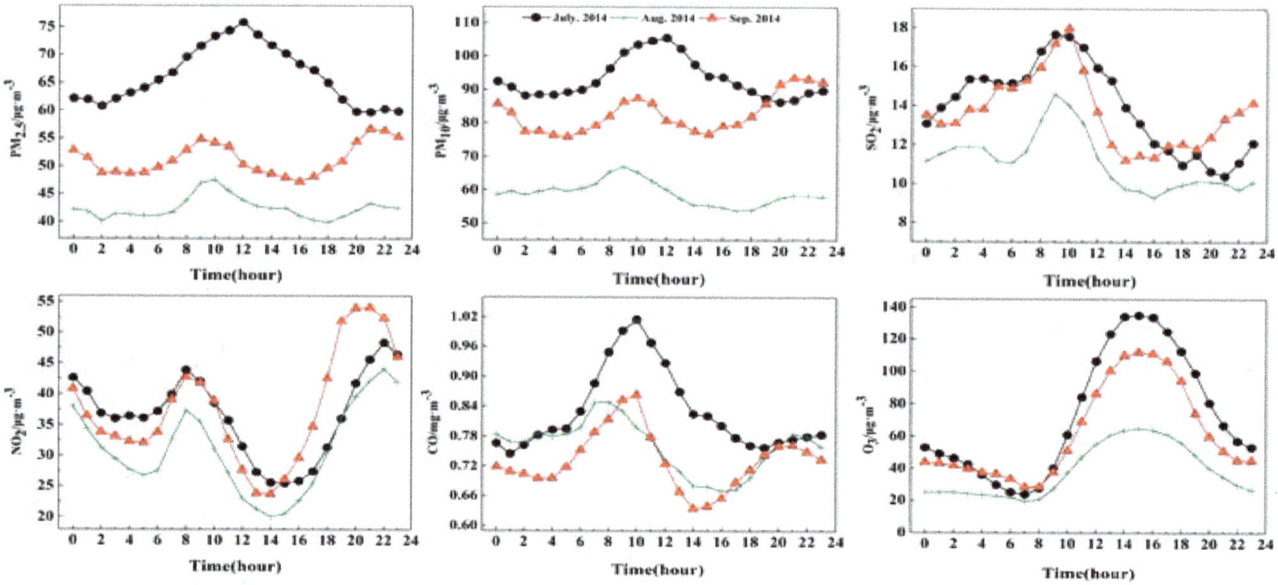

Figure 3. Diurnal variation of air pollutant concentrations in different periods.

$PM_{2.5}$ concentration was gradually reduced from midnight to 2:00–4:00, followed by a morning peak around 9:00–10:00, except for July 2014 (before the emission reduction). Concentrations then decreased until 16:00–18:00, after which they rose until 21:00–22:00. Therefore, two periods displayed a bimodal pattern with peaks between 9:00–10:00 and 21:00–22:00. The morning peak was attributed to enhanced anthropogenic activity during rush hour [16]. After sunrise, strong turbulence in the developing convective boundary layer led to lower $PM_{2.5}$ concentration. During the night, lower mixed layer height led to higher $PM_{2.5}$ concentration. The diurnal variation characteristic of PM_{10} was similar to that of $PM_{2.5}$ concentration, which reflected the positive effects of the emission reduction measures.

SO_2 concentration showed a unimodal pattern; the maximum SO_2 concentrations tended to appear around 9:00–10:00 during the day, which was similar to that of $PM_{2.5}$. The mixing layer was high and photochemical conversion was strong at this time, as SO_2 in the upper air was transferred to the ground which led to the peak [9]. The main source of NO_2 in Nanjing is automobile exhaust. Diurnal variation of NO_2 presented a double-peak curve. The first peak appeared at 08:00 after the morning peak traffic, then the concentration of NO_2 was reduced by photolysis due to the enhancement of solar radiation, and the minimum values of NO_2 concentration occurred at 13:00–14:00. With the arrival of the evening peak traffic, NO_2 concentration gradually increased from 13:00–14:00 to 21:00–22:00, and another significant peak appeared at 21:00–22:00, similar to that observed in Chengdu City [17].

The highest CO hourly concentration appeared at around 08:00–10:00 in the morning, followed by a sharply decreasing trend to the lowest value at around 14:00–15:00, except for July 2014 (before the emission reduction). A slight but continuous increase was observed from 14:00–15:00 to 21:00, and finally a second weaker peak was seen at night due to the decrease in boundary layer height, which resulted in the accumulation of pollutants. The diurnal variations of O_3 concentrations in three periods showed similar characteristics; the lowest ozone concentrations for three periods appeared at around 07:00–08:00 with values around 20–30 $\mu g \cdot m^{-3}$. With increasing temperature and solar radiation in the daytime, O_3 concentration increased from 07:00–08:00 until 15:00, when it reached its peak. The concentration then rapidly fell until midnight.

3.4. Spatial Variation of the AQI and Air Pollutant Concentrations

In order to analyze the spatial variation of air quality and air pollutants during the period of the emission reduction, the mean values and standard errors of the AQI and six air pollutant concentrations in nine different sites of Nanjing were compared with the period before the emission reduction, as shown in Figure 4. The AQI during the emission reduction was lower than that in July 2014 at all monitoring stations; the percentage of decrease of ranged from 31.54 to 44.90% on spatial variation, which indicated the effects on air quality of implementing the reduction measures.

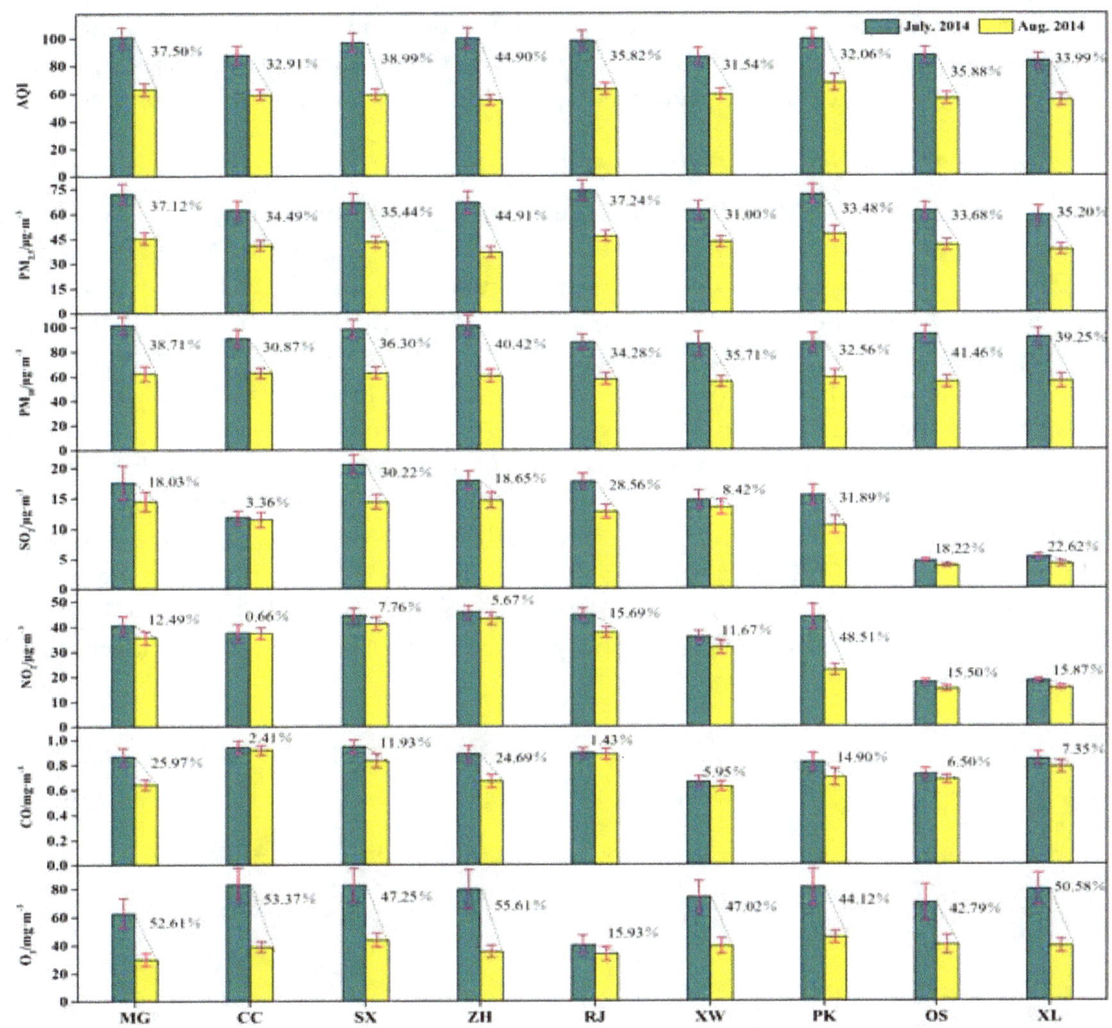

Figure 4. Spatial variation of the AQI and air pollutants at each site before and during the emission reduction.

Concentrations of $PM_{2.5}$, PM_{10} and O_3 during the emission reduction period decreased significantly at nine monitoring sites compared with those in July 2014. The percentage of decrease of $PM_{2.5}$ ranged from 31.00 to 44.91%, showing that the strict pollution control measures carried out in Nanjing could reduce the concentration of $PM_{2.5}$. The percentage of decrease of PM_{10} ranged from 30.87 to 41.46%, which was similar to $PM_{2.5}$ in the spatial variation. The percentage of decrease of O_3 ranged from 15.93 to 55.61%, which was the largest of the six air pollutants. The percentage of decrease of SO_2 ranged from 3.36 to 31.89%, which was lower than that observed in Beijing during the APEC period [9]. Most of the coal-burning factories were shut down during the period of the emission reduction in Nanjing. Thus, the decrease of SO_2 indicated the effects of the reduction of coal consumption [18]. The percentage of decrease of NO_2 ranged from 0.66 to 48.51%, which reflected the good influence of reducing motor vehicle exhaust and industrial production. The percentage of decrease of CO ranged from 1.43 to 25.97%, which was the smallest of the six air pollutants; moreover, the daily average of CO has not exceeded the Chinese national assessment standard in Nanjing in recent years.

The decrease of the AQI at the ZH site was the highest of all the sites. The decrease of $PM_{2.5}$ at the ZH site was the highest, while the decrease of $PM_{2.5}$ at the XW site was lower than the others. The decrease of PM_{10} at the OS site was higher than those at other monitoring sites, reflecting the effectiveness of control measures for PM_{10} in the main stadium of the Nanjing Youth Olympic Games. The PK site was located in the suburbs of Nanjing, and the decrease of SO_2 and NO_2 at the PK site was the highest of all the monitoring sites, indicating the effectiveness of controlling motor vehicles and industrial production. The decrease of O_3 exceeded 40% in the spatial variation, except for the RJ site, with a decrease of 55.61% at the ZH site, which was the largest of all the monitoring sites.

3.5. Variation of Air Quality in Nanjing and the Surrounding Cities

From 1 to 31 August 2014, Nanjing cooperated with 22 surrounding cities to establish an air pollution joint prevention group to guarantee the air quality in August in Nanjing. Figure 5 shows the AQI of Nanjing and 22 nearby cities in July and August 2014. In July 2014, the air quality was the worst in WX, compared with the other cities; the AQI for WX was the highest (105), followed by CZ2 and YZ. The air quality was the best in LYG, XZ and SH, where the AQI was the lowest (74). In August 2014, the maximum value of the AQI occurred in JX and the minimum value of the AQI appeared in Nanjing. We found that the difference between the AQI in July 2014 and August 2014 was very significant ($p < 0.05$). Figure 6 represents the percentage of decrease of the AQI during the emission reduction period compared with July 2014 in Nanjing and 22 nearby cities. The AQI of all cities showed a downward trend. The decrease in the AQI in Nanjing was 35.48%, which was the largest of the 23 cities, followed by YZ and HZ2, and was almost the same in ZJ, YC and WX. The decrease in the AQI in CZ2, NT and TZ were 19.15%, 18.60% and 18.18%, respectively. The decrease in the AQI in SH, HF, XZ, JX, HA and LYG were less than 10% and greater than 0%, and was the lowest in LYG with only a 2.70% reduction.

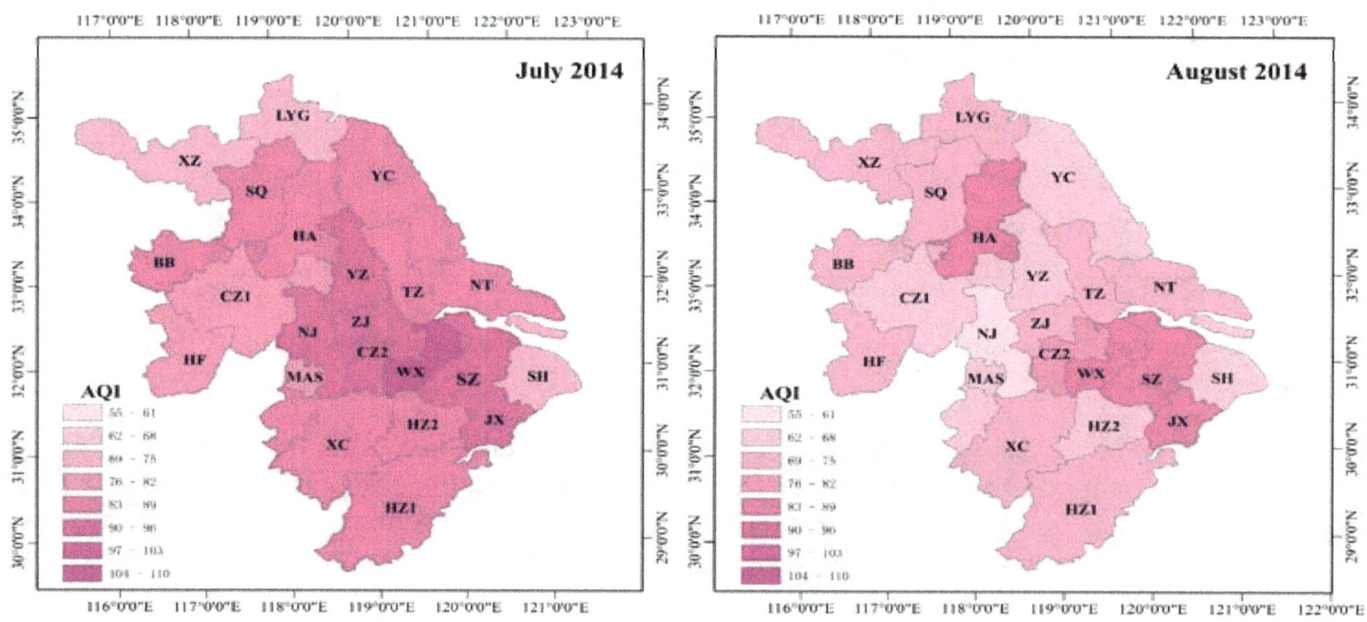

Figure 5. Variation of the AQI in Nanjing and its surrounding cities during the emission reduction period.

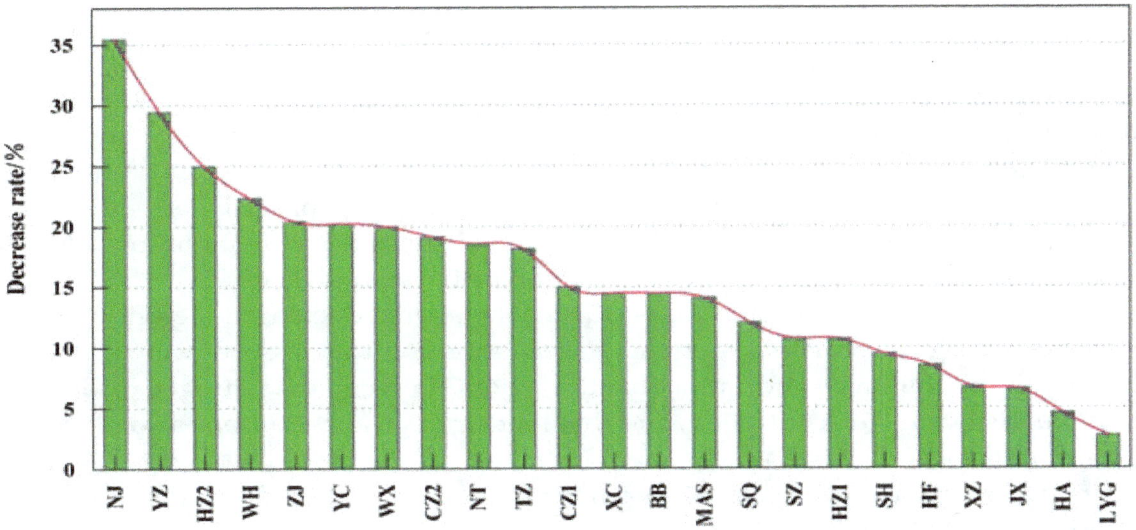

Figure 6. Decrease percentage of the AQI during the emission reduction period compared with July 2014 in Nanjing and its surrounding cities.

4. Discussion and Conclusions

Air pollutant concentrations are not only related to emissions from air pollution sources, but are also influenced by meteorological conditions. In order to eliminate the interference of meteorological conditions, we divided the meteorological data into three time periods: before, during the emission reduction and during the same period of the previous year (2013). According to the comparison of meteorological data in three periods, we found that most of the meteorological parameters in July 2014 were closer to the average level during the period of the emission reduction than that in August 2013. Therefore, we analyzed the concentrations of air pollutants before, during and after the emission reduction, which was more appropriate than the comparison with the year 2013 to evaluate the effectiveness of air pollution reduction actions.

According to our analysis and comparison, the concentrations of six air pollutants during the emission reduction period were significantly lower than those in the prophase mean before the emission reduction, especially O_3. Compared with the average levels in July 2014, the mean concentrations of $PM_{2.5}$, PM_{10}, SO_2, NO_2, CO and O_3 decreased by 35.92%, 36.75%, 20.40%, 15.05%, 8.54% and 47.15%, respectively. All these reductions in air pollutants indicated that the emission control measures were successful during the YOG, and little related to the local meteorological conditions. After the YOG, air pollutants increased significantly due to the cancellation of the emission control measures, except for CO. In addition, we also analyzed the diurnal variations of air pollutant concentrations in different periods. Overall, the diurnal variation curve of air pollutants during the period of the emission reduction was lower than that in July 2014, except for CO. In the spatial variation of pollutants, the AQI during the emission reduction was lower than that in July 2014 at all monitoring stations, the decrease of which ranged from 31.54 to 44.90% in spatial variation. Meanwhile, the decrease of $PM_{2.5}$ ranged from 31.00% to 44.91% at different sites, the decrease of PM_{10} ranged from 30.87% to 41.46%, the decrease of SO_2 ranged from 3.36% to 31.89%, the decrease of NO_2 ranged from 0.66% to 48.51%, the decrease of CO ranged from 1.43% to 25.97%, and the decrease of O_3 ranged from 15.93% to 55.61%. Our results suggested that the AQI of Nanjing and 22 nearby cities showed a downward trend when compared with July 2014, which reflected the effects on air quality of implementing the regional emission reduction measures.

In recent years, China has successfully hosted many international events. In order to improve air quality during the events, the local government has implemented some short-term periods with temporary emission control measures to reduce the emissions of pollutants. During the 2008 Olympic Games, the Beijing municipal government took many air pollution control measures, which included relocating some heavily polluting enterprises, encouraging natural gas instead of coal-fired boilers and domestic stoves, limiting the use of cars and so on [19]. During the 2010 Asian Games, the Guangzhou government made great efforts to improve the air quality, which mainly included controlling emissions from industries and transportation restrictions; vehicles could be driven only on alternate days depending on license plate numbers, and all construction activities were put on hold during the Games [20]. During the APEC period, the Chinese government took more stringent emission reduction measures in Beijing and the surrounding region, including banning heavily polluting vehicles, closing heavily polluting factories, slowing down construction activities and so on [21]. Some studies found that the mass concentration of air pollutants during the events decreased significantly, except for O_3 [9,22]. The results of our studies showed that the emission control measures were the most effective on O_3 out of the six air pollutants, indicating that the major O_3 precursors (nitrogen oxides and volatile organic compounds) were well controlled, and that meteorological conditions also played a positive role during the YOG.

The Nanjing government carried out temporary strict environmental regulations to ensure good air quality during the YOG in 2014. Other surrounding cities cooperated with Nanjing to guarantee good air quality during the Games. Approximately 2630 construction sites were halted. Heavy-industry factories, including the petrochemical industry, iron and steel industry, building materials industry and so on were required to reduce manufacturing by 20%. High-emission or yellow-labeled vehicles were not allowed to drive on the road. Open space barbecue restaurants were closed. Over 900 electric buses and 500 electric taxis have been put into operation [23]. In addition, 22 surrounding cities in the Yangtze River Delta region were asked to cooperate with Nanjing to close industries with high pollution emissions during the emission reduction period [24].

According to the analysis of this study, several suggestions have been made to improve the air quality of Nanjing. On the one hand, as regional emission reduction measures were the main factor in improving the air quality in Nanjing, regional cooperation and joint defense and control are considered the future direction in air pollution control. On the other hand, greater efforts should be exerted to control dust, coal-burning, vehicle emissions and industrial production around Nanjing, which would play a significant role in improving the air quality of Nanjing.

In order to protect human health and improve air quality, we may consider taking the following several measures and recommendations for air pollution reduction over long-term period:

1. To reduce emissions. Use more non-pollution energy, such as solar energy, wind energy and hydropower. Reform energy structure and use low-polluting energy, such as natural gas, biogas and alcohol. In addition, before the pollutants enter the atmosphere, use technology such as dust and smoke abatement to eliminate the partial pollutant in the waste gas, as it can reduce the quantity of pollutants entering the atmosphere.
2. To curb emissions and make full use of the atmosphere's self-purification ability. Due to different meteorological conditions, the efficiency of the removal of pollutants in the atmosphere is different. The pollutants concentration may be different in different regions, even when the same amount of pollutants has been released into the atmosphere. Therefore, respond to different regions at different times for the effective control of emissions.
3. Factors such as the location of factory sites, chimney designs, city planning and the location of industrial districts must be given proper attention so as to not cause redundant iterations of pollution, which can in turn result in serious local pollution events.
4. Afforestation enables more plants to absorb pollutants, thus reducing the degree of air pollution.

Acknowledgments: This work was financially supported by the National Natural Science Fund (41475108).

Author Contributions: Hui Zhao and Youfei Zheng conceived and designed the experiments as well as writing the paper; Hui Zhao analyzed the data; Ting Li helped perform the statistical analysis.

References

1. Chan, C.K.; Yao, X.H. Air pollution in mega cities in China. *Atmos. Environ.* **2008**, *42*, 1–42. [CrossRef]
2. Gao, J.J.; Tian, H.Z.; Cheng, K.; Lu, L.; Zheng, M.; Wang, S.X.; Hao, J.M.; Wang, K.; Hua, S.B.; Zhu, C.Y.; Wang, Y. The variation of chemical characteristics of $PM_{2.5}$ and PM_{10} and formation causes during two haze pollution events in urban Beijing, China. *Atmos. Environ.* **2015**, *107*, 1–8. [CrossRef]
3. Leiva, M.A.; Santibanez, D.A.; Ibarra, S.; Matus, P.; Seguel, R. A five-year study of particulate matter ($PM_{2.5}$) and cerebrovascular diseases. *Environ. Pollut.* **2013**, *181*, 1–6. [CrossRef] [PubMed]
4. Pascal, M.; Falq, G.; Wagner, V.; Chatignoux, E.; Corso, M.; Blanchard, M.; Host, S.; Pascal, L.; Larrieu, S. Short-term impacts of particulate matter (PM_{10}, $PM_{10-2.5}$, $PM_{2.5}$) on mortality in nine French cities. *Atmos. Environ.* **2014**, *95*, 175–184. [CrossRef]
5. Shankardass, K.; Jerrett, M.; Dell, S.D.; Foty, R.; Stieb, D. Spatial analysis of exposure to traffic-related air pollution at birth and childhood atopic asthma in Toronto, Ontario. *Health Place* **2015**, *34*, 287–295. [CrossRef] [PubMed]
6. Yang, Y.R.; Liu, X.G.; Qu, Y.; Wang, J.L.; An, J.L.; Zhang, Y.H.; Zhang, F. Formation mechanism of continuous extreme haze episodes in the megacity Beijing, China, in January 2013. *Atmos. Res.* **2015**, *155*, 192–203. [CrossRef]
7. Xin, J.Y.; Wang, Y.S.; Tang, G.Q.; Wang, L.L.; Sun, Y.; Wang, Y.H.; Hu, B.; Song, T.; Ji, D.S.; Wang, W.F.; et al. Variability and reduction of atmospheric pollutants in Beijing and its surrounding area during the Beijing 2008 Olympic Games. *Chin. Sci. Bull.* **2010**, *55*, 1937–1944. [CrossRef]
8. Wang, W.T.; Primbs, T.; Tao, S.; Simonich, S.M. Atmospheric Particulate Matter Pollution during the 2008 Beijing Olympics. *Environ. Sci. Technol.* **2009**, *15*, 5314–5320. [CrossRef]
9. Wang, Z.S.; Li, Y.T.; Chen, T.; Li, L.J.; Liu, B.X.; Zhang, D.W.; Sun, F.; Wei, Q.; Jiang, L.; Pan, L.B. Changes in atmospheric composition during the 2014 APEC conference in Beijing. *J. Geophys. Res. Atmos.* **2015**, *120*, 695–707. [CrossRef]
10. Hao, N.; Valks, P.; Loyola, D.; Cheng, Y.F.; Zimmer, W. Space-based measurements of air quality during the World Expo 2010 in Shanghai. *Environ. Res. Lett.* **2011**, *6*, 67–81. [CrossRef]
11. Wu, F.C.; Xie, P.H.; Li, A.; Chan, K.L.; Hartl, A.; Wang, Y.; Si, F.Q.; Zeng, Y.; Qin, M.; Xu, J.; et al. Observations of SO2 and NO2 by mobile DOAS in the Guangzhou eastern area during the Asian Games 2010. *Atmos. Meas. Tech.* **2013**, *6*, 2277–2292. [CrossRef]
12. Hu, J.L.; Ying, Q.; Wang, Y.G.; Zhang, H.L. Characterizing multi-pollutant air pollution in China: Comparison of three air quality indices. *Environ. Int.* **2015**, *84*, 17–25. [CrossRef] [PubMed]
13. Shen, G.F.; Yuan, S.Y.; Xie, Y.N.; Xia, S.J.; Li, L.; Yao, Y.K.; Qiao, Y.Z.; Zhang, J.; Zhao, Q.Y.; Ding, A.J.; et al. Ambient levels and temporal variations of PM2.5 and PM10 at a residential site in the mega-city, Nanjing, in the western Yangtze River Delta, China. *J. Environ. Sci. Health. A.* **2014**, *49*, 171–178.
14. Stella, P.; Personne, E.; Loubet, B.; Lamaud, E.; Ceschia, E.; Beziat, P.; Bonnefond, J.M.; Irvine, M.; Keravee, P.; Mascher, N.; et al. Predicting and partitioning ozone fluxes to maize crops from sowing to harvest: The Surfatm-O3 model. *Biogeosciences* **2011**, *8*, 2869–2886. [CrossRef]
15. Leung, L.R.; Gustafson, W.I. Potential regional climate change and implications to U.S. air quality. *Geophys. Res. Lett.* **2005**, *32*, 367–384. [CrossRef]
16. Zhao, X.J.; Zhang, X.L.; Xu, X.F.; Xu, J.; Meng, W.; Pu, W.W. Seasonal and diurnal variations of ambient PM2.5 concentration in urban and rural environments in Beijing. *Atmos. Environ.* **2009**, *43*, 2893–2900. [CrossRef]
17. Xie, Y.Z.; Pan, Y.P.; Ni, C.J.; Chen, Z.H.; Wei, X. Temporal and spatial variations of atmospheric pollutants in urban Chengdu during summer [in Chinese]. *Acta. Sci. Circumst.* **2015**, *35*, 975–983.
18. Apergis, N.; Loomis, D.; Payne, J.E. Are fluctuations in coal consumption transitory or permanent? Evidence from a panel of US states. *Atmos. Environ.* **2010**, *87*, 2424–2426. [CrossRef]
19. Wang, S.X.; Zhao, M.; Xing, J.; Wu, Y.; Zhou, Y.; Lei, Y.; He, K.B.; Fu, L.X.; Hao, J.M. Quantifying the air pollutants emission reduction during the 2008 Olympic Games in Beijing. *Environ. Sci. Technol.* **2010**, *44*, 2490–2496. [CrossRef] [PubMed]
20. Liu, H.; Wang, X.M.; Zhang, J.P.; He, K.B.; Wu, Y.; Xu, J.Y. Emission controls and changes in air quality in Guangzhou during the Asian Games. *Atmos. Environ.* **2013**, *76*, 81–93. [CrossRef]
21. Li, R.P.; Mao, H.J.; Wu, L.; He, J.J.; Ren, P.P.; Li, X.Y. The evaluation of emission control to PM concentration during Beijing APEC in 2014. *Atmos. Pollut. Res.* **2016**, *7*, 363–369. [CrossRef]

22. Wang, Z.S.; Li, Y.T.; Zhang, D.W.; Chen, T.; Sun, F.; Li, L.J.; Li, J.X.; Sun, N.D.; Chen, C.; Wang, B.Y. Analysis on air quality in Beijing during the 2014 APEC conference [in Chinese]. *J. Environ. Sci.* **2016**, *36*, 675–683.
23. Ding, J.; Vander, A.R.J.; Mijling, B.; Levelt, P.F.; Hao, N. NOx emission estimates during the 2014 Youth Olympic Games in Nanjing. *Atmos. Chem. Phys.* **2015**, *15*, 6337–6372. [CrossRef]
24. Li, S.W.; Li, H.B.; Luo, J.; Li, H.M.; Qian, X.; Liu, M.M.; Bi, J.; Cui, X.Y.; Ma, L.Q. Influence of pollution control on lead inhalation bioaccessibility in $PM_{2.5}$: A case study of 2014 Youth Olympic Games in Nanjing. *Environ. Int.* **2016**, *94*, 69–75. [CrossRef] [PubMed]

14

Temporal and Spatial Patterns of China's Main Air Pollutants: Years 2014 and 2015

Tiancai Zhou [1,2] , Jian Sun [2,*] and Huan Yu [1,*]

1 College of Earth Sciences, Chengdu University of Technology, Chengdu 610059, China; ztc18108279610@163.com
2 Institute of Geographic Sciences and Natural Resources Research, Chinese Academy of Sciences, Beijing 100101, China
* Correspondence: sunjian@igsnrr.ac.cn (J.S.); yuhuan0622@126.com (H.Y.)

Abstract: China faces unprecedented air pollution today. In this study, a database (SO_2, NO_2, CO, O_3, $PM_{2.5}$ (particulate matter with aerodynamic diameter less than 2.5 μm), and PM_{10} (particulate matter with aerodynamic diameter less than 10 μm) was developed from recordings in 188 cities across China in 2014 and 2015 to explore the spatial-temporal characteristics, relationships among atmospheric contaminations, and variations in these contaminants. Across China, the results indicated that the average monthly concentrations of air pollutants were higher from November to February than in other months. Further, the spatial patterns of air pollutants showed that the most polluted areas were located in Shandong, Henan, and Shanxi provinces, and the Beijing-Tianjin-Hebei region. In addition, the average daily concentrations of air pollutants were also higher in spring and winter, and significant relationships between the principal air pollutants (negative for O_3 and positive for the others) were found. Finally, the results of a generalized additive model (GAM) indicated that the concentrations of PM_{10} and O_3 fluctuate dynamically; there was a consistent increase in CO and NO_2, and $PM_{2.5}$ and SO_2 showed a sharply decreasing trend. To minimize air pollution, open biomass burning should be prohibited, the energy efficiency of coal should be improved, and the full use of clean fuels (nuclear, wind, and solar energy) for municipal heating should be encouraged from November to February. Consequently, an optimized program of urban development should be highlighted.

Keywords: air pollutants; dynamics; spatial patterns; generalized additive model; policy recommendations

---

1. Introduction

Haze is an atmospheric effect that has become a serious global issue [1,2], as it affects species diversity, global climate, and human health [3], social, and economic [4]. In general, the chemical aspects of haze have been studied—specifically its physical and chemical properties, including the elements Cd, Cr, Cu, Fe, and Mn, gaseous pollutants (O_3, NO_x, SO_2, CO_x), and inorganic aerosols (SO_4^{2-}, NO_3^-, and NH_4^+ [5–7]. Source apportioning has shown that haze is attributable primarily to dust storms, biomass burning [8], coal consumption, and vehicle exhaust [9–11]. Further, secondary inorganic and organic aerosols should not be neglected [12]. Some studies have focused on the long-range transport mechanism of haze, which is controlled by meteorological conditions [13,14]; thus, affluent moisture, warm advection in the lower troposphere, and stable atmospheric stratification favor the concentration of haze [15]. Other studies have investigated the side effects of haze, including reduced visibility [16], adverse effects on health [17], and so on.

China's air quality has deteriorated in recent years, and haze affects increasing areas [18]. Moreover, numerous studies have found that the regions of Beijing [19], North China [20], Wuhan [21], and Guangzhou [22] have excessively high concentrations of haze. Other studies have examined the formation and evolution of haze [23,24], and the role of meteorological conditions in haze transportation (e.g., wind and relative humidity) [25–27]. The urbanization is another driving force in the development of haze [28]. The varied characteristics of haze in different seasons have also been reported: for example, haze in summer in East China [29] and Beijing [30], autumn in Shanghai [31], and winter in Beijing [32,33], Shanghai [34], the China Loess Plateau [35], and the North China Plain [36], all of which have relatively higher concentrations of haze.

Meanwhile, the high concentration of organic carbon in PM (particulate matter) indicates that the main source of haze is biomass burning [37], and that the high concentration of NO_2 is closely related to vehicle emissions [34]. Moreover, high concentrations of SO_2 and NO_2 contribute significantly in the formation of secondary aerosols [38]. Thus, secondary aerosols (sulfate, nitrate, and organic matter) are produced when combined with gaseous pollutants (NO_2, SO_2, O_3, and formaldehyde) and main atmospheric particles [39]. These constitute the main chemical compositions of particulate matter with aerodynamic diameter less than 2.5 μm ($PM_{2.5}$) in Hangzhou [40], and the heterogeneous chemical processes promote and sustain the growth of haze [41,42]. Further, the spatial patterns and temporal variation of $PM_{2.5}$ have been simulated [43], and the simulated values of meteorological conditions (temperature, and wind speed and direction) agree with the values observed. The diurnal cycle of land-sea breezes among the Pearl River Delta is the primary influence in the transport of particulate matter with aerodynamic diameter less than 10 μm (PM_{10}) [44].

Many studies have indeed investigated the spatiotemporal patterns of haze across China. However, the dynamics and relationships of air pollutants during long time series (2 years) have been insufficiently discussed. Hence, a special study was performed to more fully reveal the spatiotemporal distribution of the principal air pollutants at present and their future variations in China. Within this context, the objectives of this study were to: (1) analyze the dynamic and spatiotemporal distribution of the principal air pollutants for 2014 and 2015, respectively; (2) reveal the dynamics and the relationships between the principal air pollutants in the most polluted cities, respectively; and (3) identify the principal variable trends in air pollutants over major Chinese cities based on a generalized additive model (GAM). Finally, we present strategic policy recommendation for reducing the levels of air pollutants according to specific haze characteristics in China.

2. Experiments

The main air pollutants (SO_2, NO_2, CO, O_3, $PM_{2.5}$, and PM_{10}) were monitored by the continuous air-monitoring stations (CAMS, Figure 1) covering 188 cities of China. The stations were established based on the standard "Technical regulation for selection of ambient air quality monitoring stations (HJ 664-2013)", and the data were collected through the Ministry of Environmental Protection of the People's Republic of China (Available online: http://datacenter.mep.gov.cn/). In addition, atmospheric contamination was compiled daily and monthly for 2014 and 2015.

In this study, we used the Inverse Distance Weighted (IDW) in ArcGIS 10.2 (ESRI, Inc., Redlands, CA, USA) to acquire the spatial graphs, SigmaPlot 10.0 (Systat Software, Inc., Chicago, IL, USA) was used to obtain the rose diagram, and Microsoft Office Visio (2007) (Microsoft corporation, Redmond, WA, USA) for fishbone diagrams. The Performance Analytics module of R software (R Core Development Team, R Foundation for Statistical Computing, Vienna, Austria) was employed to obtain the scatterplot matrices and conduct correlation as well as regression analyses. The two-tailed Pearson's correlations at $p = 0.05$ were employed to determine the correlations between different variables. The GAM (the non-linear relationship between the pollutants and day-of-year trends) was modeled flexibly by the log link function, and we used the *mgcv* module in R to predict air pollutants in China. The package *mgcv* (the gam fit function, gam function, family function were contained) of R was used to fit GAMs specified by presenting a symbolic description of the additive predictor as well as a description of the error distribution.

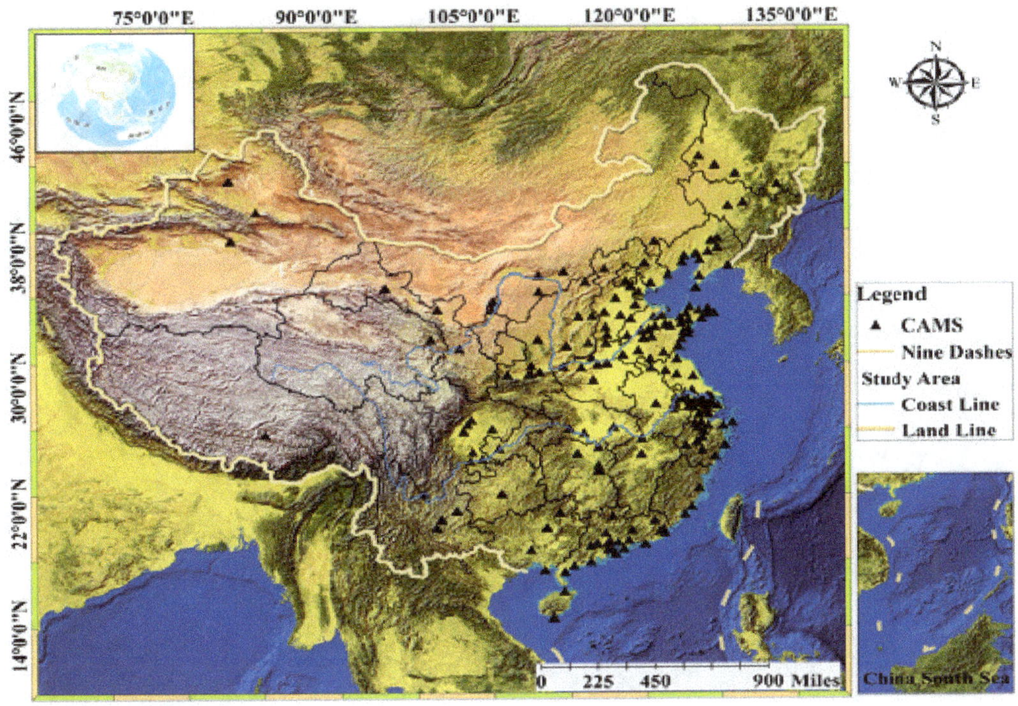

Figure 1. The study area.

3. Results

3.1. Dynamics of the Principal Air Pollutants

The average monthly concentrations of the principal air pollutants over China in 2014 are shown in Figure 2 and Table S1. The highest levels of $PM_{2.5}$, PM_{10}, CO, SO_2, and NO_2 were 106.5 $\mu g/m^{-3}$, 154.1 $\mu g/m^{-3}$, 1.69 mg/m^{-3}, 62.0 $\mu g/m^{-3}$, and 50.4 $\mu g/m^{-3}$ in January. The lowest $PM_{2.5}$ and PM_{10} levels were 39.8 $\mu g/m^{-3}$ and 69.9 $\mu g/m^{-3}$ in September, and the lowest values of CO, SO_2, and NO_2 were 0.93 mg/m^{-3}, 19.7 $\mu g/m^{-3}$, and 25.86 $\mu g/m^{-3}$ in July. In contrast, the highest and lowest levels of O_3 were 131.6 $\mu g/m^{-3}$ in July and 61.3 $\mu g/m^{-3}$ in December.

A comparison of the average monthly concentrations of $PM_{2.5}$, PM_{10}, CO, SO_2, and NO_2 indicated that the values were higher in January, February, November, and December, followed by March, April, May, and October, with lower concentrations from June to September. However, the opposite trend was observed with O_3, which had the highest average monthly levels from May to August, followed by March, April, September, and October, with lower values in January, February, November, and December. Further, the highest $PM_{2.5}$, PM_{10}, CO, SO_2, and NO_2 levels were higher than the lowest ones, respectively, indicating that the principal air pollutants were highest in January.

Similar trends were observed in 2015 (Figure 2). The highest levels of $PM_{2.5}$, PM_{10}, CO, and NO_2 were 86.8 $\mu g/m^{-3}$, 129.3 $\mu g/m^{-3}$, 1.60 mg/m^{-3}, and 48.0 $\mu g/m^{-3}$ in December, and the highest levels of SO_2 and O_3 were 53.3 $\mu g/m^{-3}$ in January and 130.8 $\mu g/m^{-3}$ in August (Table S1). The lowest $PM_{2.5}$ and PM_{10} levels were 36.4 $\mu g/m^{-3}$ and 65.4 $\mu g/m^{-3}$ in September, and the lowest values of CO, SO_2, and NO_2 were 0.83 mg/m^{-3}, 15.7 $\mu g/m^{-3}$, and 24.4 $\mu g/m^{-3}$ in July. The lowest level of O_3 was 55.7 $\mu g/m^{-3}$ in August (Table S1).

The average monthly $PM_{2.5}$, PM_{10}, CO, SO_2, and NO_2 concentrations in January and December were higher than in other months. Thus, the highest $PM_{2.5}$, PM_{10}, CO, SO_2, and NO_2 levels were 2.38, 1.98, 1.91, 3.39, 1.97, and 2.35 times greater than the lowest ones. However, the average monthly concentration of O_3 was higher from May to August, followed by March, April, September, and October, with lower values found from January to December.

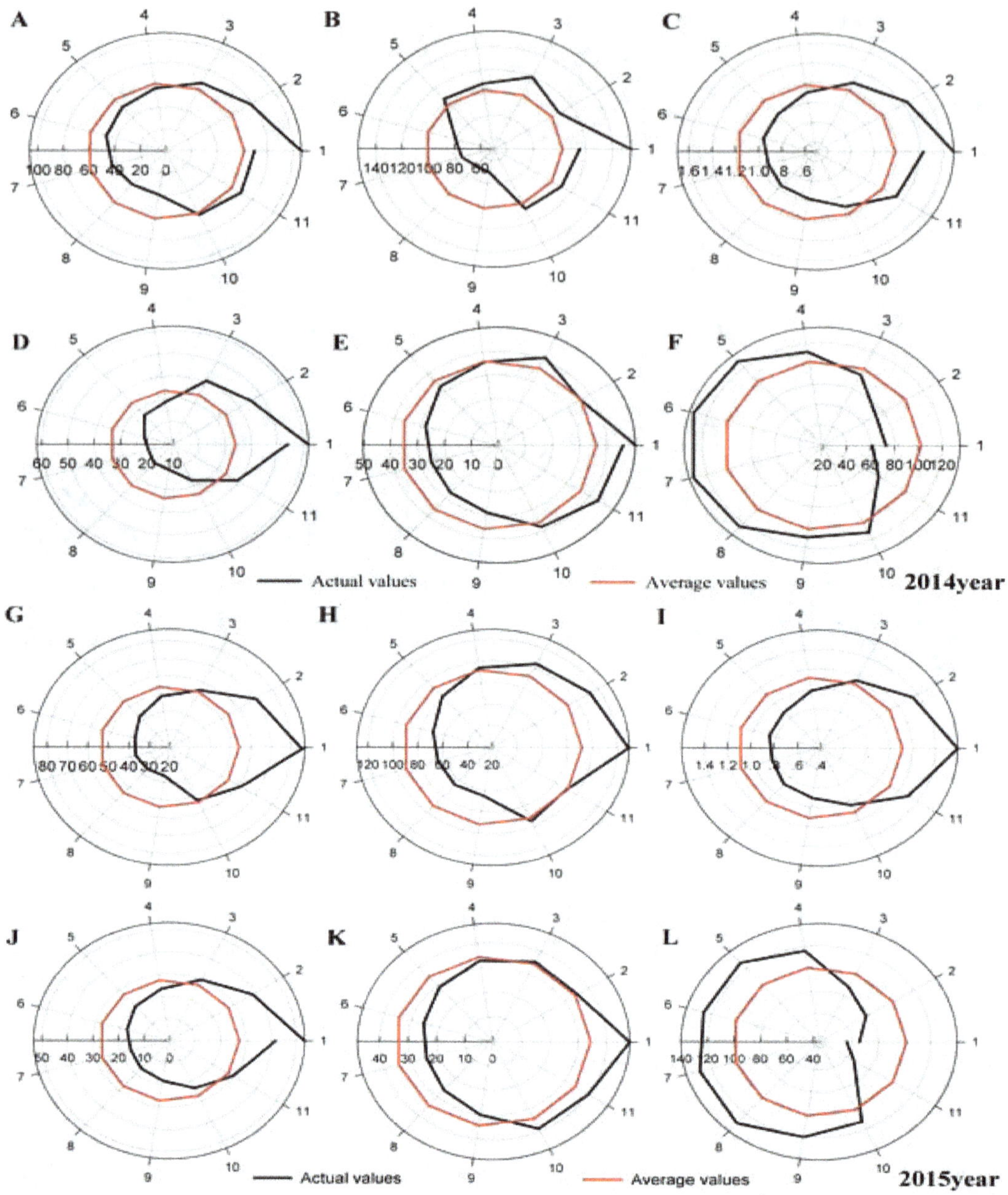

Figure 2. The monthly dynamics of the principal air pollutants over China (2014 and 2015). Graphs (**A–F**) represent the monthly concentration of $PM_{2.5}$ ($\mu g/m^{-3}$), PM_{10} ($\mu g/m^{-3}$), CO (mg/m^{-3}), SO_2 ($\mu g/m^{-3}$), NO_2 ($\mu g/m^{-3}$), and O_3 ($\mu g/m^{-3}$) in 2014, respectively; Graphs (**G–L**) represent the monthly concentration of $PM_{2.5}$ ($\mu g/m^{-3}$), PM_{10} ($\mu g/m^{-3}$), CO (mg/m^{-3}), SO_2 ($\mu g/m^{-3}$), NO_2 ($\mu g/m^{-3}$), and O_3 ($\mu g/m^{-3}$) in 2015, respectively.

3.2. Spatial Analysis of the Principal Air Pollutants

We obtained the average concentrations of the principal air pollutants during the four months (January, February, November, and December), except for O_3, the average value of which was calculated from May to August. Following the processes described above, we measured the spatial distribution of the principal air pollutants in 2014 (Figure 3). The highest and lowest levels of $PM_{2.5}$ and PM_{10} were 191.4 $\mu g/m^{-3}$ and 296.9 $\mu g/m^{-3}$ in Baoding, and 26.4 $\mu g/m^{-3}$ and 44.5 $\mu g/m^{-3}$ in Sanya (Table S2). The provinces of Shanxi, Shandong, Henan, and Shanxi, and the Beijing-Tianjin-Hebei region were the most polluted. It should be noted that most areas of China were affected by $PM_{2.5}$ and PM_{10}, including central and northern China. With respect to CO, SO_2, NO_2, and O_3, the distribution narrowed, and was more centralized in the Beijing-Tianjin-Hebei region, and Shandong and Henan provinces. The highest values of CO, SO_2, NO_2, and O_3 were 3.62 mg/m^{-3} in Baoding, 151.5 μg/m^{-3} in Yangquan, 85.2 μg/m^{-3} in Baoding, and 225.2 μg/m^{-3} in Wuhan (Table S2), respectively.

The air quality improved in 2015 in comparison to 2014 (Figure 3). The most polluted areas were primarily in the provinces of Hebei, Shandong, and Henan, with the highest levels of $PM_{2.5}$, PM_{10}, and CO (170.4 μg/m^{-3}, 254.2 μg/m^{-3}, and 3.57 mg/m^{-3}) in Baoding (Table S2). The highest values of SO_2, NO_2, and O_3 were 137.0 μg/m^{-3} in Shenyang, 77 μg/m^{-3} in Xingtai, and 184.6 μg/m^{-3} in Dezhou (Table S2), respectively. The lowest values of the principal air pollutants were observed in Hainan province.

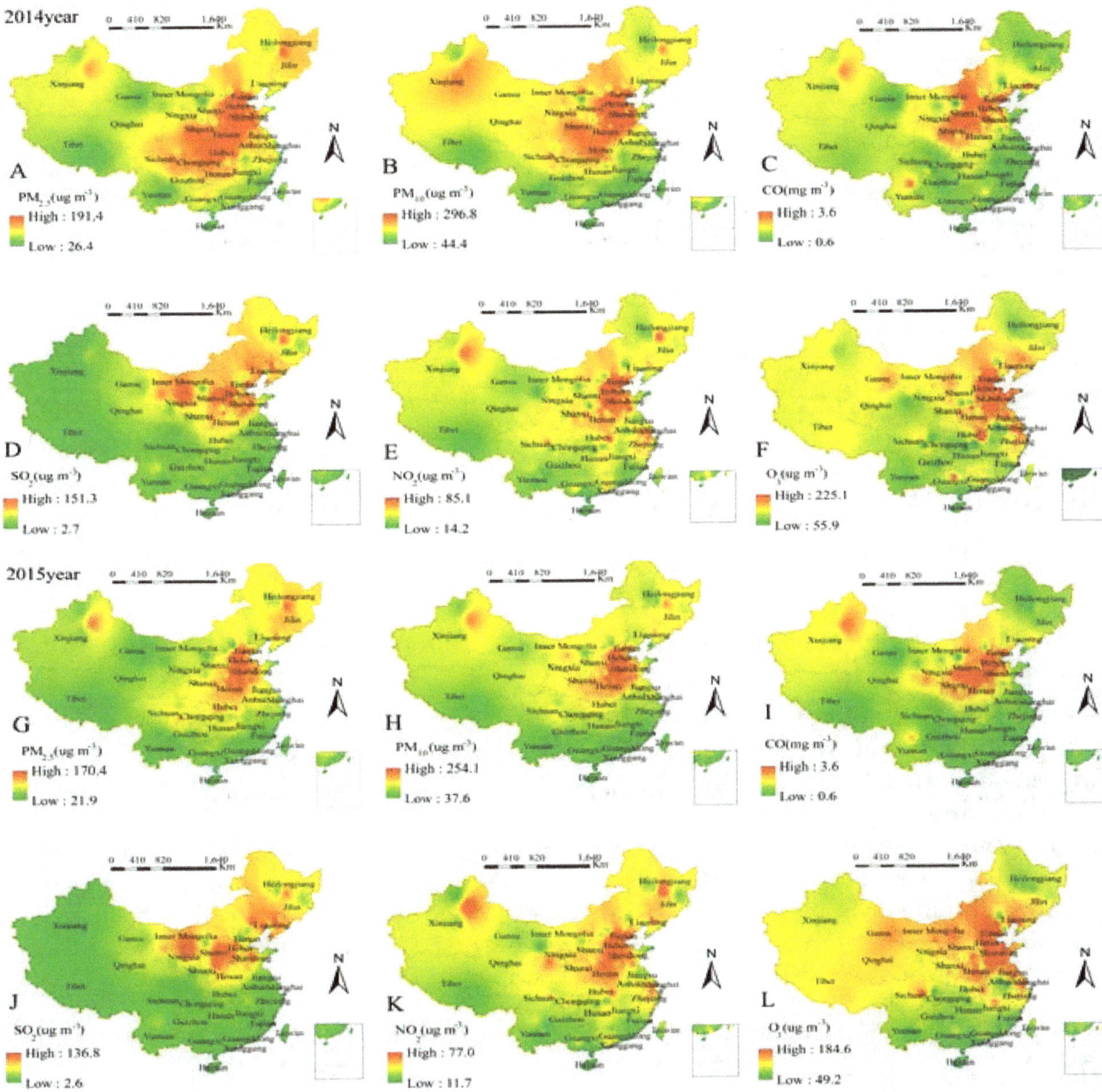

Figure 3. The spatial pattern of the mean concentrations (during January, February, November, and December) of the principal air pollutants over China (2014 and 2015). Graphs (**A–F**) represent the spatial pattern of $PM_{2.5}$, PM_{10}, CO, SO_2, NO_2, and O_3 in 2014, respectively; Graphs (**G–L**) represent the spatial pattern of $PM_{2.5}$, PM_{10}, CO, SO_2, NO_2, and O_3 in 2015, respectively.

The dynamic and spatiotemporal distribution of the principal air pollutants indicated that the cities Baoding, Xingtai, Handan, Shijiazhuang, Hengshui, Dezhou, Tangshan, Heze, Langfang, Liaocheng, Zibo, Laiwu, Anyang, Linyi, Yichang, Zhengzhou, and Pingdingshan were the most polluted regions; hence, the average daily concentrations of the principal air pollutants were analyzed in 2014 and 2015 (Figure 4). The average daily concentrations of air pollutants (SO_2, NO_2, CO, $PM_{2.5}$, and PM_{10}) were also higher in the spring and winter, except for O_3, which had the higher average daily levels in the summer. Besides, the average daily concentrations of SO_2, NO_2, CO, $PM_{2.5}$, and PM_{10} decreased at a rate of 16.2 ($\mu g/m^{-3}$)/year, 2.8 ($\mu g/m^{-3}$)/year, 0.1 (mg/m^{-3})/year, 11.8 ($\mu g/m^{-3}$)/year, and 17.8 ($\mu g/m^{-3}$)/year from 2014 to 2015, respectively. On the contrary, the average daily concentrations of O_3 increased from 102.2 ($\mu g/m^{-3}$)/year to 105.4 ($\mu g/m^{-3}$)/year between 2014 and 2015.

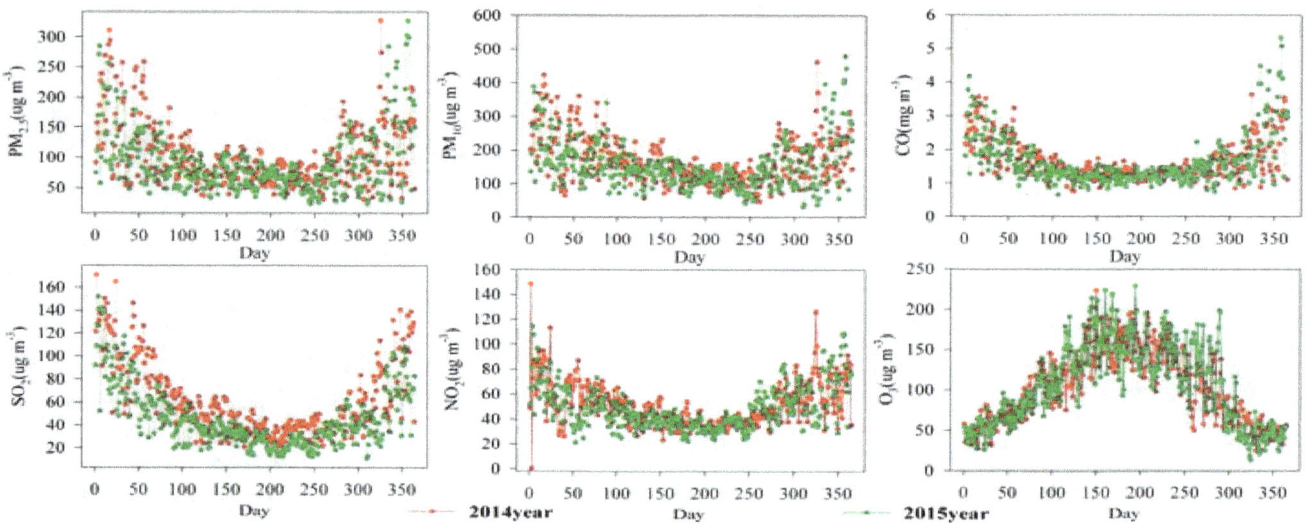

Figure 4. The daily dynamics of the principal air pollutants in the 17 most polluted cities (2014 and 2015).

3.3. The Dynamics of the Principal Air Pollutants in Beijing-Tianjin-Hebei during January, February, November, and December

As for the mosted polluteted region (Beijing-Tianjin-Hebei), Figure 5 exhibits the dynamic of the principal air pollutants during January, February, November, and December in 2014 and 2015. In Beijing, the concentrations of $PM_{2.5}$, CO, and NO_2 slightly increased; while the concentrations of PM_{10}, SO_2, and O_3 decreased at a rate of 3.5 ($\mu g/m^{-3}$)/year, 14.7 ($\mu g/m^{-3}$)/year, and 3.0 ($\mu g/m^{-3}$)/year, respectively. All the principal air pollutants in Tianjin showed a decreasing trend, especially for the SO_2, which decreased at a rate of 30.9 ($\mu g/m^{-3}$)/year. Similarly, all the principal air pollutants in Shijiazhuang also decreased, except for the CO, which increased at a rate of 0.1 (mg/m^{-3})/year. In addition, the concentration of PM_{10} obviously decreased at a rate of 83.1 ($\mu g/m^{-3}$)/year in Shijiazhuang.

3.4. The Dynamics of the Principal Air Pollutants in Shandong, Henan, and Hebei during Autumn

In terms of the major agricultural provinces Shandong, Henan, and Hebei, the dynamics of the principal air pollutants in Jinan and Zhengzhou during July and September are presented in Figure 6. In Jinan, all the principal air pollutants showed a decreasing trend, especially for SO_2, which decreased at a rate of 18.0 ($\mu g/m^{-3}$)/year. However, the average concentrations of $PM_{2.5}$, PM_{10}, NO_2, and O_3 in Zhengzhou increased at a rate of 7.0 ($\mu g/m^{-3}$)/year, 14.7 ($\mu g/m^{-3}$)/year, 9.4 ($\mu g/m^{-3}$)/year, and 38.6 ($\mu g/m^{-3}$)/year, respectively. Fortunately, the concentration of SO_2 and CO presented a decreasing trend in Zhengzhou during autumn. In addition, all the principal air pollutants in Shijiazhuang also decreased, except for the NO_2, which slightly increased at a rate of 0.4 ($\mu g/m^{-3}$)/year.

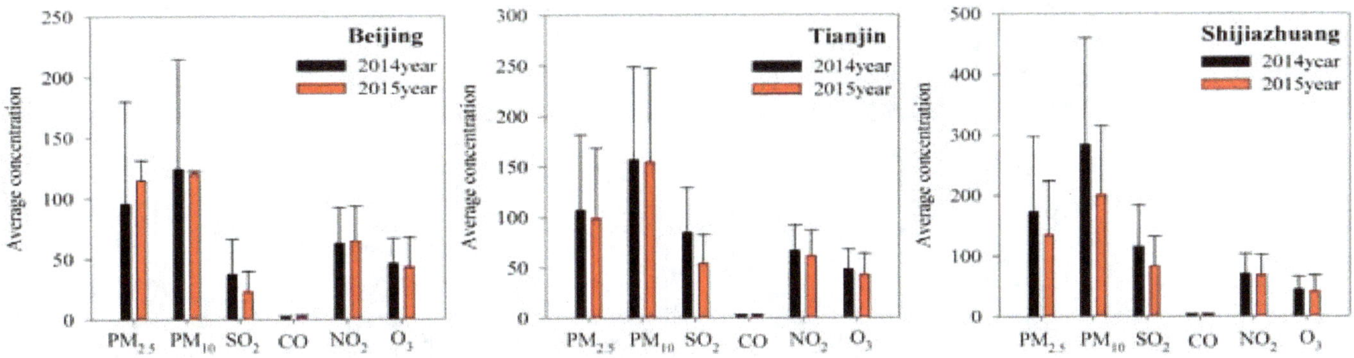

Figure 5. The average concentration of the principal air pollutants in Beijing-Tianjin-Hebei during January, February, November, and December (2014 and 2015). Units: $PM_{2.5}$ ($\mu g/m^{-3}$), PM_{10} ($\mu g/m^{-3}$), CO (mg/m^{-3}), SO_2 ($\mu g/m^{-3}$), NO_2 ($\mu g/m^{-3}$), and O_3 ($\mu g/m^{-3}$).

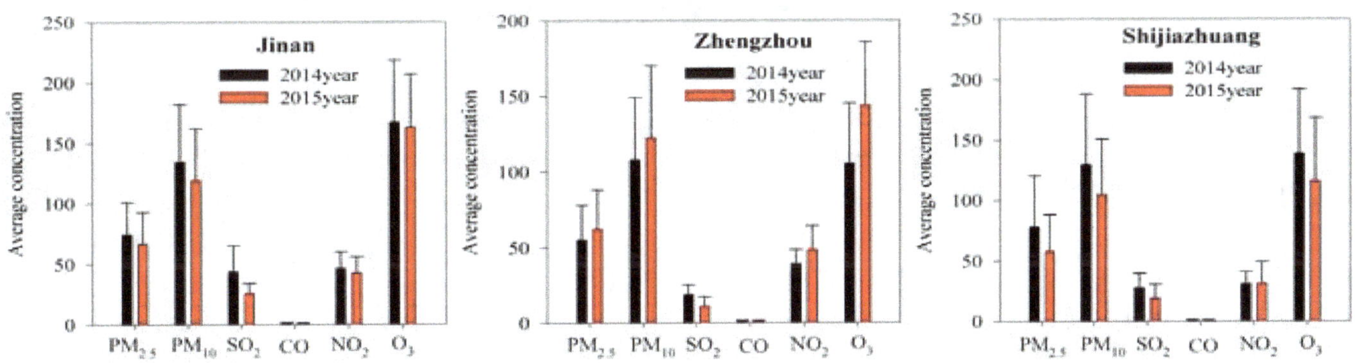

Figure 6. The average concentration of the principal air pollutants in Jinan, Zhengzhou, and Shijiazhuang in autumn (2014 and 2015). Units: $PM_{2.5}$ ($\mu g/m^{-3}$), PM_{10} ($\mu g/m^{-3}$), CO (mg/m^{-3}), SO_2 ($\mu g/m^{-3}$), NO_2 ($\mu g/m^{-3}$), and O_3 ($\mu g/m^{-3}$).

3.5. The Relationships among the Principal Air Pollutants

As Figure 7 shows, the correlation coefficients among air pollutants were significant at $p < 0.05$ in the most polluted 17 cities (2014). Our results indicated that there were positive correlations among $PM_{2.5}$, PM_{10}, CO, SO_2, and NO_2, and negative correlations for O_3. As for $PM_{2.5}$, and the correlation coefficients between PM_{10}, CO, SO_2, NO_2, and O_3 were 0.94, 0.89, 0.74, 0.82, and −0.42, respectively. High correlation coefficients among CO, SO_2, and NO_2 (0.88 for CO and SO_2, 0.89 for CO and NO_2, 0.87 for SO_2 and NO_2) were observed. We also found significant negative correlation coefficients among O_3, CO, SO_2, and NO_2 (−0.57 for O_3 and CO, −0.60 for O_3 and SO_2, and −0.56 for O_3 and NO_2).

Figure 8 shows that there were significant relationships among the principal air pollutants in 2015. The correlation coefficients between $PM_{2.5}$ and PM_{10}, CO, SO_2, NO_2, and O_3 were 0.94, 0.93, 0.73, 0.84, and −0.38. In addition, there was a slight decrease in the correlation coefficients among CO, SO_2, and NO_2 compared to those in 2014, with values of 0.77 for CO and SO_2, 0.87 for CO and NO_2, and 0.82 for SO_2 and NO_2. A similar decrease in the correlation coefficients among O_3, CO, SO_2, and NO_2 was observed (−0.54 for O_3 and CO, −0.52 for O_3 and SO_2, and −0.44 for O_3 and NO_2).

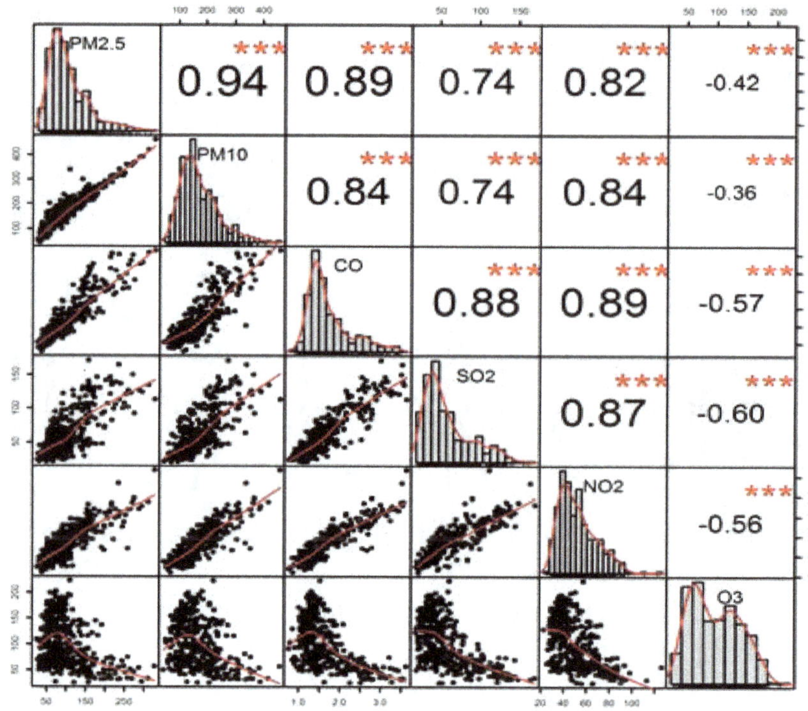

Figure 7. The relationships among the principal air pollutants in the 17 most polluted cities (2014). All the numbers and red stars represent the correlation coefficients and significant relationships among the principal air pollutants.

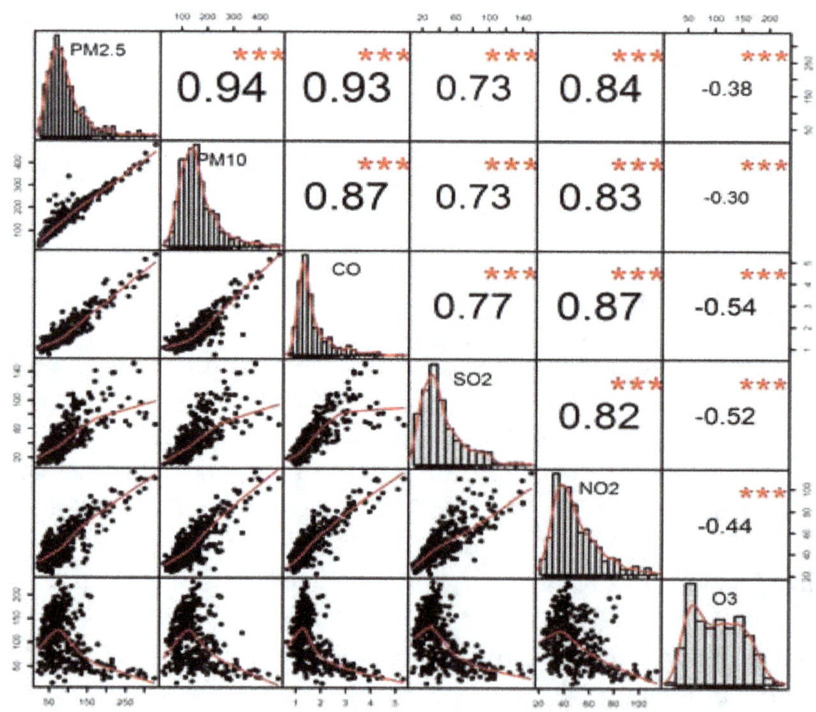

Figure 8. The relationships among the principal air pollutants in the most polluted cities (2015). All the numbers and red stars represent the correlation coefficients and significant relationships among the principal air pollutants.

3.6. GAM Prediction of the Trend for Principal Air Pollutants

The GAM was employed to reveal the future dynamics of atmospheric pollutants (Figure 9), and the non-linear relationship between the pollutants and day-of-year trends was modeled flexibly by the log link function. The results showed that although $PM_{2.5}$ and SO_2 will decrease sharply in the future, further observation is imperative. Unfortunately, a consistent increase in NO_2 and CO was predicted. Furthermore, a slight increase of PM_{10} was observed until it peaked between $100 < x < 150$, and then decreased slightly over the remaining days. Meanwhile, the estimated O_3 showed a slight increase before $x = 50$ and then decreased until approximately $x = 125$, followed by a slight upward trend during the remaining days. In summary, a small variation in PM_{10} and O_3 indicated that the simulated results were reliable and desirable.

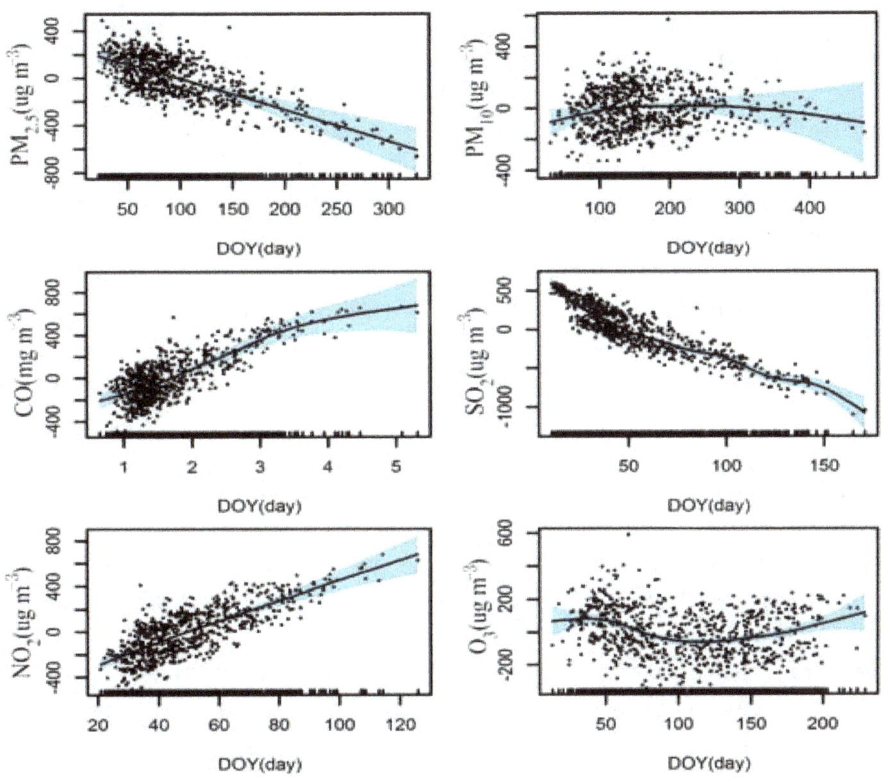

Figure 9. Predicted changes in the principal air pollutants together with temporal dynamics by generalized additive model (GAM) analysis. Rugplot on the x-axis represents the DOY (day of year), and the light blue belts indicate the credible intervals.

4. Discussion

4.1. Temporal and Spatial Patterns of China's Principal Air Pollutants

The concentration of the principal air pollutants was higher in January and December—a phenomenon also observed in Fuzhou [45] and Shanghai [46,47]. In January 2013, two severe air pollution events happened in Beijing, during which the hourly concentration of $PM_{2.5}$ rose to 680 $\mu g/m^{-3}$ and 530 $\mu g/m^{-3}$, respectively [38]. In Shanghai, the highest and lowest levels of PM_{10}, NO_2, and SO_2 were found in winter and autumn, with vehicle emissions and meteorological conditions the most probable causes [36]. In the North China Plain, high emissions of atmospheric contamination, biomass burning, and stable weather conditions contributed to the haze events in winter [48], and higher energy consumption and motor vehicle emissions occurred in winter in Guangzhou [49]. If natural gas is supplied for municipal heating in Beijing rather than coal, air pollutant emissions ($PM_{2.5}$, SO_2, NO_X, etc.) would decrease by 52% in winter [50] because the inter-transport of pollution and the secondary aerosols O_3, OC (organic carbon) and VOC (volatile organic compounds) promote pollution. Thus, reducing the levels of air pollutants during the cold season is imperative for public health [51].

The spatial patterns of the average monthly simulation values indicated that in 2014 the most polluted regions were the Beijing-Tianjin-Hebei region and the provinces of Shanxi, Shandong, Henan, Hubei, and Shanxi. Industrial and domestic sources of pollution and agricultural emissions are the chief regional contributors to pollution in these regions [52]. Compared to 2014, the distribution patterns of air pollutants show a slight shift towards a smaller area in 2015, becoming more centralized in Hebei, Shandong, and Henan provinces, where haze episodes were also more frequent. A similar situation prevailed in the North China Plain [53], Guangzhou [22], and the Yangtze River Delta [26,54]. The stricter laws protecting air quality by government may account for this decreasing trend between 2014 and 2015 to some extent.

The significant relationships among air pollutants illustrated that there could be similar or interacting sources for the pollutions. Many studies have been undertaken to explore the source of large areas of haze. Biomass open burning contributes 47% of the $PM_{2.5}$ in the Yangtze River Delta [55], is an important precursor for O_3 [56], and is a stressor on marine ecosystems [57]. The fact that the inter-transport of pollution accompanies high humidity is the dominant reason for the haze in East China [29,46] and Northern Taiwan [27]. Dust was a major source of pollution in eastern Inner Mongolia [58], local emissions and regional transport accounted for pollution in Nanjing [59], coal consumption and industry increased pollution in Beijing in winter [32,50], and industrial pollution and vehicle emissions were the dominant local contributors to the levels of NO_2 and $PM_{2.5}$ in Shanghai [31,34].

Specifically, for the most polluted region—the northeast of China—the mineral dust from the deserts of western China contributes significantly to the concentration of PM [60,61]. The long-range transport dust plumes mixed with regional pollutants aid the formation of haze episodes [53,60–62]. Thus, the local pollutants also have significant contributions to the widespread haze pollution [48]. Meanwhile, during haze episodes, the secondary inorganic pollutants evident increased in PM [36], suggesting a joint effect among them [63]. Under unfavorable meteorological conditions, the interaction of PM and the secondary inorganic pollutants produced a large amount of aerosols with the characteristic of low visibility [36].

In general, the dust, municipal heating, and vehicle and local emissions are the dominant contributors to haze in winter [31], and meteorological conditions contribute significantly in the distribution, formation, and evolution of haze: for example, higher relative humidity and weaker wind speed contribute to haze [23,25,30]. In addition, there is a positive relationship between visibility and wind speed, and a negative relationship with relative humidity [45,49,59]. Secondary inorganic ions were also positively correlated with stable weather conditions (higher relative humidity), which determined the specific chemical composition of haze [64]. In summary, the median or highest values and the distribution area of air pollutants showed a decreasing trend from 2014 to 2015, which may be explained in part by the improved energy efficiency and stricter laws protecting air quality.

4.2. Policy Recommendations

The estimated values of $PM_{2.5}$ and SO_2 showed a sharp decreasing trend. However, we observed a consistent increasing trend in CO and NO_2, while the estimated values of PM_{10} and O_3 remained stable overall. Thus, the concentration of CO and NO_2 in China will continue to increase. As we know already, vehicle emissions are a key factor in the high concentration of NO_2, which has been caused by the rapid increase in the number of vehicles and industrial parks [31,34]. Therefore, technological innovation is imperative in coal gasification, liquefaction, and storage to improve the energy efficiency of coal [65]. Indeed, replacing fossil fuels with clean fuel (natural gas, nuclear energy, etc.) for municipal heating in winter would be an effective long-term measure to mitigate the dense haze north of the Yellow River [50]. In addition, stringent measures to control particle emissions (biomass burning) should be implemented during specific periods based on meteorological conditions [29,66], while mass transit should be encouraged to avoid energy waste. Further, the central and local government should subsidize the manufacture of hybrid, flex-fuel, and electric automobiles [67,68], and remove market-entry barriers to promote competition among companies [65]. Finally, we need to improve the proportion of the tertiary industry to optimize the economic structure that uses less energy and produces fewer particle emissions [69,70].

As Figure 10 shows, this study highlighted the measures needed to decrease current sources of air pollutants. Indeed, replacing fossil fuels with clean fuels (natural gas, nuclear, wind, and solar energy) will help to address air pollution's root causes. To transform and improve air quality, restricting biomass burning during specific periods (according to meteorological conditions) is imperative, and it may be necessary to implement strict environmental policies forbidding open biomass burning when municipal heating is highest. We should also consider recycling biological feed. Equally important, the issue of vehicle emissions requires the cooperation of government, manufacturers, retailers, and consumers—particularly in the development and promotion of flex-fuel (hybrid) and electric automobiles. Further, ecosystem restoration projects such as the Three-North Shelterbelt System can be effective in defending against sandstorms while simultaneously improving the regional environment. Remote sensing should also be employed to monitor the formation and development of haze, to record, measure, and evaluate the sources of pollution, and monitor the trajectory of haze over the long-term. Indeed, predicting the values and levels of air pollutants and communicating this information by mobile messages will cement trust between individuals and governments. Further, analyzing the structure and function of cities based on local conditions to facilitate scientific urban planning and formulate reasonable population distribution policies should help to mitigate pollution. All of the aforementioned underscore the essential need to share information, provide support, and increase cooperation among different departments to achieve effective haze control. Of course, trade-offs between environmental objectives, economic growth, energy consumption, living standards, and funding are inevitable, but we believe that these can be balanced for the benefit and welfare of all.

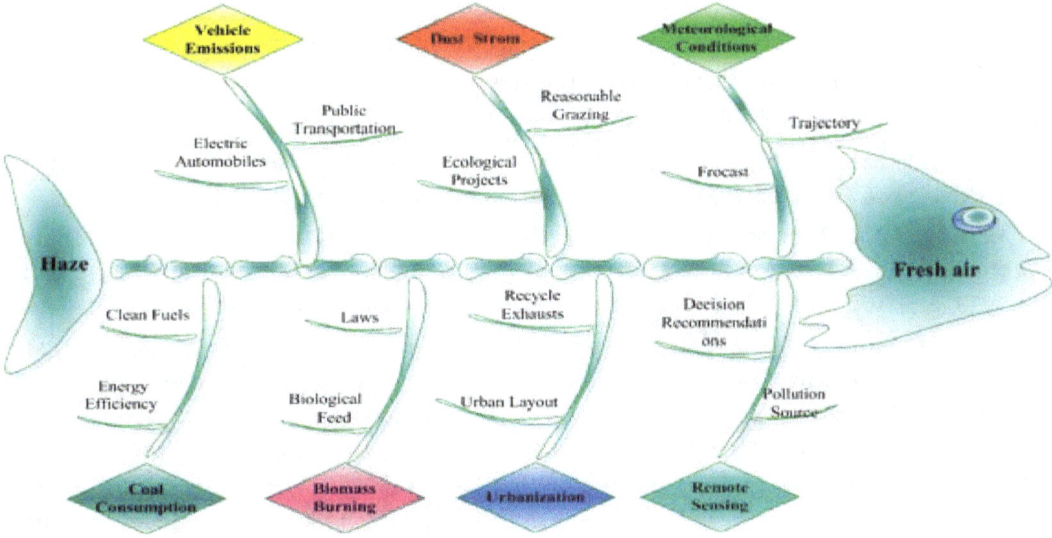

Figure 10. The mind map for decision-makers.

5. Conclusions

Because particulate haze episodes over China have increased in recent years, it is imperative that we understand pollutant pathways and propose recommendations for a policy framework. In general, November to February demonstrated the highest concentrations of air pollutants ($PM_{2.5}$, PM_{10}, CO, SO_2, and NO_2), excluding O_3 levels, which were highest from May to August. Further, the most highly polluted areas were in the provinces of Shandong, Henan, and Shanxi, and in the Beijing-Tianjin-Hebei region. Fortunately, most of the principal air pollutants presented a decreasing or relatively stable trend in Beijing-Tianjin-Hebei during January, February, November, and December. Meanwhile, most of the principal air pollutants also presented a decreasing trend in Jinan and Shijiazhuang in autumn. However, in Zhengzhou, the haze events during autumn were unoptimistic. Although the conflict between clean air and economic growth will continue, measures can be taken to mitigate the sources of air pollutants and to use resources to control, reduce, and manage air pollution.

To improve our understanding of the formation and frequency of haze, satellite observations and monitoring of high-risk areas are important subjects for future research. We need to understand the specific effects of meteorological conditions on the transport mechanism of air pollutants, and the role they play in the secondary formation process.

Acknowledgments: This work was supported by the Ministry of Environmental Protection of the People's Republic of China. The research was funded as well by the West Light Foundation of The Chinese Academy of Sciences, and the Open Fund of the Key Laboratory of Mountain Surface Processes and Eco-regulation.

Author Contributions: J.S. contributed to the study design, J.S., T.Z. and H.Y. were involved in drafting the manuscript, approving the final draft, and agree to be accountable for the work. All authors read and approved the final manuscript.

References

1. He, X.Y.; Hu, J.B.; Chen, W.; Li, X.Y. Haze removal based on advanced haze-optimized transformation (AHOT) for multispectral imagery. *Int. J. Remote Sens.* **2010**, *31*, 5331–5348. [CrossRef]
2. Han, L.J.; Zhou, W.Q.; Pickett, S.T.A.; Li, W.F.; Li, L. An optimum city size? The scaling relationship for urban population and fine particulate ($PM_{2.5}$) concentration. *Environ. Pollut.* **2016**, *208*, 96–101. [CrossRef] [PubMed]
3. Carrasco, L.R. Silver lining of Singapore's haze. *Science* **2013**, *341*, 342–343. [CrossRef] [PubMed]
4. Hao, Y.; Liu, Y.-M. The influential factors of urban $PM_{2.5}$ concentrations in China: A spatial econometric analysis. *J. Clean. Prod.* **2016**, *112*, 1443–1453. [CrossRef]
5. Malandrino, M.; Casazza, M.; Abollino, O.; Minero, C.; Maurino, V. Size resolved metal distribution in the PM matter of the city of Turin (Italy). *Chemosphere* **2016**, *147*, 477–489. [CrossRef] [PubMed]
6. Salinas, S.V.; Chew, B.N.; Miettinen, J.; Campbell, J.R.; Welton, E.J.; Reid, J.S.; Yu, L.Y.E.; Liew, S.C. Physical and optical characteristics of the October 2010 haze event over Singapore: A photometric and lidar analysis. *Atmos. Res.* **2013**, *122*, 555–570. [CrossRef]
7. Kim, K.W.; Kim, Y.J.; Bang, S.Y. Summer time haze characteristics of the urban atmosphere of Gwangju and the rural atmosphere of Anmyon, Korea. *Environ. Monit. Assess.* **2008**, *141*, 189–199. [CrossRef] [PubMed]
8. Zhao, H.M.; Zhang, X.L.; Zhang, S.C.; Chen, W.W.; Tong, D.Q.; Xiu, A.J. Effects of Agricultural Biomass Burning on Regional Haze in China: A Review. *Atmosphere* **2017**, *8*, 88. [CrossRef]
9. Szidat, S. Atmosphere. Sources of Asian Haze. *Science* **2009**, *323*, 470–471. [CrossRef] [PubMed]
10. Begum, B.A.; Hopke, P.K. Identification of haze-creating sources from fine particulate matter in Dhaka aerosol using carbon fractions. *J. Air Waste Manag.* **2013**, *63*, 1046–1057. [CrossRef]
11. Pernigotti, D.; Belis, C.A.; Spano, L. Specieurope: The European data base for PM source profiles. *Atmos. Pollut. Res.* **2016**, *7*, 307–314. [CrossRef]
12. Huang, R.J.; Zhang, Y.; Bozzetti, C.; Ho, K.F.; Cao, J.J.; Han, Y.; Daellenbach, K.R.; Slowik, J.G.; Platt, S.M.; Canonaco, F. High secondary aerosol contribution to particulate pollution during haze events in China. *Nature* **2014**, *514*, 218–222. [CrossRef] [PubMed]
13. Soleiman, A.; Ohtman, M.; Samah, A.A.; Sulaiman, N.M.; Radojevic, M. The occurrence of haze in Malaysia: A case study in an urban industrial area. *Pure Appl. Geophys.* **2003**, *160*, 221–238. [CrossRef]
14. Abdullaev, S.F.; Maslov, V.A.; Nazarov, B.I. Study of dust haze in arid zone. *Izv. Atmos. Ocean. Phys.* **2013**, *49*, 276–284. [CrossRef]

15. Zhang, R.H.; Li, Q.; Zhang, R.N. Meteorological conditions for the persistent severe fog and haze event over eastern China in January 2013. *Sci. China Earth Sci.* **2014**, *57*, 26–35.
16. Elias, T.; Haeffelin, M.; Drobinski, P.; Gomes, L.; Rangognio, J.; Bergot, T.; Chazette, P.; Raut, J.C.; Colomb, M. Particulate contribution to extinction of visible radiation: Pollution, haze, and fog. *Atmos. Res.* **2009**, *92*, 443–454. [CrossRef]
17. Sun, J.; Zhou, T.C. Health risk assessment of China's main air pollutants. *BMC Public Health* **2017**, *17*, 212. [CrossRef] [PubMed]
18. Zhang, X.Y.; Wang, Y.Q.; Niu, T.; Zhang, X.C.; Gong, S.L.; Zhang, Y.M.; Sun, J.Y. Atmospheric aerosol compositions in China: Spatial/temporal variability, chemical signature, regional haze distribution and comparisons with global aerosols. *Atmos. Chem. Phys.* **2012**, *12*, 799.
19. Wang, Y.; Li, Z.Q.; Zhang, Y.; Wang, Q.; Ma, J.Z. Impact of aerosols on radiation during a heavy haze event in Beijing. *IOP Conf. Ser. Earth Environ. Sci.* **2014**, *17*, 012012.
20. Sun, Y.L.; Jiang, Q.; Wang, Z.F.; Fu, P.Q.; Li, J.; Yang, T.; Yin, Y. Investigation of the sources and evolution processes of severe haze pollution in Beijing in January 2013. *J. Geophys. Res. Atmos.* **2014**, *119*, 4380–4398. [CrossRef]
21. Zhang, M.; Ma, Y.Y.; Gong, W.; Zhu, Z.M. Aerosol optical properties of a haze episode in Wuhan based on ground-based and satellite observations. *Atmosphere* **2014**, *5*, 699–719. [CrossRef]
22. Zhang, Z.L.; Wang, J.; Chen, L.H.; Chen, X.Y.; Sun, G.Y.; Zhong, N.S.; Kan, H.D.; Lu, W.J. Impact of haze and air pollution-related hazards on hospital admissions in Guangzhou, China. *Environ. Sci. Pollut. Res.* **2014**, *21*, 4236–4244. [CrossRef] [PubMed]
23. Liu, X.G.; Li, J.; Qu, Y.; Han, T.; Hou, L.; Gu, J.; Chen, C.; Yang, Y.; Liu, X.; Yang, T.; et al. Formation and evolution mechanism of regional haze: A case study in the megacity Beijing, China. *Atmos. Chem. Phys.* **2013**, *13*, 4501–4514. [CrossRef]
24. Quan, J.; Zhang, Q.; He, H.; Liu, J.; Huang, M.; Jin, H. Analysis of the formation of fog and haze in North China Plain (NCP). *Atmos. Chem. Phys.* **2011**, *11*, 8205–8214. [CrossRef]
25. Bei, N.F.; Xiao, B.; Meng, N.; Feng, T. Critical role of meteorological conditions in a persistent haze episode in the Guanzhong basin, China. *Sci. Total Environ.* **2016**, *550*, 273–284. [CrossRef] [PubMed]
26. Tang, L.L.; Yu, H.X.; Ding, A.J.; Zhang, Y.J.; Qin, W.; Wang, Z.; Chen, W.T.; Hua, Y.; Yang, X.X. Regional contribution to PM_1 pollution during winter haze in Yangtze River Delta, China. *Sci. Total Environ.* **2016**, *541*, 161–166. [CrossRef] [PubMed]
27. Wang, S.H.; Hung, W.T.; Chang, S.C.; Yen, M.C. Transport characteristics of Chinese haze over Northern Taiwan in winter, 2005–2014. *Atmos. Environ.* **2016**, *126*, 76–86. [CrossRef]
28. Gabrielli, M.; Fracassetti, D.; Tirelli, A. Release of phenolic compounds from cork stoppers and its effect on protein-haze. *Food Control* **2016**, *62*, 330–336. [CrossRef]
29. Tao, M.H.; Chen, L.F.; Wang, Z.F.; Tao, J.H.; Su, L. Satellite observation of abnormal yellow haze clouds over East China during summer agricultural burning season. *Atmos. Environ.* **2013**, *79*, 632–640. [CrossRef]
30. Sun, Y.; Song, T.; Tang, G.Q.; Wang, Y.S. The vertical distribution of $PM_{2.5}$ and boundary-layer structure during summer haze in Beijing. *Atmos. Environ.* **2013**, *74*, 413–421. [CrossRef]
31. Wang, Y.J.; Li, L.; Chen, C.H.; Huang, C.; Huang, H.Y.; Feng, J.L.; Wang, S.X.; Wang, H.L.; Zhang, G.; Zhou, M.; et al. Source apportionment of fine particulate matter during autumn haze episodes in Shanghai, China. *J. Geophys. Res. Atmos.* **2014**, *119*, 1903–1914. [CrossRef]
32. Lin, Y.C.; Hsu, S.C.; Chou, C.C.K.; Zhang, R.J.; Wu, Y.F.; Kao, S.J.; Luo, L.; Huang, C.H.; Lin, S.H.; Huang, Y.T. Wintertime haze deterioration in Beijing by industrial pollution deduced from trace metal fingerprints and enhanced health risk by heavy metals. *Environ. Pollut.* **2016**, *208*, 284–293. [CrossRef] [PubMed]
33. Gui, K.; Che, H.Z.; Chen, Q.L.; An, L.C.; Zeng, Z.L.; Guo, Z.Y.; Zheng, Y.; Wang, H.; Wang, Y.Q.; Yu, J.; et al. Aerosol optical properties based on ground and satellite retrievals during a serious haze episode in December 2015 over Beijing. *Atmosphere* **2016**, *7*, 70. [CrossRef]
34. Zhao, W.C.; Cheng, J.P.; Li, D.L.; Duan, Y.S.; Wei, H.P.; Ji, R.X.; Wang, W.H. Urban ambient air quality investigation and health risk assessment during haze and non-haze periods in Shanghai, China. *Atmos. Pollut. Res.* **2013**, *4*, 275–281. [CrossRef]
35. Li, W.J.; Shi, Z.B.; Zhang, D.Z.; Zhang, X.Y.; Li, P.R.; Feng, Q.J.; Yuan, Q.; Wang, W.X. Haze particles over a coal-burning region in the China Loess Plateau in winter: Three flight missions in December 2010. *J. Geophys. Res. Atmos.* **2012**, *117*. [CrossRef]

36. Zhao, X.J.; Zhao, P.S.; Xu, J.; Meng, W.; Pu, W.W.; Dong, F.; He, D.; Shi, Q.F. Analysis of a winter regional haze event and its formation mechanism in the North China Plain. *Atmos. Chem. Phys.* **2013**, *13*, 5685–5696. [CrossRef]
37. Zhu, C.S.; Cao, J.J.; Tsai, C.J.; Shen, Z.X.; Liu, S.X.; Huang, R.J.; Zhang, N.N.; Wang, P. The rural carbonaceous aerosols in coarse, fine, and ultrafine particles during haze pollution in northwestern China. *Environ. Sci. Pollut. Res.* **2016**, *23*, 4569–4575. [CrossRef] [PubMed]
38. Wang, Y.S.; Yao, L.; Wang, L.L.; Liu, Z.R.; Ji, D.S.; Tang, G.Q.; Zhang, J.K.; Sun, Y.; Hu, B.; Xin, J.Y. Mechanism for the formation of the January 2013 heavy haze pollution episode over central and eastern China. *Sci. China Earth Sci.* **2014**, *57*, 14–25. [CrossRef]
39. Zhu, T.; Shang, J.; Zhao, D.F. The roles of heterogeneous chemical processes in the formation of an air pollution complex and gray haze. *Sci. China Chem.* **2011**, *54*, 145–153. [CrossRef]
40. Jansen, R.C.; Shi, Y.; Chen, J.M.; Hu, Y.J.; Xu, C.; Hong, S.M.; Li, J.; Zhang, M. Using hourly measurements to explore the role of secondary inorganic aerosol in $PM_{2.5}$ during haze and fog in Hangzhou, China. *Adv. Atmos. Sci.* **2014**, *31*, 1427–1434. [CrossRef]
41. Shen, X.J.; Sun, J.Y.; Zhang, X.Y.; Zhang, Y.M.; Zhang, L.; Che, H.C.; Ma, Q.L.; Yu, X.M.; Yue, Y.; Zhang, Y.W. Characterization of submicron aerosols and effect on visibility during a severe haze-fog episode in Yangtze River Delta, China. *Atmos. Environ.* **2015**, *120*, 307–316. [CrossRef]
42. Lin, Y.F.; Huang, K.; Zhuang, G.S.; Fu, J.S.; Wang, Q.Z.; Liu, T.N.; Deng, C.R.; Fu, Q.Y. A multi-year evolution of aerosol chemistry impacting visibility and haze formation over an Eastern Asia megacity, Shanghai. *Atmos. Environ.* **2014**, *92*, 76–86. [CrossRef]
43. Catalano, M.; Galatioto, F.; Bell, M.; Namdeo, A.; Bergantino, A.S. Improving the prediction of air pollution peak episodes generated by urban transport networks. *Environ. Sci. Policy* **2016**, *60*, 69–83. [CrossRef]
44. Chen, X.L.; Feng, Y.R.; Li, J.N.; Lin, W.S.; Fan, S.J.; Wang, A.Y.; Fong, S.K.; Lin, H. Numerical simulations on the effect of sea-land breezes on atmospheric haze over the Pearl River Delta Region. *Environ. Model. Assess.* **2009**, *14*, 351–363.
45. Zhang, F.W.; Xu, L.L.; Chen, J.S.; Chen, X.Q.; Niu, Z.C.; Lei, T.; Li, C.M.; Zhao, J.P. Chemical characteristics of $PM_{2.5}$ during haze episodes in the urban of Fuzhou, China. *Particuology* **2013**, *11*, 264–272. [CrossRef]
46. Wang, X.M.; Chen, J.M.; Cheng, T.T.; Zhang, R.Y.; Wang, X.M. Particle number concentration, size distribution and chemical composition during haze and photochemical smog episodes in Shanghai. *J. Environ. Sci.* **2014**, *26*, 1894–1902. [CrossRef] [PubMed]
47. Wang, H.L.; Qiao, L.P.; Lou, S.R.; Zhou, M.; Ding, A.J.; Huang, H.Y.; Chen, J.M.; Wang, Q.; Tao, S.K.; Chen, C.H.; et al. Chemical composition of $PM_{2.5}$ and meteorological impact among three years in urban Shanghai, China. *J. Clean. Prod.* **2016**, *112*, 1302–1311. [CrossRef]
48. Gao, M.; Carmichael, G.R.; Wang, Y.; Saide, P.E.; Yu, M.; Xin, J.; Liu, Z.; Wang, Z. Modeling study of the 2010 regional haze event in the North China Plain. *Atmos. Chem. Phys.* **2016**, *16*, 1673–1691. [CrossRef]
49. Tan, J.H.; Duan, J.C.; Chen, D.H.; Wang, X.H.; Guo, S.J.; Bi, X.H.; Sheng, G.Y.; He, K.B.; Fu, J.M. Chemical characteristics of haze during summer and winter in Guangzhou. *Atmos. Res.* **2009**, *94*, 238–245. [CrossRef]
50. Li, H.C.; Yang, S.Y.; Zhang, J.; Qian, Y. Coal-based synthetic natural gas (SNG) for municipal heating in China: Analysis of haze pollutants and greenhouse gases (GHGs) emissions. *J. Clean. Prod.* **2016**, *112*, 1350–1359. [CrossRef]
51. Fang, Z.Q.; Zhu, H.L.; Bao, W.Z.; Preston, C.; Liu, Z.; Dai, J.Q.; Li, Y.Y.; Hu, L.B. Highly transparent paper with tunable haze for green electronics. *Energ. Environ. Sci.* **2014**, *7*, 3313–3319. [CrossRef]
52. Wang, L.T.; Wei, Z.; Yang, J.; Zhang, Y.; Zhang, F.F.; Su, J.; Meng, C.C.; Zhang, Q. The 2013 severe haze over southern Hebei, China: Model evaluation, source apportionment, and policy implications. *Atmos. Chem. Phys.* **2014**, *14*, 3151–3173. [CrossRef]
53. Zhang, X.L.; Huang, Y.B.; Zhu, W.Y.; Rao, R.Z. Aerosol characteristics during summer haze episodes from different source regions over the coast city of North China Plain. *J. Quant. Spectrosc. Radiat. Transf.* **2013**, *122*, 180–193. [CrossRef]
54. Wang, S.; Zhou, C.; Wang, Z.; Feng, K.; Hubacek, K. The characteristics and drivers of fine particulate matter ($PM_{2.5}$) distribution in China. *J. Clean. Prod.* **2017**, *142*, 1800–1809. [CrossRef]
55. Cheng, Z.; Wang, S.; Fu, X.; Watson, J.G.; Jiang, J.; Fu, Q.; Chen, C.; Xu, B.; Yu, J.; Chow, J.C.; et al. Impact of biomass burning on haze pollution in the Yangtze River delta, China: A case study in summer 2011. *Atmos. Chem. Phys.* **2014**, *14*, 4573–4585. [CrossRef]

56. Zhang, Y.W.; Zhang, X.Y.; Zhang, Y.M.; Shen, X.J.; Sun, J.Y.; Ma, Q.L.; Yu, X.M.; Zhu, J.L.; Zhang, L.; Che, H.C. Significant concentration changes of chemical components of PM_1 in the Yangtze River Delta area of China and the implications for the formation mechanism of heavy haze-fog pollution. *Sci. Total Environ.* **2015**, *538*, 7–15. [CrossRef] [PubMed]
57. Jaafar, Z.; Loh, T.L. Linking land, air and sea: Potential impacts of biomass burning and the resultant haze on marine ecosystems of Southeast Asia. *Glob. Chang. Biol.* **2014**, *20*, 2701–2707. [CrossRef] [PubMed]
58. Guo, J.P.; Zhang, X.Y.; Cao, C.X.; Che, H.Z.; Liu, H.L.; Gupta, P.; Zhang, H.; Xu, M.; Li, X.W. Monitoring haze episodes over the Yellow Sea by combining multisensor measurements. *Int. J. Remote Sens.* **2010**, *31*, 4743–4755. [CrossRef]
59. Kang, H.Q.; Zhu, B.; Su, J.F.; Wang, H.L.; Zhang, Q.C.; Wang, F. Analysis of a long-lasting haze episode in Nanjing, China. *Atmos. Res.* **2013**, *120*, 78–87. [CrossRef]
60. Tao, M.H.; Chen, L.F.; Su, L.; Tao, J.H. Satellite observation of regional haze pollution over the North China Plain. *J. Geophys. Res. Atmos.* **2012**, *117*, D12. [CrossRef]
61. Yang, T.; Wang, X.Q.; Wang, Z.F.; Sun, Y.L.; Zhang, W.; Zhang, B.; Du, Y.M. Gravity-current driven transport of haze from North China Plain to Northeast China in Winter 2010-Part I: Observations. *Sola* **2012**, *8*, 13–16. [CrossRef]
62. Ye, X.X.; Song, Y.; Cai, X.H.; Zhang, H.S. Study on the synoptic flow patterns and boundary layer process of the severe haze events over the North China Plain in January 2013. *Atmos. Environ.* **2016**, *124*, 129–145. [CrossRef]
63. Kirrane, E.; Svendsgaard, D.; Ross, M.; Buckley, B.; Davis, A.; Johns, D.; Kotchmar, D.; Long, T.C.; Luben, T.J.; Smith, G.; et al. A Comparison of risk estimates for the effect of short-term exposure to PM, NO_2 and CO on cardiovascular hospitalizations and emergency department visits: Effect size modeling of study findings. *Atmosphere* **2011**, *2*, 688–701. [CrossRef]
64. Han, B.; Zhang, R.; Yang, W.; Bai, Z.P.; Ma, Z.Q.; Zhang, W.J. Heavy haze episodes in Beijing during January 2013: Inorganic ion chemistry and source analysis using highly time-resolved measurements from an urban site. *Sci. Total Environ.* **2016**, *544*, 319–329. [CrossRef] [PubMed]
65. Zhang, X.H.; Wu, L.Q.; Zhang, R.; Deng, S.H.; Zhang, Y.Z.; Wu, J.; Li, Y.W.; Lin, L.L.; Li, L.; Wang, Y.J.; et al. Evaluating the relationships among economic growth, energy consumption, air emissions and air environmental protection investment in China. *Renew. Sustain. Energy Rev.* **2013**, *18*, 259–270. [CrossRef]
66. Mu, M.; Zhang, R.H. Addressing the issue of fog and haze: A promising perspective from meteorological science and technology. *Sci. China Earth Sci.* **2014**, *57*, 1–2. [CrossRef]
67. Sun, J.; Wang, J.N.; Wei, Y.Q.; Li, Y.R.; Liu, M. The Haze Nightmare Following the Economic Boom in China: Dilemma and Tradeoffs. *Int. J. Environ. Res. Public Health* **2016**, *13*, 402. [CrossRef] [PubMed]
68. Wu, X.; Tan, L.; Guo, J.; Wang, Y.; Liu, H.; Zhu, W. A study of allocative efficiency of $PM_{2.5}$ emission rights based on a zero sum gains data envelopment model. *J. Clean. Prod.* **2016**, *113*, 1024–1031. [CrossRef]
69. Wang, S.S.; Zhou, D.Q.; Zhou, P.; Wang, Q.W. CO_2 emissions, energy consumption and economic growth in China: A panel data analysis. *Energy Policy* **2011**, *39*, 4870–4875. [CrossRef]
70. Yuan, X.L.; Mu, R.M.; Zuo, J.; Wang, Q.S. Economic development, energy consumption, and air pollution: A critical assessment in China. *Hum. Ecol. Risk Assess.* **2015**, *21*, 781–798. [CrossRef]

PERMISSIONS

All chapters in this book were first published in MDPI; hereby published with permission under the Creative Commons Attribution License or equivalent. Every chapter published in this book has been scrutinized by our experts. Their significance has been extensively debated. The topics covered herein carry significant findings which will fuel the growth of the discipline. They may even be implemented as practical applications or may be referred to as a beginning point for another development.

The contributors of this book come from diverse backgrounds, making this book a truly international effort. This book will bring forth new frontiers with its revolutionizing research information and detailed analysis of the nascent developments around the world.

We would like to thank all the contributing authors for lending their expertise to make the book truly unique. They have played a crucial role in the development of this book. Without their invaluable contributions this book wouldn't have been possible. They have made vital efforts to compile up to date information on the varied aspects of this subject to make this book a valuable addition to the collection of many professionals and students.

This book was conceptualized with the vision of imparting up-to-date information and advanced data in this field. To ensure the same, a matchless editorial board was set up. Every individual on the board went through rigorous rounds of assessment to prove their worth. After which they invested a large part of their time researching and compiling the most relevant data for our readers.

The editorial board has been involved in producing this book since its inception. They have spent rigorous hours researching and exploring the diverse topics which have resulted in the successful publishing of this book. They have passed on their knowledge of decades through this book. To expedite this challenging task, the publisher supported the team at every step. A small team of assistant editors was also appointed to further simplify the editing procedure and attain best results for the readers.

Apart from the editorial board, the designing team has also invested a significant amount of their time in understanding the subject and creating the most relevant covers. They scrutinized every image to scout for the most suitable representation of the subject and create an appropriate cover for the book.

The publishing team has been an ardent support to the editorial, designing and production team. Their endless efforts to recruit the best for this project, has resulted in the accomplishment of this book. They are a veteran in the field of academics and their pool of knowledge is as vast as their experience in printing. Their expertise and guidance has proved useful at every step. Their uncompromising quality standards have made this book an exceptional effort. Their encouragement from time to time has been an inspiration for everyone.

The publisher and the editorial board hope that this book will prove to be a valuable piece of knowledge for researchers, students, practitioners and scholars across the globe.

LIST OF CONTRIBUTORS

Daniel-Eduard Constantin, Mirela Voiculescu, Carmelia Dragomir and Lucian Georgescu
European Center of Excellence for the Environment, Faculty of Sciences and Environment, "Dunarea de Jos" University of Galati, Str. Domneasca 111, Galati 800008, Romania

Alexis Merlaud, Francois Hendrick, Gaia Pinardi and Michel Van Roozendael
Royal Belgian Institute for Space Aeronomy, Ringlaan-3-Avenue Circulaire, B-1180 Brussels, Belgium

Samuel D. Lightstone, Fred Moshary and Barry Gross
Optical Remote Sensing Lab, City College of New York, New York, NY 10031, USA

Chengming Li, Zhaoxin Dai, Xiaoli Liu and Pengda Wu
Chinese Academy of Surveying and Mapping, Beijing 100830, China

Richard Ménard and Martin Deshaies-Jacques
Air Quality Research Division, Environment and Climate Change Canada, 2121 Transcanada Highway, Dorval, QC H9P 1J3, Canada

Debra A. Schmitz, Emma Di Stefano and John R. Froines
Department of Environmental Health Sciences, School of Public Health, University of California Los Angeles, Los Angeles, CA 90095-1772, USA

William P. Melega
Department of Molecular and Medical Pharmacology, University of California Los Angeles, Los Angeles, CA 90095-1772, USA

Arthur K. Cho
Department of Environmental Health Sciences, School of Public Health, University of California Los Angeles, Los Angeles, CA 90095-1772, USA
Department of Molecular and Medical Pharmacology, University of California Los Angeles, Los Angeles, CA 90095-1772, USA

Yasuhiro Shinkai
Environmental Biology Laboratory, Faculty of Medicine, University of Tsukuba, 1-1-1 Tennodai, Tsukuba, Ibaraki 305-8575, Japan

Arantza Eiguren-Fernandez
Aerosol Dynamics, Inc., Berkeley, CA 94710, USA

Aline Lefol Nani Guarieiro
Universidade Federal da Bahia, Instituto de Química, 40170290 Salvador-BA, Brasil

Erika M. Salinas
División de Ciencias Básicas e Ingeniería, Universidad Autónoma Metropolitana, Azcapotzalco, Av San Pablo Xalpa 150, Mexico

George M. Woodall, Ronald Williams, Kristen Benedict, Annie M. Jarabek, Michael J. Stewart, James S. Brown, Andrea L. Clements, Amanda Kaufman, Martha Keating and Bob Thompson
Environmental Protection Agency, Research Triangle Park, NC 27711, USA

Mark D. Hoover, Martin Harper and Jhy-Charm Soo
National Institute for Occupational Safety and Health, Morgantown, WV 26505, USA

Janis E. Hulla
Army Corps of Engineers, Sacramento, CA 95814, USA

Motria Caudill
Agency for Toxic Substances and Disease Registry, Atlanta, GA 30329, USA

Alison J. Parker
ORISE Fellow hosted by U.S. Environmental Protection Agency, Washington, DC 20004, USA

David Balshaw
National Institute for Environmental Health Sciences, Research Triangle Park, NC 27709, USA

Kevin Garrahan and Laureen Burton
Environmental Protection Agency, Washington, DC 20004, USA

Sheila Batka and Vijay S. Limaye
Environmental Protection Agency, Chicago, IL 60605, USA

Pertti J. Hakkinen
National Library of Medicine, Bethesda, MD 20892, USA

Barry Baker and Li Pan
NOAA Air Resources Laboratory, College Park, MD 20740, USA
Cooperative Institute for Climate and Satellites, University of Maryland at College Park, College Park, MD 20740, USA

Afifa Aslam, Muhammad Ibrahim, Abid Mahmood, Muhammad Kashif Irshad, Ghazala and Muhammad Tariq
Department of Environmental Sciences & Engineering, Government College University Faisalabad, Faisalabad 38000, Pakistan

Imran Shahid
Environmental Science Centre, Qatar University, Doha, Qatar

Muhammad Yamin
Department of Farm Machinery & Power, Faculty of Agricultural Engineering & Technology, University of Agriculture, Faisalabad 38040, Pakistan

Redmond R. Shamshiri
Leibniz Institute for Agricultural Engineering and Bioeconomy, Max-Eyth-Allee 100, 14469 Potsdam-Bornim, Germany

Casey D. Bray, William Battye, Pornpan Uttamang, Priya Pillai and Viney P. Aneja
Department of Marine, Earth, and Atmospheric Sciences, North Carolina State University, Raleigh, NC 27695, USA

Rodrigo Munoz-Alpizar, Radenko Pavlovic, Sylvain Ménard, Jacinthe Racine, Annie Duhamel, Samuel Gilbert, Paul-André Beaulieu, Hugo Landry, Didier Davignon, Sophie Cousineau and Véronique Bouchet
Canadian Meteorological Centre Operations, Environment and Climate Change Canada, 2121 Rte Transcanadienne, Montreal, QC H9P 1J3, Canada

Michael D. Moran, Jack Chen and Sylvie Gravel
Air Quality Research Division, Environment and Climate Change Canada, Toronto, ON M3H 5T4, Canada

Sarah B. Henderson
Environmental Health Services, British Columbia Centre for Disease Control, Vancouver, BC V5Z 4R4, Canada

Juanming Zhan, Minyi Wang, Yonghong Liu, Ting Gan, Li Li and Hui Ding
School of Intelligent Systems Engineering, Sun Yat-sen University, Guangzhou 510275, China
Guangdong Provincial Key Laboratory of Intelligent Transport System, Guangzhou 510275, China
Guangdong Provincial Engineering Research Center for Traffic Environmental Monitoring and Control, Guangzhou 510275, China

Chunming Feng and Ruiwen Ou
Guangzhou Haizhu District Environmental Monitoring Station, Guangzhou 510220, China

Hui Zhao
Key Laboratory for Aerosol-Cloud-Precipitation of China Meteorological Administration, Nanjing University of Information Science and Technology, Nanjing 210044, China

Ting Li
Key Laboratory of Atmospheric Environment Monitoring and Pollution Control, Collaborative Innovation Center of Atmospheric Environment and Equipment Technology, Nanjing University of Information Science and Technology, Nanjing 210044, China

Youfei Zheng
Key Laboratory for Aerosol-Cloud-Precipitation of China Meteorological Administration, Nanjing University of Information Science and Technology, Nanjing 210044, China
Key Laboratory of Atmospheric Environment Monitoring and Pollution Control, Collaborative Innovation Center of Atmospheric Environment and Equipment Technology, Nanjing University of Information Science and Technology, Nanjing 210044, China

Huan Yu
College of Earth Sciences, Chengdu University of Technology, Chengdu 610059, China

Jian Sun
Institute of Geographic Sciences and Natural Resources Research, Chinese Academy of Sciences, Beijing 100101, China

Tiancai Zhou
College of Earth Sciences, Chengdu University of Technology, Chengdu 610059, China
Institute of Geographic Sciences and Natural Resources Research, Chinese Academy of Sciences, Beijing 100101, China

Index

www.ingramcontent.com/pod-product-compliance
Lightning Source LLC
LaVergne TN
LVHW081015210926
848485LV00004B/868